Lecture Notes in Bioinformatics 12313

Subseries of Lecture Notes in Computer Science

More information about this subseries at http://www.springer.com/series/5381

Paolo Cazzaniga · Daniela Besozzi ·
Ivan Merelli · Luca Manzoni (Eds.)

Computational Intelligence Methods for Bioinformatics and Biostatistics

16th International Meeting, CIBB 2019
Bergamo, Italy, September 4–6, 2019
Revised Selected Papers

 Springer

Editors
Paolo Cazzaniga ⓘD
University of Bergamo
Bergamo, Italy

Ivan Merelli ⓘD
National Research Council
Segrate, Italy

Daniela Besozzi ⓘD
University of Milano-Bicocca
Milan, Italy

Luca Manzoni ⓘD
Università degli Studi di Trieste
Trieste, Italy

ISSN 0302-9743   ISSN 1611-3349   (electronic)
Lecture Notes in Bioinformatics
ISBN 978-3-030-63060-7   ISBN 978-3-030-63061-4   (eBook)
https://doi.org/10.1007/978-3-030-63061-4

LNCS Sublibrary: SL8 – Bioinformatics

This Springer imprint is published by the registered company Springer Nature Switzerland AG
The registered company address is: Gewerbestrasse 11, 6330 Cham, Switzerland

Preface

This volume contains the revised selected papers of the 16th edition of the international conference on Computational Intelligence Methods for Bioinformatics and Biostatistics (CIBB 2019), which provides a multidisciplinary forum for the scientific community interested in the advancements and future perspectives in bioinformatics and biostatistics.

Following its tradition and roots, CIBB 2019 brought together researchers with different background – from mathematics to computer science, from materials science to medicine and biology – interested in the application of computational intelligence methods for the solution of open problems in bioinformatics, biostatistics, systems biology, synthetic biology, and medical informatics. At CIBB 2019, current trends and future opportunities at the edge of computer and life sciences, as well as their impact on innovative medicine, were presented. Both theoretical and experimental biologists and computer scientists discussed novel challenges, and started multidisciplinary collaborations with the aim of combining theory with practice.

The conference program included only oral presentations, among which cutting-edge plenary keynote lectures were given by four prominent speakers: Malu Calle Rosingana (University of Vic - Central University of Catalonia, Spain), Ana Cvejic (Wellcome Trust Sanger Institute, UK), Uzay Kaymak (Eindhoven University of Technology, The Netherlands), and Marco Masseroli (Politecnico di Milano, Italy). The conference program was organized with a main conference track, including papers dealing with heterogeneous open problems at the forefront of current research, and four special sessions on specific themes: Modeling and Simulation Methods for Computational Biology and Systems Medicine, Machine Learning in Healthcare Informatics and Medical Biology, Algebraic and Computational Methods for the Study of RNA Behaviour, and Intelligence Methods for Molecular Characterization and Dynamics in Translational Medicine.

The conference was held at the Department of Human and Social Sciences, University of Bergamo, Italy, during September 4–6, 2019, thanks to the synergistic effort of the Organizing, Program, and Steering Committees, and to the support of the Department of Informatics, Systems and Communication, University of Milano-Bicocca, Italy, and the Institute of Biomedical Technologies of the National Research Council, Italy. The conference was attended by more than 90 participants from Europe, Asia, USA, Central and South America.

A total of 55 contributions were submitted for consideration to CIBB 2019 and, after a round of reviews handled by the members of the Program Committee, 47 papers were accepted for an oral presentation. After the conference, the authors of selected papers were invited to submit an extended version of their manuscript, and the accepted papers are collected in this volume. The volume is divided into five major sections: Computational Intelligence Methods for Bioinformatics and Biostatistics (i.e., the main conference track), Algebraic and Computational Methods for the Study of RNA

Behaviour, Intelligence Methods for Molecular Characterization and Dynamics in Translational Medicine, Machine Learning in Healthcare Informatics and Medical Biology, and Modeling and Simulation Methods for Computational Biology and Systems Medicine.

The editors warmly thank the Program Committee, the invited speakers, the authors of the papers, the reviewers, and all the participants for their contributions to the success of CIBB 2019.

October 2020

Paolo Cazzaniga
Daniela Besozzi
Ivan Merelli
Luca Manzoni

Organization

Conference Co-chairs

Paolo Cazzaniga   University of Bergamo, Italy
Daniela Besozzi   University of Milano-Bicocca, Italy
Ivan Merelli   CNR, Italy

Publicity Chairs

Stefano Beretta   University of Milano-Bicocca, Italy
Simone Spolaor   University of Milano-Bicocca, Italy

Technical Chair

Angela Serra   University of Tampere, Finland

Proceedings Chair

Luca Manzoni   University of Trieste, Italy

Local Organizers

Riccardo Dondi   University of Bergamo, Italy
Stefano Beretta   University of Milano-Bicocca, Italy
Marco S. Nobile   Eindhoven University of Technology, The Netherlands
Dario Pescini   University of Milano-Bicocca, Italy
Simone Spolaor   University of Milano-Bicocca, Italy

Steering Committee

Pierre Baldi   University of California, Irvine, USA
Elia Biganzoli   University of Milan, Italy
Clelia Di Serio   Università Vita-Salute San Raffaele, Italy
Alexandru Floares   Oncological Institute Cluj-Napoca, Romania
Jon Garibaldi   University of Nottingham, UK
Nikola Kasabov   Auckland University of Technology, New Zealand
Francesco Masulli   University of Genova, Italy, and Temple University, USA
Leif Peterson   TMHRI, USA
Roberto Tagliaferri   University of Salerno, Italy

Program Committee

Antonino Abbruzzo	University of Palermo, Italy
Claudio Angione	Teesside University, UK
Sansanee Auephanwiriyakul	Chiang Mai University, Thailand
Marco Beccuti	University of Torino, Italy
Gilles Bernot	University of Nice Sophia Antipolis, France
Andrea Bracciali	University of Stirling, UK
Giulio Caravagna	Institute of Cancer Research, UK
Mauro Castelli	Universidade NOVA de Lisboa, Portugal
Davide Chicco	Peter Munk Cardiac Centre, Canada
Angelo Ciaramella	Pathenope University of Naples, Italy
Chiara Damiani	University of Milano-Bicocca, Italy
Mohammed El-Kebir	University of Illinois at Urbana-Champaign, USA
Marco Frasca	University of Milan, Italy
Christoph M. Friedrich	Dortmund University of Applied Sciences and Arts, Germany
Enrico Formenti	University of Nice Sophia Antipolis, France
Filippo Geraci	Institute for Informatics and Telematics, CNR, Italy
Yair Goldberg	University of Haifa, Israel
Alex Graudenzi	Institute of Molecular Bioimaging and Physiology, CNR, Italy
Marco Grzegorczyk	Groningen University, The Netherlands
Anne-Christin Hauschild	Philipps-Universität Marburg, Germany
Sean Holden	University of Cambridge, UK
Giuseppe Jurman	Fondazione Bruno Kessler, Italy
Johannes Köster	University of Duisburg-Essen, Germany
Giosuè Lo Bosco	University of Palermo, Italy
Hassan Mahmoud	University of Genova, Italy
Anna Marabotti	University of Salerno, Italy
Elena Marchiori	Radboud University, The Netherlands
Tobias Marschall	Saarland University, Max Planck Institute for Informatics, Germany
Giancarlo Mauri	University of Milano-Bicocca, Italy
Mickael Mendez	University of Toronto, Canada
Emanuela Merelli	University of Camerino, Italy
Luciano Milanesi	Institute of Biomedical Technologies, CNR, Italy
Bud Mishra	New York University, USA
Marianna Pensky	University of Central Florida, USA
Joao Ribeiro Pinto	Instituto de Engenharia de Sistemas e Computadores Tecnologia e Ciencia, Portugal
Paulo Ribeiro	Universidade NOVA de Lisboa, Portugal
Davide Risso	Cornell University, USA
Riccardo Rizzo	Institute for High Performance Computing and Networking, CNR, Italy
Paolo Romano	Ospedale Policlinico San Martino, Italy

Stefano Rovetta	University of Genova, Italy
Antonino Staiano	Parthenope University of Naples, Italy
Filippo Utro	IBM, USA
Maria Grazia Valsecchi	University of Milano-Bicocca, Italy
Alfredo Vellido	Universitat Politècnica de Catalunya, Spain
Blaz Zupan	University of Ljubljana, Slovenia

Special Session Organizers

Modeling and Simulation Methods for Computational Biology and Systems Medicine

Rosalba Giugno	University of Verona, Italy
Marco Beccuti	University of Turin, Italy
Marzio Pennisi	University of Catania, Italy
Pietro Liò	University of Cambridge, UK
Marco S. Nobile	Eindhoven University of Technology, The Netherlands

Machine Learning in Healthcare Informatics and Medical Biology

Davide Chicco	Peter Munk Cardiac Centre, Canada
Anne-Christin Hauschild	Philipps-Universität Marburg, Germany
Mickael Mendez	University of Toronto, Canada
Giuseppe Jurman	Fondazione Bruno Kessler, Italy
Joao Ribeiro Pinto	Instituto de Engenharia de Sistemas e Computadores Tecnologia e Ciencia, Portugal

Algebraic and Computational Methods for the Study of RNA Behaviour

Emanuela Merelli	University of Camerino, Italy
Stefano Maestri	University of Camerino, Italy
Michela Quadrini	University of Camerino, Italy

Intelligence Methods for Molecular Characterization and Dynamics in Translational Medicine

Andrea Calabria	San Raffaele Telethon Institute for Gene Therapy, Italy

Contents

Intelligence Methods for Molecular Characterization and Dynamics in Translational Medicine

Machine Learning in Healthcare Informatics and Medical Biology

Modeling and Simulation Methods for Computational Biology and Systems Medicine

Computational Intelligence Methods for Bioinformatics and Biostatistics

A Smartphone-Based Clinical Decision Support System for Tremor Assessment

Guillaume Zamora[1], Caro Fuchs[1][(✉)] , Aurélie Degeneffe[2] ,
Pieter Kubben[3] , and Uzay Kaymak[1]

[1] Eindhoven University of Technology, Eindhoven, The Netherlands
{C.E.M.Fuchs,U.Kaymak}@tue.nl
[2] Erasme University Hospital, Brussels, Belgium
[3] Maastricht University Medical Center, Maastricht, The Netherlands

Abstract. Tremor severity assessment is an important element for the diagnosis and treatment decision-making process of patients suffering from Essential Tremor (ET). Classically, questionnaires like the ETRS and QUEST surveys have been used to assess tremor severity. Recently, attention around computerized tremor analysis has grown. In this study, we use regression trees to map the relationship between tremor data that is collected using the TREMOR12 smartphone application with ETRS and QUEST scores. We aim to develop a model that is able to automatically assess tremor severity of patients suffering from Essential Tremor without the use of more subjective questionnaires. This study shows that tremor data gathered using the TREMOR12 application is useful for constructing machine learning models that can be used to support the diagnosis and monitoring of patients who suffer from Essential Tremor.

Keywords: Essential Tremor · Tremor severity · Smartphone application · Clinical decision support

1 Introduction

Tremor, which consists of 'rhythmic and involuntary movements of any body part' [6], is a symptom resulting from a neurological disorder. Often, tremor is associated with Parkinson's disease. However, sometimes the tremor is present without a demonstrable cause. In this case, the patient is diagnosed with Essential Tremor (ET).

Tremor-related diseases affect millions of people around the world, hindering various everyday life tasks, such as holding a glass of water. Tremor severity assessment is an important element for the diagnosis and treatment decision-making process. For decades, tremor severity was measured using qualitative clinical rating scales. Examples of this are the Quality of life in Essential Tremor (QUEST) scoring scale, a survey filled in by the patient that measures the effect of the tremor on the patient's quality of life, and the Essential Tremor Rating Scale (ETRS), in which the assessor evaluates the severity of tremor in rest,

© Springer Nature Switzerland AG 2020
P. Cazzaniga et al. (Eds.): CIBB 2019, LNBI 12313, pp. 3–12, 2020.
https://doi.org/10.1007/978-3-030-63061-4_1

postural and during movement, as well as the patient's performance in line and spiral drawing.

Recently, attention around computerized tremor analysis has grown (see for example [3,12,13,15]).

While previous studies measure tremor using dedicated devices that are expensive and not practical for everyday use, we use the smartphone application called 'TREMOR12' [10], which makes data collection easier and cheaper. In this study, we aim to find the relationship between the more subjective QUEST and ETRS score and smartphone sensor data that measures the patient's tremor, using a machine learning model. When this relationship is known, the model could more objectively determine the severity of the patient's tremor and improve the inter-rater reliability among clinicians.

This work is structured as follows. First, some related studies are described in Sect. 2. Then, the method of the current study is described in Sect. 3: We show how the data is collected, how it is processed, and how the decision support models are developed. In Sect. 4 the results of the study are shown, and in Sect. 5 we draw some final conclusions.

2   Related Work

According to [8], four types of sensors are used to measure tremor for ET patients:

1. *Accelerometers* measure linear acceleration (in g) on a specific axis. Measurements collected using accelerometers are a combination of linear acceleration, gravity, and additive noise. No analytic model has been validated to separate gravity from linear acceleration. Low-pass filtering is used in most studies [8].
2. *Gyroscopes* can be used to assess how fast the angular position or orientation of an object changes with time. Therefore, gyroscopes are able to measure rotation speed (in radians per second) and rotation/orientation (in radians).
3. *Elektromyogram (EMG)* is an electrical receptor of the muscle fibers activity. To collect this data, a fine wire is inserted into a muscle of the patient. EMGs were successfully used by [1] in order to differentiate Parkinson's disease from Essential Tremor.
4. *Force sensors* measure the torque in newton meter (N.m) and angular motion in radians per second. Torque being the force applied to rotate an object around a pivot, or axis. This type of sensor is described as promising but most of these haptic devices are non-portable and expensive.

The characteristics of using each of these sensors for measuring tremor severity are summarized in Table 1. As can be observed, both the accelerometer and gyroscope are affordable sensors that measure tremors in a non-invasive manner. Therefore, in the rest of this section, we will focus on studies that use these two sensors.

In [13], a pilot study is presented to assess the feasibility of using accelerometer data to estimate the severity of symptoms and motor complications in

Table 1. Characteristics of sensors to measure tremor severity.

	Accelerometer	Gyroscope	EMG	Force sensors
Gravitational component	Yes	No	No	No
Signal to noise ratio	Low to high	High	High	High
Size of the sensor	Small	Small	Small	Large
Easy to use	Yes	Yes	Variable	No
Invasive	No	No	Yes	No
Cost	Cheap	Cheap	Cheap	Expensive

patients with Parkinson's disease. Data were gathered from 12 patients using a body sensor network consisting of 8 accelerometers: two on the lower arms (one left, one right), two on the upper arms, two on the lower legs, and two on the upper legs of each patient. Based on this data, a support vector machine classifier was trained to estimate the severity of tremor, bradykinesia, and dyskinesia. The estimates of the classifier were then compared with estimates of trained clinicians. The estimates of the classifier show an average error of 2.5%.

In [4] accelerometer data of 18 patients with Parkinson's Diseases and 5 control patients were collected. The sensors were placed on the wrists of the patients. The data was used to train six classifiers: A random forest, decision tree, nearest neighbor, Bayes, multilayer perceptron, and support vector machine classifier. In this study, the random forest and decision tree have the best performance with an accuracy of 82%.

In a similar study [3] that used data of 7 patients with Parkinson's disease, accelerometer and gyroscope measurements were used to train a least-square-estimation model to assess the severity of rest, postural, and action tremors. The measured tremor severity correlated well with the judgments of a neurologist ($r = 0.98$).

These studies show great potential for computerized tremor assessment, but all use dedicated devices to assess tremor severity. To overcome this, some studies have proposed using smartphone sensors to assess tremor severity (see for example [2,9,11]). However, these studies (and the studied smartphone applications) all focus on tremors caused by Parkinson's disease, which differ significantly from Essential Tremor as can be observed in Table 2. To our knowledge, only [15] used smartphone sensors to assess the severity of Essential Tremor. However, in this study, only the accelerometer of the smartphone is used.

In this work, we would like to replicate the results of previously mentioned studies, but this time for Essential Tremor. To do this, we use the accelerometer and gyroscope of a smartphone and the freely available TREMOR12 application [10].

3 Method

3.1 Data Collection

The data that is used in this study are gathered using the TREMOR12 application [10]. This open-source mobile app was developed by clinicians from the Maastricht University Medical Center, the Netherlands. TREMOR12 samples acceleration, rotation, rotation speed, and gravity, each along three axes and time-stamped in a frequency up 100 Hz. The raw measurement data can be exported as a comma-separated value file for further analysis. The goal of the application is to enable the evaluation of the effect of the treatment of tremor.

The data for this study are collected with a 10-ms sample rate (100 Hz) using an iPhone 6, of which two sensors are used, namely the accelerometer and the gyroscope, both along three axes (x, y, z). The extracted parameters are acceleration (in g) and rotation speed (in radians per second).

Twenty patients (11 men, 9 women) were included in this study. The patients were on average 67 years old at the time of the experiment and were on average suffering from Essential Tremor for 22 years. All these patients were diagnosed with Essential Tremor underwent Deep Brain Stimulation as treatment in the past. At the time of the experiment, these patients had ETRS scores between 10 and 68 and QUEST scores varying between 3 and 68.1.

Essential tremor is most commonly present in the upper limbs for at least 95% of the patients [14]. Therefore, the tremor data for this study is gathered by strapping the measuring devices, in this study iPhones, around the wrists of ET patients. These patients performed five different tests and data was collected on both the right and left wrist of the patient. The following experiments were conducted:

1. **Rest**: The rest tremor is being measured in a position were both forearms and hands are resting on a table. Tremor is measured on both sides for one minute.
2. **Postural 1**: The postural tremor is being measured in a position where both arms are outstretched forward with palms facing down, as depicted in Fig. 1a. Tremor is measured on both sides for one minute.
3. **Postural 2**: The postural tremor is being measured in a position where both arms are in front of the chest with palms facing down, as depicted in Fig. 1b. Tremor is measured on both sides for one minute.
4. **Glass**: The action tremor is measured by raising a cup filled with water from the table, bringing it towards the mouth, and putting the cup back on the table. This measurement was repeated three times.
5. **Finger-nose test**: The action tremor is measured by bringing the index finger to the nose and subsequently bringing the index finger towards the finger of the researcher. This movement was repeated three times.

Table 2. Characteristics of tremor caused by Parkinson's disease and Essential Tremor.

	Parkinson's disease	Essential Tremor
Tremor	Resting	Postural & kinetic
Frequency of tremor	4–6 Hz	7–12 Hz
Presence in hands	>70%	>95%

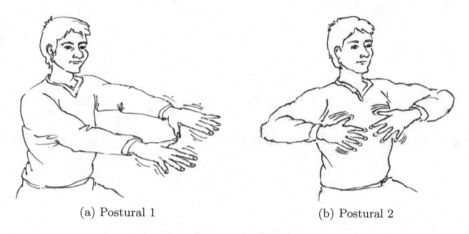

(a) Postural 1   (b) Postural 2

Fig. 1. Body positions for data measuring.

3.2   Data Preparation

Filtering Out Noise. To start and stop the recording of data, the researcher had to tap a button on the screen of the recording device. This creates noise in the beginning and end of the recording as can be seen in Fig. 2. Graphical representations of the data show that this noise typically occurs in the first and last 50 samples of the data recorded for each test. Therefore, the first and last 50 samples of each signal file were discarded.

Another source of noise is voluntary movement, as can be clearly observed in the results of the glass and finger-nose test. An example of this is shown in Fig. 3, where the unprocessed signal is indicated in blue. One can observe the big, low-frequency peaks, that originate from the patient moving his finger from his nose to the researcher's hand and vice versa. As can be seen in the graph, this movement is repeated three times. The noise originating from this voluntary movement is removed by applying an Equiripple Finite Impulse Response (FIR) filter between 7 to 12 Hz, corresponding to Essential Tremor frequencies [6]. In this way, gravitational and motion components are removed from the signals. The result of this processing is indicated in Fig. 3 with a red line.

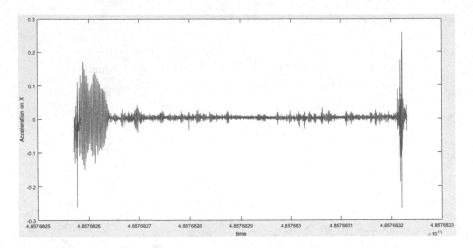

Fig. 2. Tapping the start and stop recording button on the device creates noise in the data.

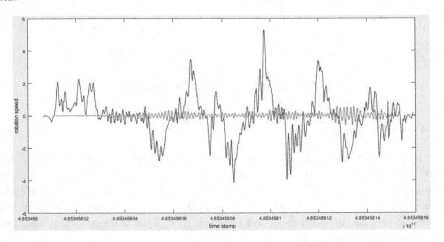

Fig. 3. Patient 19 - finger-nose test - right hand: The signal before (blue) and after (red) applying the filter. (Color figure online)

3.3 Feature Extraction and Selection

From each cleaned signal, various features are extracted in order to represent the time series in single values that can later be used as input values for modeling through machine learning.

Two time-domain features are extracted, namely, i) the root-mean-square (RMS) value of the signal, which is a measure for the signal strength, and ii) the signal period, which represents the average duration of one wave in the signal. The dominant magnitude and frequency are extracted from the frequency domain using the Welch method [16], (with Hamming 2.5 s and 50%

overlapping windows) and Daubechies 8 (D8) wavelet [5]. The extracted features give information about the highest strength peak accumulated over the frequency range and the number of oscillations per second associated. The power growth is extracted from the time-frequency domain in order to represent the power increase over time.

This results in a table with 50 variables (left and right wrist, five tests, and five features per signal), describing the tremor for every patient. Not all extracted features have to be relevant when classifying the severity of the patient's tremor. To distill the most relevant features that should be included in the model, sequential forward selection (SFS) in combination with cross-validation is used.

Missing Data. Unfortunately, not all tests were performed by all patients. The finger-nose test was added to the experiment after the data of the first four patients were already collected. In order to complete the data set, we used spline interpolation to estimate the value of the missing data points.

Modeling. In [4], six different classifiers (a random forest, a decision tree, a nearest neighbor classifier, a naive Bayes classifier, a multilayer perceptron artificial neural network, and a support vector machine) were used to classify the severity of patient's tremor. The best accuracy was obtained by using a decision tree. Therefore, in this study regression decision trees were applied. The trees were constructed using MATLAB's function *fitrtree*.

A regression tree is a machine learning algorithm where the feature dimensions are split in a binary way creating a structure visually similar to trees, with nodes (or leaves). The trees are constructed step by step, each time looking for the best split in a branch that minimizes the tree's overall classification error. The resulting tree is evaluated based on its Mean Absolute Error (MAE), Mean Absolute Percentage Error (MAPE) and Mean Absolute Scaled Error (MASE) using leave-one-out cross-validation.

4   Results

We built regression decision trees that aim to map the features extracted from the smartphone's sensor data to the patient's ETRS and QUEST score. The values for the evaluation criteria for all models can be found in Table 3.

Other studies (such as [3, 12, 13, 15]) that aim to classify tremor severity based on sensor data often rely on the Welch method to extract features from the sensor data and only extract information about the dominant frequency, dominant magnitude, and the signal RMS. To be able to compare our results with these previously constructed models, we developed Model 0 using similar settings.

For all the other models (Table 3, Model 1 - Model 4) feature selection was performed before the models were built. As can be seen in the results for Model 2, this significantly increases the model's performance. Adding the signal period and power growth to this model does not increase the model's performance substantially (Table 3, Model 3). However, using Daubechies 8 (D8) wavelet to extract features from the signal does increase the performance of the resulting models (Table 3, Model 4).

As can be observed from Table 3, the decision trees perform better when mapping to the ETRS than to the QUEST score. This can be explained by the fact that QUEST scores are self-reported by the patient and therefore more subjective, while the ETRS score is assigned based on a trained clinician's observations and should therefore be more consistent.

Table 3. Mean Absolute Error (MAE), Mean Absolute Percentage Error (MAPE) and Mean Absolute Scaled Error (MASE) of the developed models.

Model	Extraction	Features	ETRS			QUEST		
			MAE	MAPE	MASE	MAE	MAPE	MASE
0	Welch	Dominant frequency Dominant magnitude Signal RMS	10.23	34.91	0.64	18.96	119.18	1.04
1	Welch	Dominant frequency Dominant magnitude Signal RMS	7.53	27.58	0.47	9.43	80.42	0.52
2	Welch	Dominant frequency Dominant magnitude Signal RMS Signal period Power growth	7.51	27.14	0.47	8.65	44.25	0.47
3	Wavelet	Dominant frequency Dominant magnitude Signal RMS No feature selection	7.01	26.26	0.44	9.67	51.12	0.53
4	Wavelet	Dominant frequency Dominant magnitude Signal RMS Signal period Power growth	6.41	22.24	0.39	9.01	42.47	0.49

Table 4. Features selected for Model 4 using SFS.

ETRS			QUEST		
Test	Sensor	Feature	Test	Sensor	Feature
Rest	Accelerometer	Power growth	Postural 1	Rotation speed	Dominant frequency
Postural 1	Accelerometer	Dominant magnitude	Postural 2	Accelerometer	Power growth
Postural 1	Rotation speed	Signal period			
Postural 2	Accelerometer	Dominant frequency			
Postural 2	Rotation speed	Dominant magnitude			

Table 4 shows the selected features for Model 4, the best performing model. When evaluating these features, it can be observed that the data collected during postural tests is the most informative, which is congruent with the literature [6] that states that Essential Tremor can mainly be observed during postural phases of movement. Also, it can be observed that adding data from the gyroscope adds valuable information (in addition to the accelerometer as used in [15]) since the rotation speed has proven to be valuable for estimating the ETRS.

5   Concluding Remarks and Discussion

This study shows that tremor data gathered using the TREMOR12 application is useful for constructing machine learning models that can be used to support the diagnosis and monitoring of patients who suffer from Essential Tremor. TREMOR12 is freely available and runs on widely available devices (in this study iPhones), which makes it easy and relatively inexpensive to collect the data.

For this study, we posed the problem as a regression problem. However, the precision of having an exact ETRS or QUEST score might not be very relevant for diagnosing or monitoring the disease. For these purposes it is similarly useful to classify the tremor data in categories, varying from 'light tremor' to 'severe tremor'. Instead of using the ETRS and QUEST score, one might consider using the Fahn score, which divides the ET patients into five classes based on tremor severity.

Alternatively, one could model the mapping between the QUEST and ETRS score and the sensor data using a fuzzy inference system, which is better able to capture the uncertainty that comes from the subjectivity of the QUEST questionnaire. One could for example follow the approach as described in [7].

The data that was used for creating the models for this study contains data of 20 ET patients. All these patients were being treated with deep brain stimulation at the time the tests were performed. This led to relatively low ETRS and QUEST scores, which means that patients with severe tremors were not represented in this data set. To make the models useful for diagnosing new, untreated patients, the data of these patients should be included in the training data set. Also, collecting data from more patients would result in generally more robust models. However, since we were only testing the viability of training machine learning models using tremor data collected through smartphone applications, the relatively small data set of this study was sufficiently large.

Acknowledgments. We would like to thank Pieter van Gorp from the University of Technology Eindhoven and Pierre Lemaire from the Grenoble Institute of Technology for their useful input and support during the development of this research work.

References

1. Breit, S., Spieker, S., Schulz, J., Gasser, T.: Long-term EMG recordings differentiate between parkinsonian and essential tremor. J. Neurol. **255**(1), 103–111 (2008). https://doi.org/10.1007/s00415-008-0712-2

2. Carignan, B., Daneault, J.-F., Duval, C.: Measuring tremor with a smartphone. In: Rasooly, A., Herold, K.E. (eds.) Mobile Health Technologies. MMB, vol. 1256, pp. 359–374. Springer, New York (2015). https://doi.org/10.1007/978-1-4939-2172-0_24

3. Dai, H., Zhang, P., Lueth, T.: Quantitative assessment of parkinsonian tremor based on an inertial measurement unit. Sensors 15(10), 25055–25071 (2015)

4. Darnall, N.D., et al.: Application of machine learning and numerical analysis to classify tremor in patients affected with essential tremor or Parkinson's disease. Gerontechnology 10(4), 208–219 (2012)

5. Daubechies, I.: Ten Lectures on Wavelets, vol. 61. Siam (1992)

6. Elias, W.J., Shah, B.B.: Tremor. JAMA 311(9), 948–954 (2014)

7. Fuchs, C., Wilbik, A., Kaymak, U.: Towards more specific estimation of membership functions for data-driven fuzzy inference systems. In: 2018 IEEE International Conference on Fuzzy Systems (FUZZ-IEEE), pp. 1–8. IEEE (2018)

8. Grimaldi, G., Manto, M.: Neurological tremor: sensors, signal processing and emerging applications. Sensors 10(2), 1399–1422 (2010)

9. Kostikis, N., Hristu-Varsakelis, D., Arnaoutoglou, M., Kotsavasiloglou, C.: A smartphone-based tool for assessing parkinsonian hand tremor. IEEE J. BioMed. Health 19(6), 1835–1842 (2015)

10. Kubben, P.L., Kuijf, M.L., Ackermans, L.P., Leentjes, A.F., Temel, Y.: TREMOR12: an open-source mobile app for tremor quantification. Stereotact. Funct. Neurosurg. 94(3), 182–186 (2016)

11. LeMoyne, R., Mastroianni, T.: Use of smartphones and portable media devices for quantifying human movement characteristics of gait, tendon reflex response, and Parkinson's disease hand tremor. In: Rasooly, A., Herold, K.E. (eds.) Mobile Health Technologies. MMB, vol. 1256, pp. 335–358. Springer, New York (2015). https://doi.org/10.1007/978-1-4939-2172-0_23

12. Pan, D., Dhall, R., Lieberman, A., Petitti, D.B.: A mobile cloud-based Parkinson's disease assessment system for home-based monitoring. JMIR mHealth uHealth 3(1), e29 (2015)

13. Patel, S., et al.: Monitoring motor fluctuations in patients with Parkinson's disease using wearable sensors. IEEE Trans. Inf. Technol. Biomed. 13(6), 864–873 (2009)

14. Poston, K.L., Rios, E., Louis, E.D.: Action tremor of the legs in essential tremor: prevalence, clinical correlates, and comparison with age-matched controls. Parkinsonism Relat. Disord. 15(8), 602–605 (2009)

15. Senova, S., Querlioz, D., Thiriez, C., Jedynak, P., Jarraya, B., Palfi, S.: Using the accelerometers integrated in smartphones to evaluate essential tremor. Stereotact. Funct. Neurosurg. 93(2), 94–101 (2015)

16. Welch, P.: The use of fast fourier transform for the estimation of power spectra: a method based on time averaging over short, modified periodograms. IEEE Trans. Audio Electroacoust. 15(2), 70–73 (1967)

cyTRON and cyTRON/JS: Two Cytoscape-Based Applications for the Inference of Cancer Evolution Models

Lucrezia Patruno[1], Edoardo Galimberti[2,3], Daniele Ramazzotti[4,8],
Giulio Caravagna[5], Luca De Sano[1], Marco Antoniotti[1,7(✉)],
and Alex Graudenzi[1,6,7(✉)]

[1] Department of Informatics, Systems and Communication,
University of Milano-Bicocca, Milan, Italy
{marco.antoniotti,alex.graudenzi}@unimib.it
[2] Department of Computer Science, University of Turin, Turin, Italy
[3] ISI Foundation, Turin, Italy
[4] Department of Pathology, Stanford University, Stanford, CA, USA
[5] The Institute of Cancer Research - ICR, London, UK
[6] Institute of Molecular Bioimaging and Physiology - IBFM of the National Research
Council - CNR, Segrate, Milan, Italy
[7] Bicocca Bioinformatics Biostatistics and Bioimaging (B4) Center,
University of Milano-Bicocca, Milan, Italy
[8] School of Medicine and Surgery, Università degli Studi di Milano-Bicocca,
Monza, Italy

Abstract. The increasing availability of sequencing data of cancer
samples is fueling the development of algorithmic strategies to investigate tumor heterogeneity and infer reliable models of cancer evolution. We here build up on previous works on cancer progression inference from genomic alteration data, to deliver two distinct Cytoscape-
based applications, which allow to produce, visualize and manipulate
cancer evolution models, also by interacting with public genomic and
proteomics databases. In particular, we here introduce cyTRON, a stand-
alone Cytoscape app, and cyTRON/JS, a web application which employs
the functionalities of Cytoscape/JS.

cyTRON was developed in Java. cyTRON/JS was developed in JavaScript
and R.

Keywords: Cancer · Cancer evolution · Mutational graphs · Tumor
phylogenies · Cytoscape

1 Scientific Background

Cancer is a complex disease, whose development is caused by the accumulation
of alterations in the genome. Some alterations may confer a selective advantage

E. Galimberti and L. Patruno—Equal contributors.

© Springer Nature Switzerland AG 2020
P. Cazzaniga et al. (Eds.): CIBB 2019, LNBI 12313, pp. 13–18, 2020.
https://doi.org/10.1007/978-3-030-63061-4_2

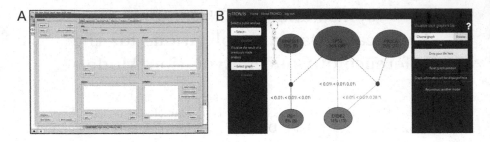

Fig. 1. We show in the figure a view of cyTRON workspace (A) and an example of output model by cyTRON/JS (B).

to cancer cells, and this may result in the expansion of cancer clones. In order to understand how cancer evolves, it is of great importance to understand how such *driver* alterations accumulate over time [5]. This goal can be pursued by reconstructing cancer evolution models, which are graphs that encode the evolutionary history of drivers and their temporal relationships. The reconstruction of such models is a complex task mainly because of two reasons: first, much of the data publicly available from initiatives such as TCGA [https://portal.gdc.cancer.gov/] comes from cross-sectional samples, and hence they lack of temporal information. The second main reason can be found in the heterogeneity of tumors [2,7].

2   Materials and Methods

In order to produce meaningful evolution models, we developed a pipeline, PICNIC [2], which includes the following steps: *i*) identification of homogeneous sample subgroups (e.g., tumor subtypes), *ii*) identification of drivers (i.e., the nodes of the output model), *iii*) identification of mutually exclusive patterns, *iv*) inference of cancer evolution models via distinct algorithms (e.g., [4,6,8]). These algorithms perform the inference of probabilistic graphical models of cancer progression in two steps; first, the assessment of statistical associations between a set of genomic variants is performed and a partially order set (poset) among them is reconstructed by considering their frequencies, i.e., variants at higher frequency can precede variants at lower frequency, and their statistical dependency, i.e., arcs are allowed only among positively dependent variants. Then, a maximum likelihood directed acyclic graph [6] or a maximum score tree is inferred within such poset [4,8].

The PICNIC pipeline was implemented within the TRONCO R suite for TRanslational ONCOlogy [1,3], which was recently employed, for instance, to analyze the largest kidney cancer cohort currently available [10].

However, TRONCO presents two practical limitations: first, it requires at least some basic programming skills due to its underlying R infrastructure; second, TRONCO is not integrated with publicly available genomic databases, hence providing a non-interactive visualization of the output graphs.

Therefore, to improve the practicality, effectiveness, interactivity and diffusion of our framework, we integrated it within Cytoscape, a user-friendly open-source platform for the visualization and manipulation of complex networks [9]. We here present cyTRON, a stand-alone Cytoscape app, and cyTRON/JS a web application which employs the functionalities of Cytoscape/JS, both of which allow to produce, visualize and manipulate cancer evolution models, also by interacting with public genomic databases. Figure 1B shows an example of the output in cyTRON/JS, which exploits Cytoscape/JS to provide an interactive visualization of the evolution model.

cyTRON and cyTRON/JS were designed for two main purposes:

- Providing an *interactive* and *user-friendly* visualization of TRONCO models: while TRONCO R-based graph display is static, cyTRON and cyTRON/JS provide interactive views, which allow to directly retrieve information about genes involved in the study, by accessing widely-used public genome databases.
- Making TRONCO *accessible* to users unfamiliar with R programming: cyTRON and cyTRON/JS provide interfaces that enable the usage of TRONCO respectively from Cytoscape and a Web browser, thus removing the need for users to execute any code in order to complete a whole analysis.

The architecture of both tools can be conceptually defined as follows:

- The *front-end* side is composed of an interface that can be used to:
  1. Select input data for the TRONCO analysis [3]: the input files should be either MAF, GISTIC or user-defined Boolean matrices that contain information about the mutations observed in each sample.
  2. Set the parameters for the inference in order to access most TRONCO capabilities. Users can indicate which driver mutations to include in the analysis, which algorithm among those implemented in TRONCO to use for the reconstruction and the algorithm's corresponding parameters.
  3. Visualize cancer evolution models and dynamically interact with the result: for instance, by clicking on the genes of the output graph, it is possible to retrieve the information available on public genomic databases. cyTRON gives access to gene information through databases such as Ensembl[1] and Entrez[2], that are accessible through the Cytoscape interface.

     In cyTRON/JS the data displayed for each node are retrieved from the Entrez database Gene[3] using E-Utils, an API provided by the National Center for Biotechnology Information.
- The *back-end* side includes the communication channel with R. For cyTRON, a Java bridge with R is built by means of rJava. Instead, for cyTRON/JS it is based on js-call-r, a Node.js package which collects the data and parameters set by the user, encodes them in JSON and sends them to R.

---

[1] http://www.ensembl.org/index.html.

[2] https://www.ncbi.nlm.nih.gov/search/.

[3] https://www.ncbi.nlm.nih.gov/gene/.

Then, R commands are transparently executed in order to perform any specific step of the analysis by TRONCO.

Figure 1 shows a view of cyTRON workspace (left) and an example of output model by cyTRON/JS (right). In order to choose between the two tools, users should take into consideration the data and the type of analysis they need to carry out. In particular, since cyTRON/JS is a web application, it is readily accessible from any device and computations are carried out on the back-end side. This feature is useful in case a user needs to carry out a computational-expensive analysis. However, cyTRON is more complete with respect to all the functionalities implemented in TRONCO: for example, it implements also the option of testing hypothesis on mutations through the algorithm Capri [6].

A Note on How cyTRON/JS *Handles Uploads.* In order to perform its analysis, cyTRON/JS requests a pro-forma registration on the site. This is only to identify the datasets subsequently loaded. cyTRON/JS does not require, or record any email or other identifying information. The datasets are uploaded by the user under his responsibility; that is, cyTRON/JS assumes that the user has all the rights to upload and share the data, which is used "as-is" and not modified by cyTRON/JS. In other words, cyTRON/JS does not claim to be GDPR-compliant or to follow the *data stewardship* guidelines put forth by FAIR [11].

3   Case Study

We present a case study on mutation data of prostate cancer, which can be downloaded from cyTRON/JS's GitHub repository[4]. This dataset was downloaded from TCGA, and it is composed of a MAF file which contains data about SNPs, Insertions and Deletions and a GISTIC file which contains data about CNVs.

For the first part of the analysis, both cyTRON and cyTRON/JS provide an interface to select the input files and, optionally, to choose which genes and which samples to consider in the analysis. The last two options can be exploited to carry out an analysis over a specific cancer subtype and to restrict the inference only to driver mutations. For this case study we selected only those samples that present a mutation on the ERG gene, and we restricted the analysis to driver mutations.

Finally, we ran the inference with CAPRI algorithm: through the user interface it is possible to indicate the heuristic search to be performed, enable the estimation of error rates through bootstrap and specify the regularization criterion for the likelihood estimation. Once the inference was performed, we visualized the result in cyTRON/JS: we present an example of how cyTRON/JS can serve as a means to identify associations among genes. Indeed, by clicking on a node of the graph users can read additional information about the corresponding gene and they can exploit this information to find which genes are involved in the same processes.

---

[4] https://github.com/lucreziaPatruno/cyTRON_JS/tree/master/examples.

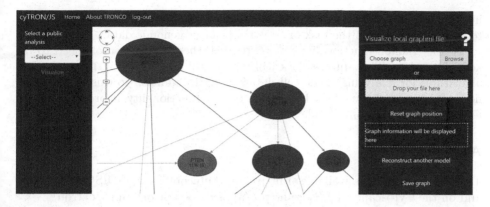

Fig. 2. Example of the visualization provided by cyTRON/JS: this figure displays a portion of the progression model reconstructed from prostate cancer mutation data.

Figure 2 displays a portion of the output graph that resulted from the case study. The observations that stood out while exploring the progression model are the following:

- There is an oriented edge which connects RB1 to TP53. The protein encoded by the former serves as a negative regulator for cell cycle, and the one encoded by the latter has a role in inducing cell cycle arrest, apoptosis and senescence.
- Another oriented edge connects RB1 to PTEN. The latter encodes a protein which is a negative regulator of the AKT/PKB signaling pathway.

Therefore through the cyTRON/JS interface it was possible to identify two connected nodes that represent genes which both have a role in cell cycle.

Even though these might be trivial associations, this is an example of how this tool can be exploited to better understand the role of genes that are connected in the graph, possibly finding interesting associations.

Given the fact that some researchers may want to access cyTRON/JS only to explore some already constructed model, the web application contains a section which allows users to access without authenticating, in order to visualize one of their local **graphml** files.

4   Conclusion and Future Work

TRONCO is an R package that implements state-of-the-art algorithms for the inference of cancer evolution models with the ultimate goal of understanding the evolutionary trajectories driving tumor evolution. In such a multidisciplinary domain, where computer scientists actively cooperate with biologists, being capable of visually understanding the data is crucial to both parties. In order to effectively allow the use of TRONCO, we here presented cyTRON and cyTRON/JS, two Cytoscape-based applications which translate many of TRONCO functionalities into `Cytoscape`. Our effort aims at designing user-friendly and accessible tools to support the user in the task of exploring cancer genomic data.

As the `TRONCO` functionalities are extended, these two new tools need to be kept up to date. Thus, future work will be focused on integrating `TRONCO`'s new algorithms for analyzing single cell datasets [8]. In addition to this, `cyTRON/JS` needs to be extended with the hypothesis testing functionality, in order to enable users to carry out more complex analysis.

Availability

The code for `cyTRON` is available at https://github.com/BIMIB-DISCo/cyTRON and on the Cytoscape App Store http://apps.cytoscape.org/apps/cytron.

The source code for `cyTRON/JS` is available at https://github.com/BIMIB-DISCo/cyTRON-js and the tool itself is accessible from https://bimib.disco.unimib.it/cytronjs/welcome.

Acknowledgments. `cyTRON` and `cyTRON/JS` were developed within the Google summer of Code program, in collaboration with the National Resource for Network Biology (NRNB – www.nrnb.org); we thank Alexander Pico and Kristina Hanspers of Gladstone Institutes of San Francisco, CA, USA, for their support. The authors declare no conflicts of interest.

References

1. Antoniotti, M., et al.: Design of the TRONCO bioconductor package for translational oncology. R J. **8**(2), 39–59 (2016)
2. Caravagna, G., et al.: Algorithmic methods to infer the evolutionary trajectories in cancer progression. Proc. Nat. Acad. Sci. **113**(28), E4025–E4034 (2016)
3. De Sano, L., et al.: TRONCO: an R package for the inference of cancer progression models from heterogeneous genomic data. Bioinformatics **32**(12), 1911–1913 (2016)
4. Loohuis, L.O., et al.: Inferring tree causal models of cancer progression with probability raising. PLoS ONE **9**(10), e108358 (2014)
5. Nowell, P.C.: The clonal evolution of tumor cell populations. Science **194**(4260), 23–28 (1976)
6. Ramazzotti, D., et al.: CAPRI: efficient inference of cancer progression models from cross-sectional data. Bioinformatics **31**(18), 3016–3026 (2015)
7. Ramazzotti, D., Graudenzi, A., Caravagna, G., Antoniotti, M.: Modeling cumulative biological phenomena with Suppes-Bayes causal networks. Evol. Bioinform. **14**, 1176934318785167 (2018)
8. Ramazzotti, D., Graudenzi, A., De Sano, L., Antoniotti, M., Caravagna, G.: Learning mutational graphs of individual tumour evolution from single-cell and multiregion sequencing data. BMC Bioinformatics **20**(1), 210 (2019)
9. Shannon, P., et al.: Cytoscape: a software environment for integrated models of biomolecular interaction networks. Genome Res. **13**(11), 2498–2504 (2003)
10. Turajlic, S., et al.: Deterministic evolutionary trajectories influence primary tumor growth: TRACERx renal. Cell **173**(3), 595–610 (2018)
11. Wilkinson, M.D., et al.: The FAIR Guiding Principles for scientific data management and stewardship. Sci. Data **3**(1), 160018 (2016)

Effective Use of Evolutionary Computation to Parameterise an Epidemiological Model

Ryan Mitchell[1], David Cairns[1] iD, Dalila Hamami[2] iD, Kevin Pollock[3] iD, and Carron Shankland[1]([⊠]) iD

[1] Computing Science and Mathematics, University of Stirling, Stirling, Scotland
carron.shankland@stir.ac.uk
[2] Université Abdelhamid Ibn Badis Mostaganem, Mostaganem, Algeria
[3] Infection Prevention and Control, Inverness College,
University of the Highlands and Islands, Inverness, UK

Abstract. Predictive epidemiological models are able to be used most effectively when they have first been shown to fit historical data. Finding the right parameters settings for a model is complex: the system is likely to be noisy, the data points may be sparse, and there may be many inter-related parameters. We apply computational intelligence and data mining techniques in novel ways to investigate this significant problem.

We construct an original computational model of human papilloma virus and cervical intraepithelial neoplasia with the ultimate aim of predicting the outcomes of varying control techniques (e.g. vaccination, screening, treatment, quarantine). Two computational intelligence techniques (genetic algorithms and particle swarm optimisation) are used over one-stage and two-stage optimisations for eight real-valued model parameters. Rigorous comparison over a variety of quantitative measures demonstrates the explorative nature of the genetic algorithm (useful in this parameter space to support the modeller). Correlations between parameters are drawn out that might otherwise be missed. Clustering highlights the uniformity of the best genetic algorithm results.

Prediction of gender-neutral vaccination with the tuned model suggests elimination of the virus across vaccinated and cross-protected strains, supporting recent Scottish government policy. This preliminary study lays the foundation for more widespread use of computational intelligence techniques in epidemiological modelling.

Keywords: Genetic algorithm · Particle swarm optimisation · Epidemiology · Human papilloma virus · K-means clustering

1 Scientific Background

Computational models are increasingly used to investigate complex biological systems such as disease spread and the effects of interventions such as vaccination and treatment. Choosing parameters for these models can be challenging.

© Springer Nature Switzerland AG 2020
P. Cazzaniga et al. (Eds.): CIBB 2019, LNBI 12313, pp. 19–32, 2020.
https://doi.org/10.1007/978-3-030-63061-4_3

A small change to a parameter value or the model structure can be the difference between a large outbreak, endemic disease, or elimination of the disease. Computational intelligence approaches are highly suitable tools to support modellers to fit a model to historical data, as shown in our previous work [1], and provide a base for future predictions. We compare and contrast two such tools (genetic algorithm (GA) and particle swarm optimisation (PSO)) in a realistic case study of international importance. In addition, we extend our use of clustering techniques with process algebra [2] to investigate the quality of the results of the optimisation.

The Human Papillomavirus (HPV) is a sexually transmitted infection which affects the skin and moist membranes of humans. There are over 200 strains of this virus which are split into two subcategories known as oncogenic (high-risk) and non-oncogenic (low-risk). Oncogenic strains of this virus play a pivotal role in the development of various cell abnormalities including tumors [3]. Almost all cases of cervical cancer in women are attributed to high-risk HPV strains [4]. The World Health Organisation has made a global call to eliminate cervical cancer worldwide [5], driven through better understanding of HPV control measures.

This paper contributes to that understanding by presenting computational intelligence and data mining approaches to optimising a model of HPV and pre-cancerous stages. The model is fitted to Scottish data to demonstrate the techniques. According to Kavanagh et al. [6] 80 vaccine against those strains since September 1st 2008 in Scotland. This programme has reduced HPV 16 & 18 prevalence from 30% to 4% in females [6]. The bivalent vaccine provides cross-protection against strains HPV 31, 33 & 45; however, that protection may reduce over time. Predictive modelling is used to investigate equal vaccination policies and the effect of waning cross-protective immunity.

As a starting point, we use a real-valued Genetic Algorithm (GA) as a general-purpose heuristic solver and Particle Swarm Optimisation (PSO) as an alternative solver that is useful for searching continuous parameter spaces. The model has eight parameters, which is a modest challenge for the chosen optimisation techniques but has produced some unexpected observations of the correlations between the parameters. The relationships between the optimal solutions within and between each approach are further explored through the use of data mining (clustering).

We use the optimised model to predict future outcomes of variants of the vaccination programme. We predict that a gender-neutral vaccination program can eradicate high-risk oncogenic strains of HPV even where immunity to those strains wanes. This work establishes a framework for applying computational intelligence to process algebra modelling, with effective and rigorous approaches to evaluating the resulting solution set.

2   Materials and Methods

2.1   Tools

The Evolving Process Algebra toolkit (EPA) [1] was developed with the goal of bridging the gap between evolutionary computation and the formal modelling domain, specifically using process algebra for modelling due to its parsimony and suitability for biological systems of interacting components. EPA is built on the ECJ Evolutionary Computation toolkit and the Bio-PEPA (deterministic or stochastic) simulation engine [7]. For this study, EPA was extended with a parameter sweep component and a particle swarm optimisation (PSO) module, complementing the existing GA approach to optimisation of numeric model parameters. EPA is able to optimise model structure and numeric parameters [1]. Here, the core aspects of the model are standard and only numeric parameters need to be optimised.

2.2   Model

The HPV model is a classic SEIR (Susceptible, Exposed, Infected, Recovered (Immune)) compartmental model [8] extended with binary gender and the potential precursor stages of cervical cancer (three stages of cervical intraepithelial neoplasia (CIN). Epidemiological studies [6,9] informed model development. Figure 1 shows the compartments of the model and the available transitions between compartments. The Bio-PEPA code for the model is archived online [10].

There are over 200 strains of HPV. For this investigation our model groups together the two oncogenic strains HPV16 and HPV18 which are targeted by the bivalent vaccine. The cross-protected strains HPV 31, 33 and 45 (some level of protection is conferred by the bivalent vaccine) are identified as "other" in the model. Further strains of HPV have been omitted.

Vaccination has been added in a staged manner via a rolling vaccination programme for girls aged 12–13 and a targetted catch-up programme for older girls (14–18) (who may have already been exposed to HPV strains, so vaccination may be unsuccessful). The catch-up programme ran for 3 years; in 2014, the eligible routine age for vaccinations was amended to 11–12 year olds for logistical purposes. The model gradually increases both vaccine takeup and vaccine efficacy to capture these features. The timescale is one simulation tick per day, therefore rate constants are expressed as daily rates. The population is open, with births, deaths and immigration. The birth rate is chosen to balance up the death rate in the Scottish population (keeping the population constant for the duration of the study). Immigration (average number of imported infections per year) is given by the standard formula $0.02\sqrt{N}$, where N is the total population [13]. For optimisation, the simulation time is seven years, matching the data of Kavanagh et al. [6]. The vaccine is introduced after the first simulation year.

HPV is a sexually-transmitted infection. Couplings are simplified as being female-female, female-male, male-female, and male-male, in proportions reflecting sexual orientation demographics in Scotland [11]. Coupling preference is not

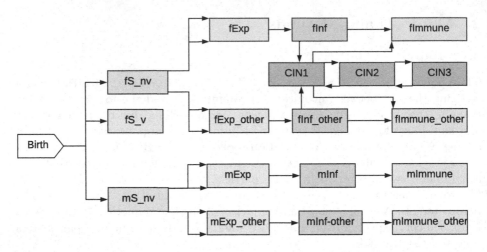

Fig. 1. Schematic of the HPV model showing compartments and routes between compartments. Birth is a source of new individuals. The compartments S, Exp, Inf, Immune are prefixed f (females) and m (males). The suffix _v or _nv relates to vaccinated or not vaccinated. HPV 16 and 18 are the Exp → Inf → Immune strands, and HPV 31, 33 and 45 have the suffix other. Three pre-cancerous stages CIN1, CIN2, CIN3 affect females only.

fixed, reflecting fluidity of orientation and behaviours. This simplification is a pragmatic modelling choice suiting the Markov chain semantics of Bio-PEPA. The two routes from susceptible to exposed in Fig. 1, e.g. fS_nv to fExp reflect that the infection can come from a female or a male partner.

HPV can be described as using frequency-dependent transmission: the number of infectious contacts an individual can make in a single day is proportional to SI/N, where S and I are the number of susceptible and infected individuals respectively and N is the total population. We assume an average of 3.5 days between exposure and becoming infectious [12].

2.3 Optimisation

The rates of progression and regression between the CIN stages is unknown, therefore these were targeted for optimisation together with the contact rate (Cr) and the rate of developing immunity to strains 16 & 18 (Immunity). Domain knowledge was used to set appropriate upper and lower bounds for these rates. The target data was extracted from Kavanagh et al. [6] (HPV, 7 years population sampling) and Pollock et al. [9] (CIN, 5 years population sampling).

Four separate experiments were carried out:

1. two-stage optimisation where infection parameters Cr, Immunity were obtained by a parameter sweep followed by CIN parameters CIN1, CIN1r, CIN2, CIN2r CIN3, CIN3r obtained using a GA,
2. two-stage optimisation as 1. but using PSO,

3. one-stage optimisation of all eight unknown parameters using a GA,
4. one-stage optimisation as 3. using PSO.

Our goal is to explore the potential use of GA and PSO with algorithm parameters that allow for a balance of exploration versus exploitation and that could be completed on our local compute cluster in a reasonable time (e.g. 1 day). The GA used a population of 100 individuals over 100 generations, with tournament selection (5% tournament pool), 1% elitism, 1-point crossover (100%), 5% mutation rate, and generational replacement. The PSO used 100 epochs, 100 particles, 0.728844 inertia, 1.49 cognitive component and 1.49 social component. For this comparison standard parameter settings were used. Ideally we would adjust the GA/PSO parameters to make the search process more efficient. We have moderated for this by using long runs and repeating the runs. Given the number of independent valid solutions we generated that converged to similar fitness scores, we can be reasonably confident that our solution set is appropriate. Optimising the algorithm parameters first may have resulted in quicker convergence but this had to be balanced with the time taken to find these best-case algorithm parameters. If our solution set had been more diverse and had not been a good fit to the data then we would have refined our algorithm parameters.

For both the GA and the PSO, inputs to be optimised were internally represented by real values as used by the HPV model. Prevalence data for HPV in Scotland [6] provides relevant target values at fixed points over time. For both optimisers, fitness was calculated as the Euclidean distance score between the data and the equivalent predicted model values. This was chosen as a straightforward standard linear measure for this data. Fixed data points in years 2009, 2010, 2012, 2013 and 2015 were used. See Figs. 2 and 3 for the fixed point values and a comparison trace produced by a high fitness simulation. For each algorithm, results were collated over 100 independent runs (i.e. 10,000 solutions), producing 100 top solutions. From this collated subset of 100 solutions, the top 20 were analysed.

3   Results

3.1   Parameter Optimisation

To illustrate optimisation results against historical data, the best match trace is shown in Figs. 2 and 3 (GA two-stage results). Mean, median and standard deviation for all eight parameters across twenty best-fit solutions are shown in Table 1, comparing GA and PSO results, and two- and one-stage processes.

Results are largely equivalent in terms of fit but there is considerable variation in some parameter values in these solutions. Distribution and correlation information (Figs. 4, 5, 6 and 7) draw out the model's susceptibility to variance in these parameters. Where there is a strong correlation, the model produces a consistent response and is sensitive to a parameter's value. Where there is little to no correlation, the model is insensitive to the parameter with respect to overall fitness. Figs. 4, 5, 6 and 7 illustrate the relationship between fitness and the other variables for GA: two-stage, GA: one-stage, PSO: two-stage and PSO: one-stage respectively. These figures are quite complex, so we explain the format here, and repeat the main features in the captions to aid the reader.

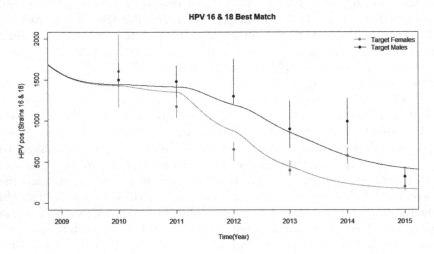

Fig. 2. A single, best-match stochastic simulation of HPV 16 & 18 infections parameterised from the two-stage process showing the data points and 95% confidence intervals of the original data

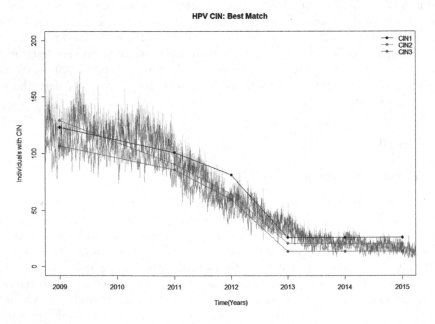

Fig. 3. A single, best-match stochastic simulation of CIN stages parameterised from the two-stage process

Table 1. Optimised parameter values with mean and standard deviation, comparing two- and one-stage, and GA and PSO (see also Figs. 4, 5, 6 and 7 diagonals).

	Bounds		Two Stage			
Parameter	Lower	Upper	GA		PSO	
Cr	0.03	0.06	0.0420	±0.00045	0.0420	±0.00045
Immunity	0.00014	0.0204	0.0028	±0.00003	0.0028	±0.00003
CIN1	0.01	0.90	0.65	±0.206	0.88	±0.019
CIN2	0.01	0.90	0.02	±0.013	0.54	±0.192
CIN3	0.01	0.90	0.45	±0.288	0.79	±0.036
CIN1r	0.01	0.90	0.39	±0.263	0.47	±0.286
CIN2r	0.01	0.90	0.39	±0.212	0.79	±0.037
CIN3r	0.01	0.90	0.46	±0.254	0.50	±0.286

	Bounds		One Stage			
Parameter	Lower	Upper	GA		PSO	
Cr	0.03	0.06	0.0500	±0.00005	0.049	±0.00065
Immunity	0.00014	0.0204	0.0015	±0.00045	0.003	±0.00004
CIN1	0.01	0.90	0.90	±0.127	0.90	±0.001
CIN2	0.01	0.90	0.01	±0.007	0.34	±0.322
CIN3	0.01	0.90	0.51	±0.238	0.90	±0.002
CIN1r	0.01	0.90	0.46	±0.224	0.59	±0.323
CIN2r	0.01	0.90	0.49	±0.223	0.80	±0.227
CIN3r	0.01	0.90	0.58	±0.268	0.53	±0.265

The x-axes labels alternate on the top and the bottom, while the y-axes labels alternate on the left and the right. The variables under optimisation are labelled along the diagonal from top left to bottom right. Each position on this diagonal includes a histogram of the named variable's distribution of values in the solution set of top twenty results. For example, the distribution of CIN1r is skewed to the left in Fig. 4 but to the right in Figs. 5, 6 and 7.

The bottom left section of the figures show scatter plots of pairs of variables. Track vertically and horizontally to locate the relevant pair: for example, in Fig. 4 the scatter plot second row from top, first column from left shows strong correlation of fitness against CIN1.

The upper right side of the figure shows the corresponding correlation values for the scatter plots from the lower left section. For example, in Fig. 4 the plot of CIN2 against CIN2r (5th row, 4th column) shows a strong correlation, and this matches a correlation value of 1.00 (4th row, 5th column).

Figures 4, 5, 6 and 7 highlight robustness, i.e. likely model response to small changes in parameter values. While both the GA and PSO approaches have some variables with low pairwise correlations, the GA had slightly better variable to fitness correlation values. Results were broadly similar for the GA across the 1-stage and 2-stage tasks: adding extra variables did not significantly vary the results. In contrast, the PSO seems more sensitive to the Cr and Immunity parameters in the 2-stage process.

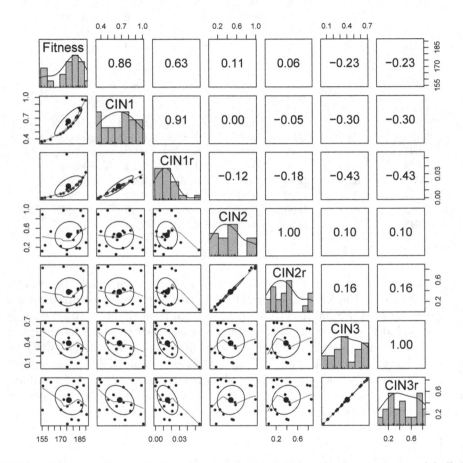

Fig. 4. GA parameter correlations: two-stage. Variables under optimisation are labelled along the diagonal from top left to bottom right. Each position on this diagonal includes a histogram of the named variable's distribution of values. The bottom left section of the figure shows scatter plots of pairs of variables. The upper right side of the figure shows the corresponding correlation values for those scatter plots. The x-axes labels alternate on the top and the bottom, while the y-axes labels alternate on the left and the right.

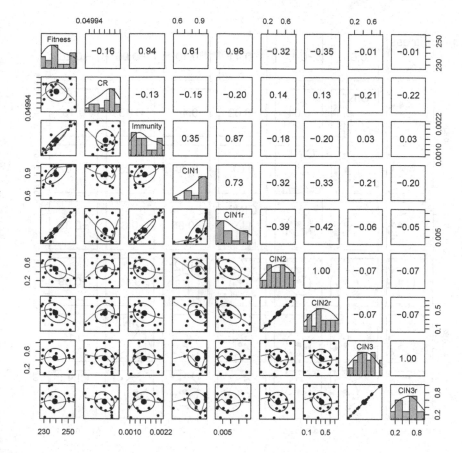

Fig. 5. GA parameter correlations: one-stage. Variables under optimisation along the diagonal. Bottom left shows scatter plots of pairs of variables. Upper right side shows the corresponding correlation values. X-axes labels alternate top and bottom. Y-axes labels alternate on left and right.

3.2   Clustering the Optimal Solutions

Several experiments were carried out using k-means clustering to further investigate the relationships between the produced optimal solutions. Clustering was applied to analyse the stability of the GA results only, and of the PSO results only. Finally, clustering was applied to GA and PSO results across the 1-stage or 2-stage process to demonstrate which technique produced more stable solutions. Values of k from 2–5 were tried. Euclidean distance was used as the distance function, with k = 3 producing optimal clusters. Distribution of data points across clusters were monitored to provide insight into how k-means splits the data for each k and for each optimisation approach. Cluster sizes are more

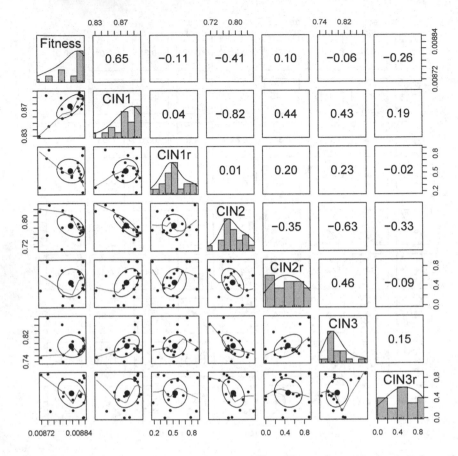

Fig. 6. PSO parameter correlations: two-stage. Variables under optimisation along the diagonal. Bottom left shows scatter plots of pairs of variables. Upper right side shows the corresponding correlation values. X-axes labels alternate top and bottom. Y-axes labels alternate on left and right.

unequal for k = 3. For the other values 2, 4 and 5 the number of instances are more equally distributed across clusters and therefore do not provide any additional information.

Table 2 shows the results of using k-means clustering (k = 3) on the combined GA and PSO results for 1-stage or 2-stage optimisation. Clustering across the approaches highlights the similarity of solutions produced within each method. All 20 GA results are in cluster 0 (1-stage optimisation). All 20 PSO instances are in cluster 0 (2-stage optimisation), highlighting sensitivity to Cr, Immunity.

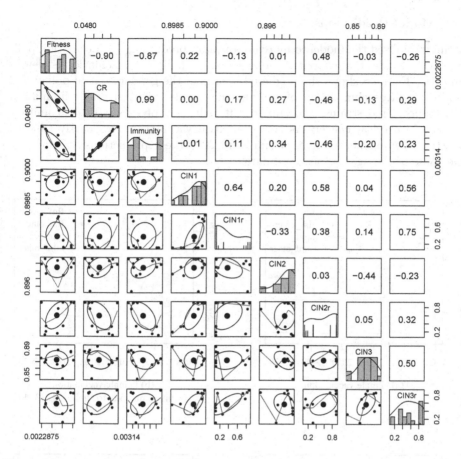

Fig. 7. PSO parameter correlations: one-stage. Variables under optimisation along the diagonal. Bottom left shows scatter plots of pairs of variables. Upper right side shows the corresponding correlation values. X-axes labels alternate top and bottom. Y-axes labels alternate on left and right.

Table 2. Clustering of GA and PSO solutions (1-stage left, 2-stage right).

	Centroids							
	1-stage GA&PSO solutions				2-stage GA&PSO solutions			
Attribute	All	Cluster0	Cluster1	Cluster2	All	Cluster0	Cluster1	Cluster2
Cr	0.0494	0.05	0.0484	0.0496	0.042	0.042	0.042	0.042
Immunity	0.0023	0.0015	0.0032	0.0032	0.0028	0.0028	0.0028	0.0028
CIN1	0.90	0.90	0.90	0.90	0.76	0.56	0.81	0.88
CIN1r	0.17	0.0083	0.31	0.38	0.27	0.0092	0.0266	0.51
CIN2	0.71	0.51	0.90	0.90	0.62	0.41	0.49	0.79
CIN2r	0.53	0.46	0.75	0.37	0.43	0.35	0.40	0.49
CIN3	0.64	0.49	0.75	0.87	0.58	0.47	0.16	0.78
CIN3r	0.56	0.58	0.46	0.64	0.48	0.57	0.19	0.51

3.3 Model Predictions

Having optimised the model to match data [6,9], predictions of vaccination policies can be made. For example, a gender-neutral policy of vaccination is clearly
equitable. Figure 8 shows the predicted results of the gender-neutral policy and
how it produces elimination of the vaccine-specific virus strains, elimination
of cross-protected strains, and subsequent reduction of CIN with concomitant
decrease in cervical cancer. As gender-neutral vaccination was partly (due to
COVID-19) implemented in Scotland from 2019 these predictions will soon be
able to be tested against observed data.

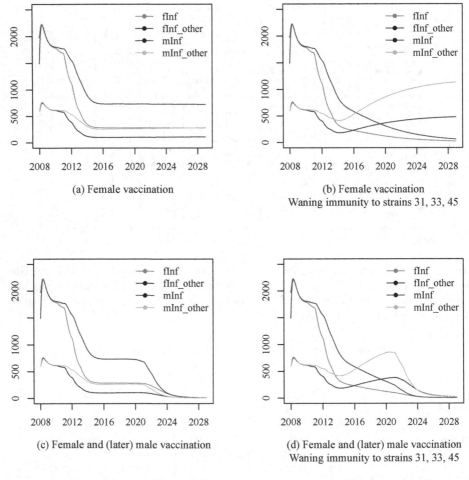

(a) Female vaccination

(b) Female vaccination
Waning immunity to strains 31, 33, 45

(c) Female and (later) male vaccination

(d) Female and (later) male vaccination
Waning immunity to strains 31, 33, 45

Fig. 8. Predicted impact on HPV 16 & 18, and cross-protected strains HPV 31, 33 &
45 generated by Ordinary Differential Equation-based simulation. Female vaccination
from 2008 (all plots). Equal vaccination from 2019 (c, d). Waning immunity to cross-
protected strains (b, d)

With no vaccination, the model suggests half of the population will be infected with HPV strains 16 & 18, and 31, 33 & 45. Female-only vaccination, shown in Fig. 8(a), reduces HPV 16 & 18 prevalence to 6% (female) and 15% (male) of the population. In contrast, Fig. 8 (c) shows these strains are virtually eliminated with equal vaccination. Figure 8(a, c) shows that strains HPV 31, 33 & 45 also reduce due to cross-protection from the bivalent vaccine. If that cross-protection wanes over time, e.g. Fig. 8(b) shows 9.4 years protection, then strains 31, 33 & 45 become dominant in a female-only vaccination regime with 10% females and 23% males infected. Equal vaccination eliminates these strains, shown in Fig. 8(d).

4   Conclusion

Two computational intelligence techniques were used to refine a hand-built model of HPV and subsequent CIN stages. We originally took a two-stage approach with two different optimisation algorithms to cross-compare results and parameter variance. A further one-stage 8-parameter optimisation run for the GA and PSO approaches was performed at the end of the study to investigate if this would lead to similar results. Clustering was used to provide further insight into the results generated.

Comparing GA performance to PSO performance, we see, as expected, the GA is more explorative (slightly wider range of fit results in Table 1 and Figs. 4 and 5) and a good general-purpose heuristic. The one-stage process provides strong correlations between CIN progression and regression parameters. The GA pulled out an interesting and unexpected strong positive correlation between Immunity and CIN1r: the state fImmune can be reached either directly from fInf, or via CIN1. Clustering shows the similarity of top GA results for the one-stage process.

For PSO the solutions (Figs. 6 and 7) showed no clear correlations between parameters except Cr and Immunity in the one-stage optimisation. Variance was lower overall in PSO results between one-stage and two-stage experiments. Clustering shows that the top PSO results are similar for the two-stage process: PSO gives a more uniform solution set when additional constraints are applied.

In future work, a range of algorithm parameters could be explored to improve GA or PSO performance. GA performance could most obviously be improved by running for longer, and decreasing mutation as the number of generations increases. The current work sets a baseline for such experiments.

The refined model can now be used to make computational predictions from health care data sets relating to gender policies on the bivalent vaccine. Varying vaccine efficacy or uptake would determine potential critical vaccination thresholds to ensure long-term eradication of relevant cancers. To calibrate our model we have used data relating to HPV in Scotland; however, due to a worldwide effort to eliminate cervical cancer, several data sets are available to validate such models. Future work will develop our model to apply to a range of data sets, testing predictions across vaccination scenarios, allowing more robust statistical analysis of the results. This would provide a contrast with the results of Simms et al. [14].

Much further predictive work is possible if the model is extended, for example, with additional HPV strains and different vaccines. In addition, there is evidence that HPV vaccination for males can reduce head and neck cancers. Given appropriate data, it would be of value to show the impact of vaccination and screening on both males and females with respect to a variety of cancers.

References

1. Marco, D., Cairns, D., Shankland, C.: Optimisation of process algebra models using evolutionary computation. In: 2011 IEEE Congress of Evolutionary Computation (CEC), pp. 1296–1301 (2011)
2. Hamami, D., et al.: Improving process algebra model structure and parameters in infectious disease epidemiology through data mining. J. Intell. Inf. Syst. **52**(3), 477–499 (2019). https://doi.org/10.1007/s10844-017-0476-1
3. Faridi, R., et al.: Oncogenic potential of Human Papillomavirus (HPV) and its relation with cervical cancer. Virol. J. **8**, 269 (2011). BioMed Central
4. Bosch, F.X., Lorincz, A., Munoz, N., Meijer, C.J.L.M., Shah, K.V.: The causal relation between human papillomavirus and cervical cancer. J. Clin. Pathol. **55**, 244–265 (2002). https://doi.org/10.1136/jcp.55.4.244
5. Brisson, M., Drolet, M.: Global elimination of cervical cancer as a public health problem. Lancet Oncol. **20**(3), 319–321 (2019)
6. Kavanagh, K., Pollock, K.G., et al.: Changes in the prevalence of human papillomavirus following a national bivalent human papillomavirus vaccination programme in Scotland: a 7-year cross-sectional study. Lancet Infect. Dis. **17**(12), 1293–1302 (2017)
7. Ciocchetta, F., Duguid, A., Gilmore, S., Guerriero, M.L., Hillston, J.: The Bio-PEPA Tool Suite (2009). http://homepages.inf.ed.ac.uk/jeh/Bio-PEPA/
8. Anderson, R., May, R.: Population biology of infectious-diseases. Nature **280**, 361–367 (1979)
9. Pollock, K.G., Kavanagh, K., et al.: Reduction of low- and high-grade cervical cancer abnormalities associated with high uptake of the HPV bivalent vaccine in Scotland. BJC **111**, 1824–1830 (2014)
10. System Dynamics model archive. www.cs.stir.ac.uk/SystemDynamics/models/
11. Sexual Orientation Demographics. Scottish Government, December 2018
12. Schiller, J.T., et al.: Current understanding of the mechanism of HPV infection. Gynecol. Oncol. **118**(1 Suppl), S12–7 (2010)
13. Finkenstädt, B.F., Keeling, M., Grenfell, B.T.: Patterns of density dependence in measles dynamics. Proc. R. Soc. B **265**, 753–762 (1998)
14. Simms, K., et al.: Impact of scaled up human papillomavirus vaccination and cervical screening and the potential for global elimination of cervical cancer in 181 countries, 2020–99: a modelling study. Lancet Oncol. **20**(3), 394–407 (2019). https://doi.org/10.1016/S1470-2045(18)30836-2

Extending Knowledge on Genomic Data and Metadata of Cancer by Exploiting Taxonomy-Based Relaxed Queries on Domain-Specific Ontologies

Eleonora Cappelli[1], Emanuel Weitschek[2], and Fabio Cumbo[3,4]

[1] Department of Engineering, University of Roma Tre, Via della Vasca Navale 79/81, 00146 Rome, Italy
`eleonora.cappelli@uniroma3.it`
[2] Department of Engineering, Uninettuno University, 00186 Rome, Italy
`emanuel.weitschek@uninettunouniversity.net`
[3] Institute for System Analysis and Computer Science "Antonio Ruberti", National Research Council of Italy, Via dei Taurini 19, 00185 Rome, Italy
[4] Department of Cellular, Computational, and Integrative Biology (CIBIO), University of Trento, Via Sommarive 9, Povo, 38123 Trento, Italy
`fabio.cumbo@unitn.it`

Abstract. The advent of Next Generation Sequencing (NGS) technologies and the reduction of sequencing costs, characterized the last decades by a massive production of experimental data. These data cover a wide range of biological experiments derived from several sequencing strategies, producing a big amount of heterogeneous data. They are often linked to a set of related metadata that are essential to describe experiments and the analyzed samples, with also information about patients from which samples have been collected. Nowadays, browsing all these data and retrieving significant insights from them is a big challenge that has been already faced with different techniques in order to facilitate their accessibility and interoperability. In this work, we focus on genomic data of cancer and related metadata exploiting domain-specific ontologies in order to allow executing taxonomy-based relaxed queries. In particular, we apply the upward and downward query extension methods to obtain a finer or coarser granularity of the requested information. We define diverse use cases with which a user can perform a query specifying particular attributes related to metadata or genomic data, even if they are not available in the considered repository. Thus, we are able to extract the requested data through the use of domain-specific ontologies of The Open Biological and Biomedical Ontology (OBO) Foundry. Finally, we propose a new ontological software layer, which allows users to interact with experimental data and metadata without knowledge about their representation schema.

Keywords: Domain-specific ontologies · Cancer · Taxonomies · Relaxed queries · The cancer genome atlas · TCGA · Genomic data commons · GDC

© Springer Nature Switzerland AG 2020
P. Cazzaniga et al. (Eds.): CIBB 2019, LNBI 12313, pp. 33–43, 2020.
https://doi.org/10.1007/978-3-030-63061-4_4

1   Scientific Background

The advent and the subsequent advancements in the technology adopted by Next Generetion Sequencing (NGS) machines [16] marked a new era in biomedicine. It moved the attention of the international research community to new challenging topics. Most of all were born from the problem of managing a huge amount of data produced by NGS machines. Many open-access repositories were released to provide this massive amount of data publicly, where the biological information is collected and annotated with all the additional data needed to understand its functionalities [10]. These databases allow to better understand the biological field, to manage, and to store different kind of information such as nucleotide/amino acid sequences, protein structures, gene expression profiles, patient/specimen metadata, and bibliographies.

A database to refer for the study of cancer is The Cancer Genome Atlas (TCGA) [15]. This project aims to catalog the genetic mutations responsible for cancer, applying NGS techniques to improve the ability to diagnose, treat, and prevent cancer through a better understanding of the genetic basis of this disease. This database contains the genomic characterization and analysis of 33 types of tumors, including 10 rare tumors. Patient samples are processed through different types of techniques such as gene expression, methylated DNA and microRNA profiling, and exon sequencing of at least 1,200 genes. Additionally TCGA collects and analyzes high quality cancer samples and makes the following data available to the research community: i) clinical information on the patients participating in the program; ii) samples metadata (for example the weight of the sample, etc.); iii) histopathological images of portions of the sample. On July 15th 2016, the TCGA data portal was officially closed, which since 2006 made available to researchers numerous genomic and clinical data of affected patients. With the NGS technologies and the increase of data, the TCGA turned out to be inefficient in storing methods and in data extraction methods by researchers. Thus the GDC (Genomic Data Commons) [11] project was created to collect and standardize all data produced not only by the TCGA project, but also by other research institutes. The GDC is born from an initiative of the National Cancer Institute (NCI) with the aim of creating a unified data system that can promote the sharing of genomic and clinical data among researchers. The GDC allows access to a high quality set of data derived from programs supported by NCI, and recommends guidelines for organizations providing personal data sets. High quality data is guaranteed by a list of procedure that the GDC strictly observes:

– **Maintenance of high quality samples of tissues.** The GDC gets most of the data from the previously listed NCI programs; these guarantee high quality because the only accepted tissues that have annotated sources and that have been subjected to rigorous quality controls throughout the entire course of processing. For organizations not supported by the NCI, on the other hand, the GDC provides recommended collection strategies, and before accepting the data, submits the samples to examination to make sure they adhere to the high quality standards used by the BCR (Biospecimen Core Resource).

- **Implementation of data validation procedures.** Data validation is performed both on data imported from NCI programs and on data sent by external organizations; the data is made available by the GDC Portal only if they pass the validation.
- **Ensure the production of reliable and harmonized derivative data.** The GDC uses the genomic sequence data available to create derived data such as somatic DNA mutations, tumor gene expression, and copy number variation. The bioinformatics pipelines described in the GDC Data Harmonization are developed with the continuous contribution of experts from the cancer genomics community. Pipelines are implemented using techniques that make them reproducible, interoperable on multiple platforms, and shareable with all interested members of the community. GDC receives all the pipeline suggestions, and keeps them constantly updated by replacing old tools and technologies to keep up with new discoveries.

The GDC distinguishes between open access data and controlled access data; open access data do not require authorization and are generally high-level genomic data that cannot be individually identified, therefore aggregated data both clinical and biological samples; data with controlled access require authorization to access it and are generally individually identifiable data such as genomic sequencing data, and some clinical data.

Because of the continuous growth of biomedical repositories and of the production of new experimental data, the problem of efficiently manage data and metadata still represents an open issue. A lot of different novel methodologies that involve the most modern technological innovations in data management have been proposed, e.g., [3, 4, 12].

These foster the development of advanced software solutions able to efficiently query genomic and clinical data [7] in order to retrieve potentially relevant insights. Unfortunately, state of the art query languages often require a deep knowledge about (i) the techniques with which the experiments are produced and (ii) the technical terminology adopted by domain-experts used to define the metadata associated to the experimental data. Additionally, we want to highlight two different requirements of those query languages: on one hand a deep knowledge about the specific biological and experimental domain is required, on the other hand knowledge about how data are modeled is also necessary in order to request and extract potentially relevant information. This last issue occurs when the structure of data is not formally defined usually due to the heterogeneous nature of the data. The metadata represent the most significant case of data variability on structure and content.

2   Materials and Methods

In this work, we propose a new query language and a novel software that relies on our previous framework called OpenOmics [3], which is a data model, a search method, and a smart access technology for GDC data. OpenOmics is a recent software framework solution that allows to easily manage through Application Programming Interfaces (APIs) all the public available experimental data and metadata of GDC (previously extended and standardized through the use of external software tools [1,5]) exploiting MongoDB as a no-SQL data management system. It is also characterized by a smart mechanism of data redundancy reduction, which results in an effective solution in terms of data modelling and storage space reduction. The authors implemented a set of application programming interfaces (APIs) to provide an easy way to interact with all these data by querying them by URLs. The APIs are released open-source, this is one of the main reasons why we focused on them. The main aim of this work is to extend their query functionalities with more powerful features able to facilitate the accessibility of experimental data and metadata even with a limited knowledge about how these data have been modelled. The new method is based on the use of domain-specific ontologies (i.e., focusing on the biological domain) able to describe the data provided by GDC and modelled by OpenOmics. We implement this feature by revising the concept of taxonomy-based relaxed query [13] applied to domain-specific ontologies.

2.1   Data Overview

The service, on which we relied on (i.e., OpenOmics), provides a programmatic access to the underlying data through a set of public accessible APIs. The data were originally retrieved from OpenGDC [1], which provides an easy access to experimental data and metadata of different research programs about cancer-related studies originally hosted on GDC. At the time of writing, OpenOmics provides access to the post-processed, extended, and standardized data of TCGA. They consists in 33 different types of tumors with thousand of experimental data each, which are the result of the application of six different experimental strategies (i.e., Gene-expression, Isoform-expression, and miRNA-expression quantification, DNA-Methylation, Masked Somatic Mutation, and Copy Number Variation).

The GDC genomic data were originally integrated with the GENECODE [9], HUGO Gene Nomenclature Committee (HGNC) [6], and miRBase [8] databases to expand the informative content and produce a BED standardized version of the original data.

According to the OpenGDC data model, each sample is identified by a unique identifier and is described by genomic regions, i.e., portions of the genome defined by genomic coordinates (chr, left, right, strand) and other arbitrary fields that depend on the considered experiment. The genomic regions are represented through attributes defined by a schema and are unique for each type of experimental data. On the other side, metadata present a different structure which is particularly simpler in terms of data modeling but very efficient to query. They are represented as attribute-value pairs on which a smart mechanism of redundancy detection and refactor of their informative content has been applied. In particular, metadata contain information about the patients (e.g., their race, gender, age), and the experimental strategies adopted to analyze their samples, usually retrieved by biopsy.

This information are extremely relevant for the identification of the domain-specific ontologies as it will be discussed in the next chapter.

2.2 Taxonomy-Based Relaxed Queries

We based our method accordingly to the *upward* and *downward* theoretical concepts of taxonomy-based relaxed queries [2,13]. Formally, starting with a taxonomy T, a set of levels $L = l_1, ..., l_n$ of T, a dataset S, and an attribute a of S, if a is provided at a level l_i, T can be used to extend the dataset to a level l_j with:

- $l_j > l_i$: moving to a high level results in a coarser data granularity;
- $l_j < l_i$: traversing the taxonomy from the top to the bottom to achieve a lower level results in a finer granularity of the data.

In particular, we refer to the concept of *upward extension* with the first case in which $l_j > l_i$. The main goal is to store new information in the original dataset providing a higher level in the taxonomy, i.e., at a less fine granularity than the one that is available. On the other hand, the second case in which $l_j < l_i$ is called *downward extension*. The goal here is the same of the *upward extension* (i.e., extend the original dataset with new information) but by providing a more detailed level, i.e., at a finer granularity than that available before.

3 Results

Starting from the analysis of metadata present in the OpenOmics framework, we exploited the *metadata* set of endpoints in order to extract a set of attributes that can represent a specific biological topic. In particular, we investigate the metadata attributes together with all their possible associated values (i.e., the data dictionary), and identify a list of metadata that can be described through domain-specific ontologies.

Here we reported some of the selected metadata attributes:

- *clinical__acc_shared__metastatic_neoplasm_initial_diagnosis_anatomic_site*:   it describes the anatomical site in which a neoplasm is identified for the first time;
- *clinical__rx__drug_name*: it is the name of the pharmaceutical drug dispensed to a patient affected by a specific type of cancer;
- *gdc__center__name*: it represents the name of the clinic or hospital in which a particular sample extracted by tumoral-affected patients has been sequenced and analyzed;
- *gdc__demographic__ethnicity*: it describes the ethnicity of the patients;
- *gdc__demographic__race*: it is the race of the involved patients (e.g., asian, white, black or african american);
- *gdc__diagnoses__primary_diagnosis*: it is the first diagnosis, including the type of cancer.

We repeated the same process for the experimental data, where we identified the *gene* names and related synonyms as the most atomic descriptive concept that characterize these kind of data.

Each of them represent a concept that can be modelled through a specific ontology. We focused on The Open Biological and Biomedical Ontology (OBO) Foundry [14], which is a repository of ontologies that explain different sides of the biological world domain. In particular, we extracted from this repository five distinct Web Onotlogy Language (OWL) standardized ontologies:

- the *Geographical Entity Ontology* (GEO), able to describe geographical relations between regions and countries (for the identification of a particular sample sequencing site, or the original country of the patients involved in the experiments);
- the *Uberon multi-species Anatomy Ontology* (UAO), essential to identify the anatomical regions of some pathologies;
- the *Drug Ontology* (DO), able to identify the chemical components of pharmaceutical drugs dispensed to the patients followed during the whole process of disease development;
- the *Human Disease Ontology* (HDO), for the definition of causes and effects of a series of diseases that affect humans (included different cancer types);
- the *Gene Ontology* (GO), able to describe the function of genes and gene products.

We adopted GraphDB, a no-SQL graph-oriented database management system, to organize the selected ontologies in the same environment, merging them together in a single graph ontological representation. Through GraphDB, we were able to query the produced graph exploiting the integrated SPARQL query engine. SPARQL is for *Simple Protocol and Resource Description Framework (RDF) Query Language*, which is a SQL-like query language able to efficiently search this kind of relational data. This produced a new application layer in the OpenOmics framework, which is accessible from http://bioinformatics.iasi.cnr.it/openomics/api/routes, as shown in Fig. 1. We strongly suggest the reader to refer to the OpenOmics paper [3] for a deep explanation about how to interact with the APIs.

According to the previously described concept of Taxonomy-based relaxation queries, here we report some promising results based on a couple of characterizing use cases, which make use of two of the selected ontologies (i.e.: the *Human Disease Ontology* and the *Gene Ontology*). We exploit the SPARQL query language to interact with both of them. It is worth noting that no SPARQL queries have to be written by the users in order to interact with the selected ontologies. A set of implemented queries are defined in a new software layer operating after the user request through URLs and directly before the interaction with the OpenOmics endpoints. Here we report a couple of these SPARQL query schemas in order to show a practice use case of both the *upward* and *downward* extension concepts. Finally, we conclude the section with a real use case scenario to highlight the potentiality of merging both query extension approaches.

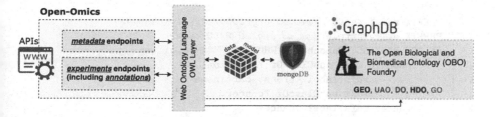

Fig. 1. The Ontological (OWL) Layer in the OpenOmics Framework is marked with the yellow area. We adopted GraphDB for the management and querying of the OWL-defined ontologies retrieved from The OBO Foundry (i.e., the Geographical Entity, Uberon multi-species Anatomy, Drug, Human Disease, and Gene ontologies). The new application layer is a bridge between the the endpoints and the data model with which data are organized in MongoDB. (Color figure online)

3.1 Upward Extension

This scenario shows an example of upward-extended SPARQL query. Here we interact with the Human Disease Ontology and the Uberon multi-species Anatomy Ontology entities through a upward-extended SPARQL query that is shown in Listing 1. In this case, we are searching for the anatomical region related to a specific type of cancer (i.e. the Breast Invasive Carcinoma (BRCA), called *breast cancer* in the query). We used on the result of this query to retrieve all tumors located in a particular anatomical region (the same of BRCA) and focus on tumor-related experiments exploiting the set of endpoints provided by the OpenOmics APIs. In particular, we extended the *metadata* endpoints with this feature that is automatically used everytime the attributes specified into the URL request are not known metadata attributes.

3.2 Downward Extension

This scenario shows an example of downward-extended SPARQL query. Here we interacted with the Gene Ontology (GO) to retrieve the properties of a specific gene in the ontological graph. In particular, as shown in Listing 2, we extracted the GO IDs of the gene functions, and their related domains, called namespaces (i.e., biological process, molecular function, and cellular component) starting from the gene-product (i.e., protein name) retrieved through querying the UniProt SPARQL endpoint with a specific gene name. This kind of query retrieves the biological function related to a gene of interest. Thus, the same query can be used reversing the arguments order, i.e. retrieving a set of genes involved in a particular biological function. These information is powerful to deeply analyze the extracted group of genes and investigate their properties exploiting the set of *experiments* endpoints of the OpenOmics framework (e.g., retrieving their expression values, or obtaining information about the methylated sites that occur in their genomic regions).

```
1   PREFIX rdfs: <http://www.w3.org/2000/01/rdf-schema#>
2   PREFIX owl: <http://www.w3.org/2002/07/owl#>
3   SELECT DISTINCT ?obouberon WHERE {
4       ?obodoid rdfs:subClassOf ?relation;
5                rdfs:label 'breast carcinoma';
6                rdfs:subClassOf ?cancer.
7       ?cancer rdfs:label 'cancer'.
8       ?relation owl:onProperty ?property;
9                 owl:someValuesFrom ?obouberon.
10      ?property rdfs:label 'located_in'.
11      ?obouberon rdfs:label ?property_value.
12  }
```

Listing 1: Upward-extended SPARQL query on the Human Disease Ontology and the Uberon multi-species Anatomy Ontology to retrieve the anatomical region (i.e. *located_in*) in which a specific type of tumor (i.e. *breast cancer*) occurs on.

```
1   PREFIX up: <http://purl.uniprot.org/core/>
2   PREFIX rdfs: <http://www.w3.org/2000/01/rdf-schema#>
3   PREFIX skos: <http://www.w3.org/2004/02/skos/core#>
4   PREFIX oboInOwl: <http://www.geneontology.org/formats/oboInOwl#>
5   SELECT DISTINCT ?goid ?namespace WHERE {
6       SERVICE <https://sparql.uniprot.org/> {
7           ?protein a up:Protein .
8           ?protein up:encodedBy ?gene ;
9                    up:classifiedWith ?go.
10          ?gene skos:prefLabel 'EAF3' .
11          ?go rdfs:label ?golabel.
12      }
13      ?go oboInOwl:id ?goid;
14          oboInOwl:hasOBONamespace ?namespace.
15  }
```

Listing 2: Downward-extended SPARQL query on the Gene Ontology and the UniProt SPARQL endpoint to retrieve GO functions of a specific gene name (i.e., *EAF3* in this case).

The same logic can be used to address a wide range of different problems. Another scenario, where the power of the downward extension concept emerges, consists in retrieving data about a particular cancer and all other tumors that occur in the same anatomical site. In particular, in Listing 3, we built a SPARQL downward-relaxed query to retrieve data about the *thoracic cancer*. In this case, the thorax is the anatomical site involved in cancer, which is the main site of metastasis also for other diseases like the *breast invasive carcinoma*. For this reason, if data about other diseases related to the specified one exist in the database, they will be reported in the results.

```
1    PREFIX rdfs: <http://www.w3.org/2000/01/rdf-schema#>
2    select DISTINCT ?label where {
3        ?o rdfs:label 'thoracic cancer'.
4        ?subclass rdfs:subClassOf ?o.
5        ?subclass rdfs:label ?label.
6    }
```

Listing 3: Downward-extended SPARQL query for the selection of experimental data about a specific cancer and all the related types of tumors (i.e., in this case *thoracic cancer*, which is linked to the *breast invasive carcinoma* by the *thorax* anatomical site).

3.3   Use Case Scenario

The previous examples of upward and downward query extensions demonstrated the ability of our approach to discover hidden relations among TCGA data. We are indeed able to group multiple experiments according to information that are not actually known at all, and can not emerge without the support of domain-specific ontologies as additional resources.

In this section, we show the capability of the proposed approach by combining both upward and downward extensions to answer a real biological question. Consider a use case where we want to check for significant variations in the expressions of specific genes among tumors of a specific anatomical area. Starting from a tumor it is possible to query for data that is not present in the OpenOmics knowledge base, like information about the anatomical site where the tumor develops.

A query like http://bioinformatics.iasi.cnr.it/openomics/metadata/tumor/ tcga-brca/attribute/located_in/list to the metadata endpoint will report the anatomical site where the BRCA tumor develops (i.e., the thorax). It is now possible to retrieve a list of tumors that share the same anatomical area with a query to the metadata endpoint like http://bioinformatics.iasi.cnr.it/ openomics/metadata/site/thorax/attribute/tumor/list. OpenOmics will return a list of two tumor types with *thorax* as common anatomical site: BRCA and LUSC.

The same logic can be applied to retrieve a set of genes that are involved in a common pathway by searching for a single specific gene of interest, like the onco-suppressor *BRCA1* gene. In this case, OpenOmics will retrieve the molecular function of the gene by querying the annotation endpoint with the following URL http://bioinformatics.iasi.cnr.it/openomics/annotation/ geneexpression/BRCA1. That function can be used in the opposite way to get the list of genes that are involved in the same pathway. It is finally possible to query the experiment endpoint to retrieve the expression level of the selected genes in all the tumor types occurring in the thorax area. This simple and efficient use case can be generalised to extract data about tumors that occur in a specific anatomical area. Thus, looking at the variation in the expression of a set of genes is very simple.

4   Conclusion

In this work, we presented a new method to query a set of data and metadata with no a priori knowledge about how they are modelled and organized. We focused on the biological domain, in particular on data derived by experiments aimed at investigating cancer development in humans, showing real case studies and obtaining promising results. Our solution is characterized by a high flexibility so that it can be possibly exploited with all data for which an ontology is available. This leads to many advantages. One of these is the ability to unlock the potential of easily grouping data according to attributes and to criteria that are not originally available in the dataset. Additionally, extending the dataset by adding new information produces a series of positive side-effects (e.g., the creation of a new more informative dataset that can be analyzed through machine learning algorithms for better classifying data according to new potentially relevant features). With a deep awareness about the capabilities of the method here proposed, we will further investigate the adoption of other domain-specific ontologies to improve The Ontological (OWL) Layer that is in the middle between the user interface and the conceptual schema with which data are modelled.

References

1. Cappelli, E., Cumbo, F., Bernasconi, A., Masseroli, M., Weitschek, E.: OpenGDC: standardizing, extending, and integrating genomics data of cancer. In: ESCS 2018: 8th European Student Council Symposium, International Society for Computational Biology (ISCB), p. 1 (2018)
2. Cappelli, E., Weitschek, E.: Extending the genomic data model and the genometric query language with domain taxonomies. In: Cabot, J., De Virgilio, R., Torlone, R. (eds.) ICWE 2017. LNCS, vol. 10360, pp. 567–574. Springer, Cham (2017). https://doi.org/10.1007/978-3-319-60131-1_44
3. Cappelli, E., Weitschek, E., Cumbo, F.: Smart persistence and accessibility of genomic and clinical data. In: Anderst-Kotsis, G., et al. (eds.) DEXA 2019. CCIS, vol. 1062, pp. 8–14. Springer, Cham (2019). https://doi.org/10.1007/978-3-030-27684-3_2
4. Ceri, S., et al.: Overview of GeCo: a project for exploring and integrating signals from the genome. In: Kalinichenko, L., Manolopoulos, Y., Malkov, O., Skvortsov, N., Stupnikov, S., Sukhomlin, V. (eds.) DAMDID/RCDL 2017. CCIS, vol. 822, pp. 46–57. Springer, Cham (2018). https://doi.org/10.1007/978-3-319-96553-6_4
5. Cumbo, F., Fiscon, G., Ceri, S., Masseroli, M., Weitschek, E.: TCGA2BEd: extracting, extending, integrating, and querying the cancer genome atlas. BMC Bioinformatics **18**(1), 6 (2017)

6. Eyre, T.A., Ducluzeau, F., Sneddon, T.P., Povey, S., Bruford, E.A., Lush, M.J.: The HUGO gene nomenclature database, 2006 updates. Nucleic Acids Res. **34**(suppl_1), D319–D321 (2006)

7. Fernandez, J.D., Lenzerini, M., Masseroli, M., Venco, F., Ceri, S.: Ontology-based search of genomic metadata. IEEE/ACM Trans. Comput. Biol. Bioinf. **13**(2), 233–247 (2015)

8. Griffiths-Jones, S., Saini, H.K., van Dongen, S., Enright, A.J.: miRBase: tools for microRNA genomics. Nucleic Acids Res. **36**(suppl_1), D154–D158 (2007)

9. Harrow, J., et al.: GENCODE: the reference human genome annotation for the encode project. Genome Res. **22**(9), 1760–1774 (2012)

10. Howe, D., et al.: Big data: the future of biocuration. Nature **455**(7209), 47 (2008)

11. Jensen, M.A., Ferretti, V., Grossman, R.L., Staudt, L.M.: The NCI genomic data commons as an engine for precision medicine. Blood **130**(4), 453–459 (2017)

12. Kaitoua, A., Gulino, A., Masseroli, M., Pinoli, P., Ceri, S.: Scalable genomic data management system on the cloud. In: 2017 International Conference on High Performance Computing & Simulation (HPCS), pp. 58–63. IEEE (2017)

13. Martinenghi, D., Torlone, R.: Taxonomy-based relaxation of query answering in relational databases. VLDB J. **23**(5), 747–769 (2014)

14. Smith, B., et al.: The OBO foundry: coordinated evolution of ontologies to support biomedical data integration. Nat. Biotechnol. **25**(11), 1251 (2007)

15. Weinstein, J.N., et al.: The Cancer Genome Atlas Pan-Cancer analysis project. Nat. Genet. **45**(10), 1113 (2013)

16. Weitschek, E., Santoni, D., Fiscon, G., De Cola, M.C., Bertolazzi, P., Felici, G.: Next generation sequencing reads comparison with an alignment-free distance. BMC Res. Notes **7**(1), 869 (2014)

GAN-Based Multiple Adjacent Brain MRI Slice Reconstruction for Unsupervised Alzheimer's Disease Diagnosis

Changhee Han[1,2,3(✉)], Leonardo Rundo[4,5], Kohei Murao[2],
Zoltán Ádám Milacski[6], Kazuki Umemoto[7], Evis Sala[4,5], Hideki Nakayama[3,8],
and Shin'ichi Satoh[2]

[1] LPIXEL Inc., Tokyo, Japan
han@lpixel.net
[2] Research Center for Medical Big Data,
National Institute of Informatics, Tokyo, Japan
[3] Graduate School of Information Science and Technology,
The University of Tokyo, Tokyo, Japan
[4] Department of Radiology, University of Cambridge, Cambridge, UK
[5] Cancer Research UK Cambridge Centre, Cambridge, UK
[6] Department of Artificial Intelligence,
ELTE Eötvös Loránd University, Budapest, Hungary
[7] Department of Rehabilitation Medicine,
Juntendo University School of Medicine, Tokyo, Japan
[8] International Research Center for Neurointelligence (WPI-IRCN), The University
of Tokyo Institutes for Advanced Study, The University of Tokyo, Tokyo, Japan

Abstract. Unsupervised learning can discover various diseases, relying on large-scale unannotated medical images of healthy subjects. Towards this, unsupervised methods reconstruct a single medical image to detect outliers either in the learned feature space or from high reconstruction loss. However, without considering continuity between multiple adjacent slices, they cannot directly discriminate diseases composed of the accumulation of subtle anatomical anomalies, such as Alzheimer's Disease (AD). Moreover, no study has shown how unsupervised anomaly detection is associated with disease stages. Therefore, we propose a two-step method using Generative Adversarial Network-based multiple adjacent brain MRI slice reconstruction to detect AD at various stages: (*Reconstruction*) Wasserstein loss with Gradient Penalty + ℓ_1 loss—trained on 3 healthy slices to reconstruct the next 3 ones—reconstructs unseen healthy/AD cases; (*Diagnosis*) Average/Maximum loss (e.g., ℓ_2 loss) per scan discriminates them, comparing the reconstructed/ground truth images. The results show that we can reliably detect AD at a very early stage with Receiver Operating Characteristics-Area Under the Curve (ROC-AUC) 0.780 while also detecting AD at a late stage much more accurately with ROC-AUC 0.917; since our method is fully unsupervised, it should also discover and alert any anomalies including rare disease.

© Springer Nature Switzerland AG 2020
P. Cazzaniga et al. (Eds.): CIBB 2019, LNBI 12313, pp. 44–54, 2020.
https://doi.org/10.1007/978-3-030-63061-4_5

Keywords: Generative adversarial networks · Alzheimer's disease diagnosis · Unsupervised anomaly detection · Brain MRI reconstruction.

1 Introduction

Deep Learning can achieve accurate computer-assisted diagnosis when large-scale annotated training samples are available. In medical imaging, unfortunately, preparing such massive annotated datasets is often unfeasible; to tackle this important problem, researchers have proposed various data augmentation techniques, including Generative Adversarial Network (GAN)-based ones [1–5]. However, even exploiting these techniques, supervised learning still requires many images with pathological features, even for rare disease, to make a reliable diagnosis; nevertheless, it can only detect already-learned specific pathologies. In this regard, as physicians notice previously unseen anomaly examples using prior information on healthy body structure, unsupervised anomaly detection methods leveraging only large-scale healthy images can discover and alert the presence of the disease when their generalization fails.

Towards this, researchers reconstructed a single medical image *via* GANs [6], AutoEncoders (AEs) [7], or combining them, since GANs can generate realistic images and AEs, especially Variational AEs, can directly map data onto its latent representation [8]; then, unseen images were scored by comparing them with reconstructed ones to discriminate a pathological image distribution (i.e., outliers either in the learned feature space or from high reconstruction loss). However, those single image reconstruction methods mainly target diseases easy-to-detect from a single image even for non-expert human observers, such as glioblastoma on Magnetic Resonance Imaging (MRI) [8] and lung cancer on Computed Tomography images [7]. Without considering continuity between multiple adjacent images, they cannot directly discriminate diseases composed of the accumulation of subtle anatomical anomalies, such as Alzheimer's Disease (AD). Moreover, no study has shown so far how unsupervised anomaly detection is associated with disease stages. We thus propose a two-step method using GAN-based multiple adjacent brain MRI slice reconstruction to detect AD at various stages (Fig. 1): (*Reconstruction*) Wasserstein loss with Gradient Penalty (WGAN-GP) [9,10] + ℓ_1 loss—trained on 3 healthy brain axial MRI slices to reconstruct the next 3 ones—reconstructs unseen healthy/AD cases; (*Diagnosis*) Average/Maximum loss (e.g., ℓ_2 loss) per scan discriminates them, comparing the reconstructed and ground truth images.

Contributions. Our main contributions are as follows:

- **MRI Slice Reconstruction:** This first multiple MRI slice reconstruction approach can predict the next 3 brain MRI slices from the previous 3 ones only for unseen images similar to training data by combining WGAN-GP and ℓ_1 loss.

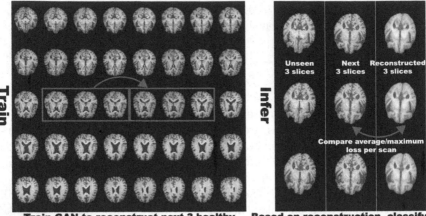

Train GAN to reconstruct next 3 healthy   Based on reconstruction, classify MRI
MRI slices from previous 3 ones   scans into healthy/Alzheimer's disease

Fig. 1. Unsupervised AD diagnosis framework: we train WGAN-GP + ℓ_1 loss on 3 healthy brain axial MRI slices to reconstruct the next 3 ones, and test it on both unseen healthy and AD cases to classify them based on average/maximum loss (e.g., ℓ_2 loss) per scan.

- **Unsupervised Anomaly Detection:** This first unsupervised anomaly detection across different disease stages reveals that, like physicians' way of diagnosis, massive healthy data can reliably aid early diagnosis, such as of Mild Cognitive Impairment (MCI), while also detecting late-stage disease much more accurately by discriminating with ℓ_2 loss.
- **Alzheimer's Disease Diagnosis:** This first unsupervised AD diagnosis study can reliably detect AD and also other diseases.

The remainder of the manuscript is organized as follows: Sect. 2 outlines the state-of-the-art of automated AD diagnosis; Sect. 3 describes the analyzed MRI dataset, as well as the proposed GAN-based unsupervised AD diagnosis framework; experimental results are shown and discussed in Sect. 4; finally, Sect. 5 provides conclusive remarks and future work.

2   Automated Alzheimer's Disease Diagnosis

Despite the clinical, social, and economic significance of early AD diagnosis—primarily associated with MCI detection—it generally relies on subjective assessment by physicians (e.g., neurologists, geriatricians, and psychiatrists); to tackle this open challenge, researchers have used classic supervised Machine Learning based on hand-crafted features [11,12]. More recently, Deep Learning has attracted great attentions owing to its more abstract and descriptive embedding based on multiple non-linear transformations: Liu *et al.* used a semi-supervised CNN to significantly reduce the need for labeled training data [13]; for clinical decision-making, Suk *et al.* integrated multiple sparse regression models (namely, Deep Ensemble Sparse Regression Network) [14]; Spasov *et al.* devised a parameter-efficient CNN for 3D separable convolutions, combining dual learning and a specific layer to predict the conversion from MCI to AD within 3 years [15]; instead of exploiting the CNNs, Parisot used a semi-supervised Graph Convolutional Network trained on a sub-set of labeled nodes with diagnostic outcomes to represent sparse clinical data [16].

To the best of our knowledge, no existing work has conducted fully unsupervised anomaly detection for AD diagnosis since capturing subtle anatomical differences between MCI and AD is challenging. Therefore, without requiring any labeled data for training, the proposed GAN-based unsupervised approach might provide new insights into AD research.

3   Materials and Methods

3.1   OASIS-3 Dataset

We use a longitudinal dataset of $176 \times 240/176 \times 256$ T1-weighted (T1w) 3T brain axial MRI slices containing both normal aging subjects/AD patients extracted from the Open Access Series of Imaging Studies-3 (OASIS-3) [17]. The 176×240 slices are zero-padded to reach 176×256 pixels. Relying on Clinical Dementia Rating (CDR) [18], common clinical scale for the staging of dementia, the subjects are comprised of:

- Unchanged CDR $= 0$: Cognitively healthy population;
- CDR $= 0.5$: Very mild dementia ($\sim$ MCI);
- CDR $= 1$: Mild dementia;
- CDR $= 2$: Moderate dementia.

Since our dataset is longitudinal and the same subject's CDRs may vary (e.g., CDR $= 0$ to CDR $= 0.5$), we only use scans with unchanged CDR $= 0$ to assure certainly healthy scans. As CDRs and MRI scans are not always simultaneously acquired, we label MRI scans with CDRs at the closest date. We only select brain MRI slices including hippocampus/amygdala/ventricles among whole 256 axial slices per scan to avoid over-fitting from AD-irrelevant information; the atrophy of the hippocampus/amygdala/cerebral cortex, and enlarged ventricles are strongly associated with AD, and thus they mainly affect the AD classification performance of Machine Learning [19]. Moreover, we discard low-quality MRI slices. The remaining dataset is divided as follows:

- Training set: Unchanged CDR $= 0$ (408 subjects/$1,133$ scans/$57,834$ slices);
- Validation set: Unchanged CDR $= 0$ (55 subjects/155 scans/$8,080$ slices),
  CDR $= 0.5$ (53 subjects/85 scans/$4,607$ slices),
  CDR $= 1$ (29 subjects/45 scans/$2,518$ slices),
  CDR $= 2$ (2 subjects/4 scans/160 slices);
- Test set: Unchanged CDR $= 0$ (113 subjects/318 scans/$16,198$ slices),
  CDR $= 0.5$ (99 subjects/168 scans/$9,206$ slices),
  CDR $= 1$ (61 subjects/90 scans/$5,014$ slices),
  CDR $= 2$ (4 subjects/6 scans/340 slices).

The same subject's scans are included in the same dataset. The datasets are strongly biased towards healthy scans similarly to MRI inspection in the clinical routine. During training for reconstruction, we only use the training set containing healthy slices to conduct unsupervised learning.

3.2   GAN-Based Multiple Adjacent Brain MRI Slice Reconstruction

To model the strong consistency of healthy brain anatomy (Fig. 1), in each scan, we reconstruct the next 3 MRI slices from the previous 3 ones using an image-to-image GAN (e.g., if a scan includes 40 slices s_i for $i = 1, \ldots, 40$, we reconstruct all possible 35 setups: $(s_i)_{i \in \{1,2,3\}} \mapsto (s_i)_{i \in \{4,5,6\}}$; $(s_i)_{i \in \{2,3,4\}} \mapsto (s_i)_{i \in \{5,6,7\}}$; $\ldots$; $(s_i)_{i \in \{35,36,37\}} \mapsto (s_i)_{i \in \{38,39,40\}}$). We concatenate adjacent 3 grayscale slices into 3 channels, such as in RGB images. The GAN uses: (*i*) a U-Net-like [20,21] generator with 4 convolutional and 4 deconvolutional layers in encoders and decoders, respectively, with skip connections; (*ii*) a discriminator with 3 decoders. We apply batch normalization to both convolution with Leaky Rectified Linear Unit (ReLU) and deconvolution with ReLU. To confirm how reconstructed images' realism and anatomical continuity affect anomaly detection, we compare the GAN models with different loss functions, namely: (*i*) Dice loss (i.e., a plain U-Net without the discriminator); (*ii*) WGAN-GP loss; (*iii*) WGAN-GP loss + 100 ℓ_1 loss. Among 8 losses comparing ground truth/reconstructon, average ℓ_2 loss per scan always outperforms the other losses during validation for U-Net and WGAN-GP without/with ℓ_1 loss, and thus we use this loss for testing.

Considering its computational speed, U-Net training lasts for 600,000 steps with a batch size of 64 and both GAN trainings last for 300,000 steps with a batch size of 32. We use 2.0×10^{-4} learning rate for the Adam optimizer [22]. The framework is implemented on Keras with TensorFlow as backend.

3.3   Unsupervised Alzheimer's Disease Diagnosis

During validation, we compare the following average/maximum losses per scan (i.e., 8 losses) between reconstructed/ground truth 3 slices (Fig. 1): (*i*) ℓ_1 loss; (*ii*) ℓ_2 loss; (*iii*) Dice loss; (*iv*) Structural Similarity loss. For each model's testing, we separately pick the loss showing the highest Receiver Operating Characteristics-Area Under the Curve (ROC-AUC) between CDR = 0 (i.e., healthy population) *vs* all the other CDRs (i.e., dementia) during validation. As a result, we pick the average ℓ_2 loss per scan for all models since squared error is sensitive to outliers and it always outperforms the others. To evaluate its unsupervised AD diagnosis performance for test sets, we show ROC and Precision-Recall (PR) curves, along with their AUCs, between CDR = 0 *vs* (*i*) all the other CDRs; (*ii*) CDR = 0.5; (*iii*) CDR = 1; (*iv*) CDR = 2. We visualize ℓ_2 loss distributions of CDR = 0/0.5/1/2 to know how disease stages affect its discrimination.

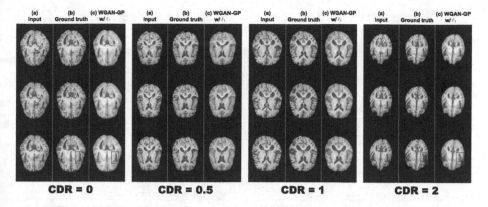

Fig. 2. Example brain MRI slices with CDR = 0/0.5/1/2 from test sets: (a) Input 3 real slices; (b) Ground truth next 3 real slices; (c) Next 3 slices reconstructed by U-Net; (d), (e) Next 3 slices reconstructed by WGAN-GP without/with ℓ_1 loss.

4 Results

4.1 Reconstructed Brain MRI Slices

Figure 2 illustrates example real MRI slices from test sets and their reconstruction by U-Net and WGAN-GP without/with ℓ_1 loss. The WGAN-GP + ℓ_1 loss can successfully capture T1w-specific appearance and anatomical changes from the previous 3 slices more smoothly than the U-Net and in more detail than the WGAN-GP without ℓ_1 loss. Since the models are trained only on healthy slices, reconstructing slices with higher CDRs tends to comparatively fail, especially around hippocampus, amygdala, cerebral cortex, and ventricles due to their insufficient atrophy after reconstruction.

4.2 Unsupervised AD Diagnosis Results

Figures 3 and 4 show ROC and PR curves, respectively—along with their AUCs—of unsupervised anomaly detection. We do not show confidence intervals since the diagnosis stage is non-trainable. Since brains with higher CDRs accompany stronger anatomical atrophy from healthy brains, their ROC-AUCs between unchanged CDR = 0 remarkably increase as CDRs increase. Clearly outperforming the other methods in every condition, WGAN-GP + ℓ_1 loss achieves excellent ROC-AUCs, especially for higher CDRs—it obtains ROC-AUC = 0.780/0.833/0.917 for CDR = 0 *vs* CDR = 0.5/1/2, respectively; this experimental finding derives from ℓ_1 loss' good realism sacrificing diversity (i.e., generalizing well only for unseen images with a similar distribution to training

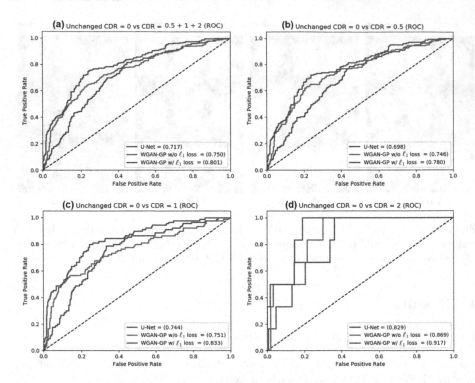

Fig. 3. Unsupervised anomaly detection results using average ℓ_2 loss per scan on reconstructed brain MRI slices (ROC curves and ROC-AUCs): unchanged CDR = 0 (i.e., cognitively healthy population) is compared with (a) all the other CDRs (i.e., dementia); (b) CDR = 0.5 (i.e., very mild dementia); (c) CDR = 1 (i.e., mild dementia); (d) CDR = 2 (i.e., moderate dementia).

images) and WGAN-GP loss' ability to capture recognizable structure. Figure 5 indicates its good discrimination ability even between healthy subjects *vs* MCI patients (i.e., CDR = 0 *vs* CDR = 0.5), which is extremely difficult even in a supervised manner [19]. Interestingly, unlike our visual expectation, WGAN-GP without ℓ_1 loss outperforms plain U-Net regardless of its very blurred reconstruction, showing the superiority of GAN-based reconstruction for diagnosis.

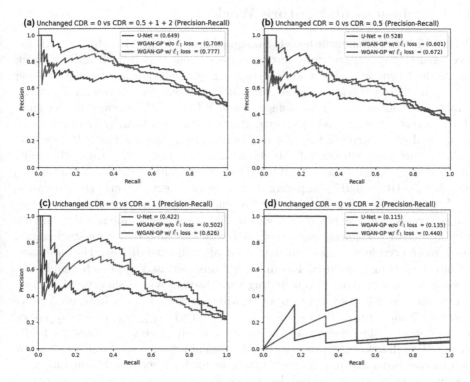

Fig. 4. Unsupervised anomaly detection results using average ℓ_2 loss per scan on reconstructed brain MRI slices (PR curves and PR-AUCs): unchanged CDR = 0 is compared with (a) all the other CDRs; (b) CDR = 0.5; (c) CDR = 1; (d) CDR = 2.

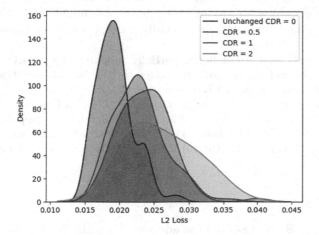

Fig. 5. Unsupervised anomaly detection resultsDistributions of average ℓ_2 loss per scan evaluated on brain MRI slices with CDR = 0/0.5/1/2 reconstructed by WGAN-GP + ℓ_1 loss.

5   Conclusions and Future Work

Using a massive amount of healthy images, our GAN-based multiple MRI slice reconstruction can successfully discriminate AD patients from healthy subjects for the first time in an unsupervised manner; our solution leverages a two-step approach: (*Reconstruction*) ℓ_1 loss generalizes well only for unseen images with a similar distribution to training images while WGAN-GP loss captures recognizable structure; (*Diagnosis*) ℓ_2 loss clearly discriminates healthy/abnormal data as squared error becomes huge for outliers. Using 1,133 healthy MRI scans for training, our approach can reliably detect AD at a very early stage, MCI, with ROC-AUC = 0.780 while detecting AD at a late stage much more accurately with ROC-AUC = 0.917—implying its ability to detect various other diseases.

Accordingly, this first unsupervised anomaly detection across different disease stages reveals that, like physicians' way of diagnosis, large-scale healthy data can reliably aid early diagnosis, such as of MCI, while also detecting late-stage disease much more accurately. Since our method deals well with diseases that are hard-to-detect even in supervised learning, this unsupervised approach should also discover/alert any anomalies including rare disease, where supervised learning is inapplicable [23]. As future work, we will reconstruct slices from both previous/next 3 slices (e.g., slices s_i for $i = 1, \ldots, 9$, $(s_i)_{i \in \{1,2,3,7,8,9\}} \mapsto (s_i)_{i \in \{4,5,6\}}$) for robustness, also optimizing the number of slices (e.g., 3 slices to 1 or 5 slices). We will investigate more reconstruction networks (e.g., GANs with attention mechanisms) and multiple loss functions for both reconstruction/diagnosis. Lastly, we plan to detect and locate various diseases, including cancer [24] and rare diseases—this work only uses brain MRI slices including hippocampus/amygdala/ventricles for AD diagnosis, but we may have to use all or most brain MRI slices to also detect anomalies appearing in other anatomical locations within the brain. Integrating multimodal imaging data, such as Positron Emission Tomography with specific radiotracers [25], might further improve AD diagnosis [26], even when analyzed modalities are partially unavailable [27].

Acknowledgment. This research was partially supported by AMED Grant Number JP18lk1010028, and also partially supported by The Mark Foundation for Cancer Research and Cancer Research UK Cambridge Centre [C9685/A25177]. Additional support has been provided by the National Institute of Health Research (NIHR) Cambridge Biomedical Research Centre. Zoltán Ádám Milacski was supported by Grant Number VEKOP-2.2.1-16-2017-00006. The OASIS-3 dataset has Grant Numbers P50 AG05681, P01 AG03991, R01 AG021910, P50 MH071616, U24 RR021382, and R01 MH56584.

References

1. Goodfellow, I., S., et al.: Generative adversarial nets. In: Proceedings of Advances in Neural Information Processing Systems (NIPS), pp. 2672–2680 (2014)
2. Frid-Adar, M., Diamant, I., Klang, E., Amitai, M., Goldberger, J., Greenspan, H.: GAN-based synthetic medical image augmentation for increased CNN performance in liver lesion classification. Neurocomputing **321**, 321–331 (2018)

3. Han, C., Rundo, L., Araki, R., Nagano, Y., Furukawa, Y., et al.: Combining noise-to-image and image-to-image GANs: brain MR image augmentation for tumor detection. IEEE Access **7**(1), 156966–156977 (2019)
4. Han, C., et al.: Synthesizing diverse lung nodules wherever massively: 3D multi-conditional GAN-based CT image augmentation for object detection. In: Proceedings of the International Conference on 3D Vision (3DV), pp. 729–737 (2019)
5. Han, C., Murao, K., Noguchi, T., et al.: Learning more with less: conditional PGGAN-based data augmentation for brain metastases detection using highly-rough annotation on MR images. In: Proceedings of the ACM International Conference on Information and Knowledge Management (CIKM), pp. 119–127 (2019)
6. Schlegl, T., Seeböck, P., Waldstein, S.M., Langs, G., Schmidt-Erfurth, U.: f-AnoGAN: fast unsupervised anomaly detection with generative adversarial networks. Med. Image Anal. **54**, 30–44 (2019)
7. Uzunova, H., Schultz, S., Handels, H., Ehrhardt, J.: Unsupervised pathology detection in medical images using conditional variational autoencoders. Int. J. Comput. Assist. Radiol. Surg. **14**(3), 451–461 (2018)
8. Chen, X., Konukoglu, E.: Unsupervised detection of lesions in brain MRI using constrained adversarial auto-encoders. In: International Conference on Medical Imaging with Deep Learning (MIDL) (2018). arXiv preprint arXiv:1806.04972
9. Gulrajani, I., Ahmed, F., Arjovsky, M., Dumoulin, V., Courville, A.C.: Improved training of Wasserstein GANs. In: Advances in Neural Information Processing Systems, pp. 5769–5779 (2017)
10. Han, C., et al.: GAN-based synthetic brain MR image generation. In: Proceedings of the International Symposium on Biomedical Imaging (ISBI), pp. 734–738. IEEE (2018)
11. Salvatore, C., Cerasa, A., Battista, P., Gilardi, M.C., Quattrone, A., Castiglioni, I.: Magnetic resonance imaging biomarkers for the early diagnosis of Alzheimer's disease: a machine learning approach. Front. Neurosci. **9**, 307 (2015)
12. Nanni, L., Brahnam, S., Salvatore, C., Castiglioni, I.: Texture descriptors and voxels for the early diagnosis of Alzheimer's disease. Artif. Intell. Med. **97**, 19–26 (2019)
13. Liu, S., Liu, S., Cai, W., Pujol, S., Kikinis, R., Feng, D.: Early diagnosis of Alzheimer's disease with deep learning. In: Proceedings of the International Symposium on Biomedical Imaging (ISBI), pp. 1015–1018. IEEE (2014)
14. Suk, H.I., Lee, S.W., Shen, D.: Deep ensemble learning of sparse regression models for brain disease diagnosis. Med. Image Anal. **37**, 101–113 (2017)
15. Spasov, S., et al.: A parameter-efficient deep learning approach to predict conversion from mild cognitive impairment to Alzheimer's disease. NeuroImage **189**, 276–287 (2019)
16. Parisot, S., et al.: Disease prediction using graph convolutional networks: application to autism spectrum disorder and Alzheimer's disease. Med. Image Anal. **48**, 117–130 (2018)
17. LaMontagne, P.J., Keefe, S., Lauren, W., et al.: OASIS-3: longitudinal neuroimaging, clinical, and cognitive dataset for normal aging and Alzheimer's disease. Alzheimers Dement. **14**(7), P1097 (2018)
18. Morris, J.C.: The clinical dementia rating (CDR): current version and scoring rules. Neurology **43**(11), 2412–2414 (1993)
19. Ledig, C., Schuh, A., Guerrero, R., Heckemann, R.A., Rueckert, D.: Structural brain imaging in Alzheimer's disease and mild cognitive impairment: biomarker analysis and shared morphometry database. Sci. Rep. **8**, 11258 (2018)

20. Ronneberger, O., Fischer, P., Brox, T.: U-Net: convolutional networks for biomedical image segmentation. In: Navab, N., Hornegger, J., Wells, W.M., Frangi, A.F. (eds.) MICCAI 2015. LNCS, vol. 9351, pp. 234–241. Springer, Cham (2015). https://doi.org/10.1007/978-3-319-24574-4_28

21. Rundo, L., Han, C., Nagano, Y., et al.: USE-Net: incorporating squeeze-and-excitation blocks into U-Net for prostate zonal segmentation of multi-institutional MRI datasets. Neurocomputing **365**, 31–43 (2019)

22. Kingma, D.P., Ba, J.: Adam: A method for stochastic optimization. arXiv preprint arXiv:1412.6980 (2014)

23. Han, C., Rundo, L., Murao, K., Nemoto, T., Nakayama, H., Satoh, S.: Bridging the gap between AI and healthcare sides: towards developing clinically relevant AI-powered diagnosis systems. In: Proceedings of the International Conference on Artificial Intelligence Applications and Innovations (AIAI) (2020)

24. Rundo, L., Militello, C., Russo, G., Vitabile, S., Gilardi, M.C., Mauri, G.: GTVcut for neuro-radiosurgery treatment planning: an MRI brain cancer seeded image segmentation method based on a cellular automata model. Nat. Comput. **17**, 521–536 (2018)

25. Rundo, L., et al.: A fully automatic approach for multimodal PET and MR image segmentation in Gamma Knife treatment planning. Comput. Methods Programs Biomed. **144**, 77–96 (2017)

26. Brier, M.R., et al.: Tau and Aβ imaging, CSF measures, and cognition in Alzheimer's disease. Sci. Trans. Med. **8**(338), 338ra66–338ra66 (2016)

27. Li, R., et al.: Deep learning based imaging data completion for improved brain disease diagnosis. In: Golland, P., Hata, N., Barillot, C., Hornegger, J., Howe, R. (eds.) MICCAI 2014. LNCS, vol. 8675, pp. 305–312. Springer, Cham (2014). https://doi.org/10.1007/978-3-319-10443-0_39

Improving the Fusion of Outbreak Detection Methods with Supervised Learning

Moritz Kulessa[1](✉) ⓘ, Eneldo Loza Mencía[1] ⓘ, and Johannes Fürnkranz[2] ⓘ

[1] Technische Universität Darmstadt, Darmstadt, Germany
{mkulessa,eneldo}@ke.tu-darmstadt.de
[2] Johannes Kepler Universität Linz, Linz, Austria
juffi@faw.jku.at

Abstract. Epidemiologists use a variety of statistical algorithms for the early detection of outbreaks. The practical usefulness of such methods highly depends on the trade-off between the detection rate of outbreaks and the chances of raising a false alarm. Recent research has shown that the use of machine learning for the fusion of multiple statistical algorithms improves outbreak detection. Instead of relying only on the binary outputs (*alarm* or *no alarm*) of the statistical algorithms, we propose to make use of their p-values for training a fusion classifier. In addition, we also show that adding contextual features and adapting the labeling of an epidemic period may further improve performance. For comparison and evaluation, a new measure is introduced which captures the performance of an outbreak detection method with respect to a low rate of false alarms more precisely than previous works. We have performed experiments on synthetic data to evaluate our proposed approach and the adaptations in a controlled setting and used the reported cases for the disease *Salmonella* and *Campylobacter* from 2001 until 2018 all over Germany to evaluate on real data. The experimental results show a substantial improvement on the synthetic data when p-values are used for learning. The results on real data are less clear. Inconsistencies in the data appearing under real conditions make it more challenging for the learning approach to identify valuable patterns for outbreak detection.

Keywords: Outbreak detection · Fusion methods · Stacking · Syndromic surveillance

1 Scientific Background

The early detection of outbreaks of infectious diseases is of great significance for public health. In particular, the spread of such outbreaks can be diminished tremendously by applying control measures as early as possible, which indeed can save lives and reduce suffering. For that purpose, statistical algorithms have been developed to automate and improve outbreak detection. Such methods raise

ⓒ Springer Nature Switzerland AG 2020
P. Cazzaniga et al. (Eds.): CIBB 2019, LNBI 12313, pp. 55–66, 2020.
https://doi.org/10.1007/978-3-030-63061-4_6

alarms in cases where an unusually high number of infections is detected, which results in a further investigation by an epidemiologist. However, if not chosen wisely or configured properly, such statistical methods may also raise many false alarms. In particular for large surveillance systems, where many time series for different diseases and different locations are monitored simultaneously, the false alarm rate is a major concern and therefore highly determines the practical usefulness of an outbreak detection method [8]. However, regulating the false alarm rate usually has an impact on the ability to detect outbreaks. To find a good trade-off between the detection rate and the false alarm rate is one of the major challenges in outbreak detection [6].

Traditional outbreak detection methods rely on historic data to fit a parametric distribution which is then used to check the statistical significance of the observation. Choosing the significance level for the statistical method beforehand makes the evaluation difficult. In line with [3], we propose a method which uses the p-values of the statistical methods in order to evaluate their performance. In particular, we propose a variant of receiver operating characteristic (ROC) curves, which shows the false alarm rate on the x-axis and the detection rate— in contrast to the true positive rate—on the y-axis. By using the area under a *partial* ROC curve [5], we obtain a measure for the performance of an algorithm that satisfies a given constraint on the false alarm rate (e.g., less than 1% false alarms). This criterion serves as the main measure for our evaluations and enables us to analyze the trade-off between the false alarm rate and the detection rate of outbreak detection methods precisely.

An interesting alternative avenue to early outbreak detection is the use of machine learning (ML), in particular of supervised learning techniques. Prior work mainly focuses on forecasting the number of infections for a disease. However, a comparably lower amount of research has been devoted to improving statistical algorithms to raise alarms and which are applied to univariate time series. In particular, [9] has used the ML technique *hierarchical mixture of experts* [2] to combine the output of the methods from EARS. However, the authors note that all algorithms rely on the assumption of a Gaussian distribution, which limits their diversity. In contrast, [1] has used a variety of classification algorithms (*logistic regression, CART* and *Bayesian networks*) for the fusion of outbreak detection methods. As underlying statistical algorithms, they have used the cumulative sum algorithm, two exponential weighted moving average algorithms, the EARS methods (C1, C2, and C3) and the Farrington algorithm [6]. In general, the results indicate that ML improves the ability to detect outbreaks while simple voting schemes (e.g. weighted voting and majority vote) did not perform well. Moreover, the algorithms have not been evaluated with respect to data which include seasonality and trend.

However, the examined approaches only rely on the binary output (*alarm* or *no alarm*) of the underlying statistical methods for the fusion which limits the information about a particular observation. Prior research in the area of ML has shown that more precise information of the underlying models improves the overall performance of the fusion [10]. Therefore, we propose an approach for the fusion of outbreak detection methods which uses the p-values of the underlying statistical methods. Moreover, one can also incorporate different information for the outbreak detection (e.g., weather data, holidays, statistics about the data, ...) by augmenting the data with additional attributes. In addition, the way outbreaks are labeled in the data also has a major influence on the learnability of outbreak detectors. Thus, we propose adaptions for the labeling of outbreaks in order to maximize the detection rate of ML algorithms. As a first step, we put our focus on improving the performance of outbreak detection methods using a univariate time series as the only source of information.

2 Materials and Methods

The key idea of our approach is to learn to combine predictions of commonly used statistical outbreak detection models with a trainable ML method. Thus, we need to generate a series of aligned predictions, one for each method, which we can then use for training the ML model. We have chosen to base our work on the methods of the *early aberration reporting system* (EARS C1, C2, and C3), the Bayes method, and the RKI (*Robert Koch-Institut*) method, which are implemented in the R package *surveillance* [7]. They all require comparably little historic data on their own, allowing us to generate longer sequences for the ML algorithm. These methods rely on a sliding window approach which uses the previous observed counts as reference values for fitting a parametric distribution. Given an observed count c_t and the fitted distribution $p(x)$, a one-tailed significance test is performed in order to identify suspicious spikes. Therefore, a p-value is computed as the probability $\int_{c_t}^{\infty} p(x)dx$ of observing c_t or higher counts. An alarm is triggered if the p-value is inferior to the pre-defined significance level α, regulating the sensitivity of raising alarms.

2.1 Evaluation Measures

Instead of manually adjusting the α parameter and examining the results individually, which is mostly done in previous works, we propose to evaluate the p-value of the statistical approaches. In particular, the p-value can be interpreted as a score which sorts examples according to their degree to which they indicate an alarm. This allows us to analyze an algorithm with ROC curves. A ROC curve can be used to examine the trade-off between the true positive rate (i.e., the probability of raising an alarm in case of an actual outbreak) and the false alarm rate (i.e., the probability of falsely raising an alarm when no outbreak is ongoing). Following [3] and [1], we use a ROC curve-like representation with the detection rate (i.e., the proportion of outbreaks in which at least one alarm is raised during their activity) on the y-axis instead of the usual true positive rate. This allows us to only measure the ability to detect outbreaks since alarms raised in cases when the epidemic has already been detected are typically not so important. To only focus on high specificity results (e.g., with a false alarm rate below 1%), which is of major importance for many medical applications, we only consider *partial ROC curves*. By using the partial area under the ROC curve as proposed in [5] we obtain a simple measure to evaluate the performance of an

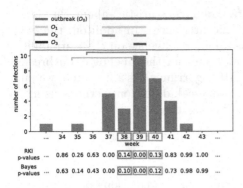

index week t	augmented features			p-values		target
	mean$_t$	prev. p-values		RKI$_t$	Bayes$_t$	outbreak$_t$
		RKI$_{t-1}$	Bayes$_{t-1}$			
...	...	...	...	...	...	...
34	1.00	0.59	0.63	0.86	0.63	no
35	0.50	0.86	0.63	0.26	0.14	no
36	0.50	0.26	0.14	0.63	0.43	no
37	0.50	0.63	0.43	0.00	0.00	yes
38	1.50	0.00	0.00	0.14	0.10	yes
39	2.25	0.14	0.10	0.00	0.00	yes
40	4.50	0.00	0.00	0.13	0.12	yes
41	6.25	0.13	0.12	0.83	0.73	yes
42	6.00	0.83	0.73	0.99	0.98	yes
43	5.50	0.99	0.98	1.00	0.99	no
...	...	...	...	...	...	...

Fig. 1. Example for the creation of training data for the learning algorithm using the statistical algorithms Bayes and RKI. The chart on the left shows the number of infections with a particular disease over time. Each time point t is also annotated with the computed p-values of the statistical algorithms (underneath) and the label indicating an outbreak (red line above the chart represents the epidemic period). Using this information, the data instances are created as shown on the right: Each particular time point is represented by one training instance, labeled according to the original targets O_0. The columns RKI$_t$ and Bayes$_t$ represent the computed p-values for the current observation while the other columns (mean$_t$, RKI$_{t-1}$ and Bayes$_{t-1}$) represent additional features, which are the mean over the previous four counts (as indicated with the brackets on top of the counts) and the previous output of the statistical algorithms. (Color figure online)

outbreak detection method, which we define as $dAUC_\tau = \frac{1}{\tau} \int_0^\tau dROC(f) \, df$, where $dROC(f)$ denotes the detection rate at a given false alarm rate of f and the parameter τ defines the maximum allowed false alarm rate to be considered.

2.2   Training Data for the ML-based Fusion

Fusion of p-Values Using Stacking. A straightforward way for combining the predictions of multiple outbreak detection methods is to simply vote and follow the majority prediction. A more sophisticated approach consists of training a classifier that uses the predictions of the detection methods as input, and is trained on the desired output, a technique that is known in ML as *stacking* [11]. In this work, we propose to use the computed p-values of the statistical methods as input features for the fusion classifier in order to include information about the certainty of an alarm for the ML algorithm. Figure 1 illustrates how the data for the learning algorithm are created.

Additional Features. The use of a trainable fusion method allows us to include additional information that can help to decide whether a given alarm should be raised or not. As additional features, we propose to include the mean of the counts over the last time points, which can give us evidence about the relia-bility of the statistical algorithms. Under the assumption that a time series is stationary an unusually high mean can also be a good indicator for an outbreak, especially in the case that the outbreak arises slowly over time. Finally, we also include the output of the statistical methods for previous time points in a window of a user-defined size as additional features.

Modelling the Output Labels. A major challenge for ML algorithms is that the duration of an outbreak period is not clearly defined [8]. A simple strategy— which we refer to as O_0—is to label all time points as positive as long as cases for the particular epidemic are reported. In this case, the goal of the learning algorithm is to predict most time points in an ongoing epidemic as positive, regardless of their timestamp. Indeed, our early results indicate that the predictor learns to recognize the fading-out of an outbreak (e.g., weeks 40 to 42 in Fig. 1). However, this also increases the number of false alarms as the ML algorithm learns to raise alarms when the count is decreasing outside an epidemic period. To avoid this, we propose three methods to adapt the basic labeling O_0 of the outbreaks: O_1 denotes the adaption which labels all time points until the peak (i.e., the point with the maximum number of counts during the period) as positive, O_2 instead skips the time points where there is a decrease in counts (i.e., it labels all increasing counts until reaching the peak), and O_3 labels only the peak of the outbreak as positive. Figure 1 visualizes the different options of labeling on the top left.

2.3 Evaluation Data

Synthetic Data: For the generation of synthetic data, we use the data generator proposed in [6]. In total, 42 different settings to generate time series (*test cases*) are proposed which reflect a wide range of application scenarios allowing to explicitly analyze the effects of trend (T), seasonality (S1) and biannual seasonality (S2). For each test case, 100 time series are created, using the first 575 weeks of all 100 time series to train an ML model and evaluated the created model on the last 49 weeks of the time series. The parameter k of the data generator, used to estimate the number of cases per outbreak, is randomly drawn from the range $1 \ldots 10$ for each outbreak. Instead of reporting averaged $dAUC_{1\%}$ scores, which could have different scales for different test cases, we determined a ranking over the methods for each considered test case and afterward computed each method's average ranking, 1 being the best rank.

For the parameter optimization on synthetic data, we created new time series for each test case by using a different random seed and used these to evaluate all parameter combinations. We wanted to optimize the performance across all test cases and, therefore, picked the parameter combination with the best average rank.

Real Data: For the evaluation on real data, we rely on the reported cases for the diseases *Salmonella* (SAL) and *Campylobacter* (CAM) in Germany, which are captured by the *Robert Koch Institute*. The reported cases are aggregated by the public health offices with respect to the 401 districts (*Landkreise* and

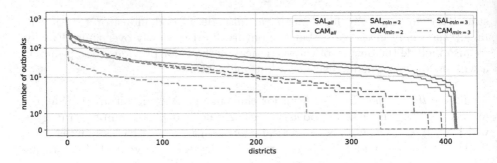

Fig. 2. Number of outbreaks for each application scenario and each district (ordered according to number of outbreaks) in Germany from 2001 until 2018.

Stadtkreise) in Germany and range from 2001 to 2018. Only for Berlin we obtain a finer granularity (12 sub-districts), resulting in a total of 412 time series for each disease. In addition, each reported case can be associated with a specific outbreak. The outbreaks were labeled in a retrospective manner by grouping all cases which can be assigned to a specific outbreak reason. However, not always all cases relating to a specific outbreak are reported since not all people visit the doctor nor is it possible to identify the source for catching the disease. Due to this, a lot of outbreaks only contain a few cases which are hard to detect. Therefore, we have generated three application scenarios: (1) Use all outbreaks (*all*), (2) use the outbreaks consisting of at least two cases (*min=2*) and (3) use the outbreaks which contain at least three cases (*min=3*). We obtain the outbreak labeling by labeling all weeks positive as long as cases are reported for a particular outbreak. In Fig. 2 the number of outbreaks for the six data configurations over all 18 years for each district is visualized.

We use the first 884 weeks (17 years) of all districts to train the ML model and evaluate on the last 52 weeks of all districts with the trained model. For parameter optimization, we split each time series into 18 folds $\{f_1, \ldots, f_{18}\}$ and performed for each parameter combination five evaluations $i \in \{1, \ldots, 5\}$, using the weeks of the folds f_1 to f_{11+i} of all districts as training data and evaluated on the weeks of fold f_{12+i} of all districts. Afterwards, we average the result for the measure $dAUC_{1\%}$ across the five evaluations and then choose the parameter configuration maximizing this score.

3   Results

The key aspect of our experimental evaluation is to demonstrate that the fusion of *p*-values leads to a further improvement in performance compared to only using the binary output of the statistical algorithms. For a deeper understanding of our proposed approaches, we first performed experiments on synthetic data to evaluate the influence of our adaptions on stacking and to compare it with the underlying statistical algorithms in a controlled environment. Afterwards, we test stacking on real data in order to underline its practical utility.

3.1   Experimental Setup

Baselines: As an implementation baseline for the statistical methods, we have used the R package *surveillance* [7] and adapted the implementation of the methods EARS (C1, C2, and C3), Bayes, and RKI so that they also return p-values. All methods use the previous seven time points as reference values, which is the standard configuration. We have evaluated the underlying statistical methods itself which serve as a baseline to which the stacking approaches are compared. In addition, we also evaluated the fusion method which only combines the binary outputs of the statistical methods as proposed in [1,9] and which we refer to as *standard fusion*.

Measures: For all evaluations, we focus on the evaluation measure $dAUC_{1\%}$ proposed above. In addition, we evaluate the conventional area under partial ROC-curve $AUC_{1\%}$ to further investigate the effect of the labeling on the true positive rate.

Parameter Optimization: We have evaluated three ML algorithms: (1) Random Forest with the different values for the minimum number of samples per leaf $\{5, 10, 20, 30\}$, (2) logistic regression with different values for the regularization parameter $\{2^{-3}, 2^{-2}, 2^{-1}, 1, 2^1, 2^2, 2^3\}$, and (3) k-nearest-neighbours with different values for the number of considered neighbours $\{1, 3, 5, 7, 9\}$. For the labeling, we have the parameters $\{O_0, O_1, O_2, O_3\}$, $\{False, True\}$ for the use of mean, and $\{0, 1, 2, 3, 4, 6, 9\}$ for the windowing where 0 represents no windowing. For each algorithm we performed a grid search in which we evaluate all possible parameter combinations. For the standard fusion approach, we have set the threshold of each single statistical method beforehand on the validation data so that it has a false alarm rate of 1% and then performed the parameter optimization. The reported results can always be attributed to one of the random forest configurations since this approach always achieved the best results in the parameter optimization, on the synthetic as well as on the real data.

3.2   Evaluation on Synthetic Data

Comparison to the Baselines: Considering the results of the parameter optimization, we evaluated both fusion approaches with the adaption of the labeling O_3 and including the mean. For the p-value fusion a window size of two and for the standard fusion a window size of four is used. The results in Table 1 clearly show that p-value fusion performs best across all test cases, with average ranks close to 1. In line with [9] and [1], the results show an improvement of the standard fusion approach on the time series without trend and seasonality. However, this improvement is not consistent for all compared test cases, resulting only in an average rank of 3.143. On the other test cases, the fusion of binary outputs obtains often a low rank, making it often worse than the underlying statistical algorithms.

Table 1. Each column shows the average ranks of the baseline algorithms, the standard fusion, and the p-value fusion considering different test case combinations: *overall* denotes all 42 test cases, the other columns cases (not) containing trend ($[\neg]T$), annual ($[\neg]S1$) or biannual ($[\neg]S2$) seasonality.

Approach	Overall	$\{\neg T, \neg S1, \neg S2\}$	$\{\neg T, S1, \neg S2\}$	$\{\neg T, S1, S2\}$	$\{T, \neg S1, \neg S2\}$	$\{T, S1, \neg S2\}$	$\{T, S1, S2\}$
C1	5.738	6.571	6.000	4.571	6.857	5.428	5.000
C2	4.905	4.000	4.429	5.143	4.714	5.429	5.714
C3	4.857	5.143	5.143	4.143	5.571	5.000	4.143
Bayes	2.881	4.143	3.143	3.429	2.000	2.000	2.571
RKI	4.024	4.142	3.571	3.857	4.714	3.571	4.286
Standard fusion	4.500	3.143	4.857	5.286	3.143	5.429	5.143
p-value fusion	**1.381**	**1.143**	**1.143**	**1.857**	**1.286**	**1.429**	**1.429**

Effect of Additional Features and Adaption of the Labeling: From the results of the parameter optimization, we can also observe that including the mean is always beneficial and that a window size between two and three performed best for both fusion approaches. In addition, by narrowing the labeling of the outbreak on particular events (i.e., O_1, O_2 or O_3) a better performance can be achieved. In particular, learning only the peaks (O_3) achieved the best results for both fusion approaches. A more detailed evaluation can be found in [4].

Evaluation with Respect to the Outbreak Size: Furthermore, we evaluated the approaches with respect to the number of cases per outbreak. The results for the measure $dAUC_{1\%}$ across the 42 test cases with a fixed value for the parameter k is visualized as box plots, representing minimum, first quantile, mean, third quantile and maximum, in Fig. 3. In addition to $dAUC_{1\%}$, we include the analysis of the $AUC_{1\%}$ measure and compare to the original labeling O_0 in order to further investigate the effect of the labeling on detection rate and true positive rate.

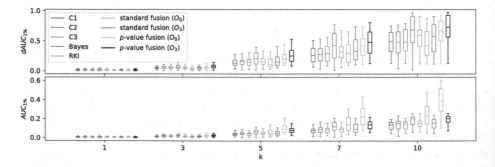

Fig. 3. Results for the measures $dAUC_{1\%}$ and $pAUC_{1\%}$. Each box plot represents the distribution of measure values for a particular method computed over all 42 test cases for a fixed outbreak size defined by the parameter k).

As the outbreak size increases all methods are more likely to detect the outbreak. While the C1, C2, C3, and RKI method achieve similar results across all outbreak sizes, it is surprising to observe that the Bayes method has a better performance in case of larger outbreaks. Regarding the p-value fusion approaches, the results confirm the better overall performance across all outbreak sizes while the performance of the standard fusion approach gets worse compared to the other methods with an increasing number of cases per outbreak. Moreover, as discussed in Sect. 2.2, we can observe that we yield a slightly worse result for the measure $dAUC_{1\%}$ using the basic labeling (O_0) than with adapting the labeling (O_3). A closer examination of the graphs for the measures $dAUC_{1\%}$ and $AUC_{1\%}$ reveals the difference between the adaption of the labeling for learning. In particular, without adaption, the ML algorithm achieves a tremendous better performance for the trade-off between the true positive rate and the false alarm rate. However, this also has an effect on the ability to detect outbreaks as discussed in Sect. 2.2, yielding a slightly worse result for the measure $dAUC_{1\%}$ than with adapting the labeling.

3.3 Evaluation on Real Data

Comparison to the Statistical Surveillance Baselines: Using the optimal parameters found by the parameter optimization, we obtain the results which are shown in Table 2. Only considering the results of the statistical methods, we can observe that the *C2* method achieves the best results for *SAL* and the worst results for *CAM*. This shows the diversity of the data but also highlights the difficulty in choosing a suitable statistical algorithm for a particular disease. Regarding our fusion approaches, we are able to achieve better results compared to the underlying statistical algorithms for the configurations *all* and *min=2*. However, the results for the configuration *min=3* are indicating a somehow unexpected behavior: While the statistical methods can double the $dAUC_{1\%}$ score compared to configuration *min=2*, the stacking approaches are unable to achieve such an improvement. Therefore, we further investigated the results to understand these effects.

Table 2. Results for the measure $dAUC_{1\%}$ evaluated on real data using optimized parameters.

Approach	SAL_{all}	$SAL_{min=2}$	$SAL_{min=3}$	CAM_{all}	$CAM_{min=2}$	$CAM_{min=3}$
C1	0.1545	0.1763	0.3392	0.0473	0.0478	0.0799
C2	0.1599	**0.1867**	**0.3692**	0.0389	0.0383	0.0780
C3	0.1444	0.1668	0.3294	0.0442	0.0441	0.0894
Bayes	0.1212	0.1314	0.2811	0.0398	0.0428	0.0921
RKI	0.1495	0.1730	0.3303	0.0506	0.0526	**0.1159**
Standard fusion	0.1591	0.1869	0.3300	0.1459	0.1457	0.0892
p-value fusion	**0.1732**	**0.1901**	0.3341	**0.1589**	**0.1516**	**0.1300**

Table 3. Results for the measure $dAUC_{1\%}$ evaluated on the training data.

Approach	SAL_{all}	$SAL_{min=2}$	$SAL_{min=3}$	CAM_{all}	$CAM_{min=2}$	$CAM_{min=3}$
Standard fusion	0.1982	0.2013	0.2693	0.1369	0.1404	0.1808
p-value fusion	0.2745	0.3006	0.5668	0.7208	0.7359	0.7548

Inconsistencies: From a machine learning point of view, an important reason for low performance of learning algorithms are inconsistencies in the data. Roughly speaking, inconsistent data points can not or only hardly be discriminated by the learning algorithm, but are associated with different outcomes (e.g., *outbreak* yes/no). Our hypothesis is that such inconsistencies are the main cause for the observed results. They can be identified by analyzing how well the model can adapt to the observed data. Table 3 shows the training set performance of random forests, which are generally capable of memorizing observed data well [12]. The low values for SAL indicate that the dataset contains a high ratio of inconsistencies from the perspective of the learner. Especially small outbreaks are difficult to differentiate from ordinary cases.

Heterogeneous Data Sources: As already mentioned, in addition to undiagnosed patients, our dataset may contain cases which were erroneously not attributed to an epidemic outbreak, or even undetected outbreaks. Furthermore, the unequal distribution of outbreaks (cf. Fig. 2) indicates that there may be some heterogeneity in the districts' policies for reporting of cases and labelling of outbreaks. To support this assumption, Fig. 4 shows the results for CAM_{all} for two districts. The alarms which would have been triggered if the false alarm rate on all evaluated districts was set to 1%, respectively for each approach separately, are visualized on top of the graphs. Furthermore, the red lines indicate the outbreaks, which are as we see sometimes overlapping. The stacking approaches achieve excellent results for district A while for district B the predictions do not fit at all. Since a global model is learned, the ML algorithm is not able to differentiate between the districts. Therefore, patterns are learned which work best across all districts, even though the predictions for particular

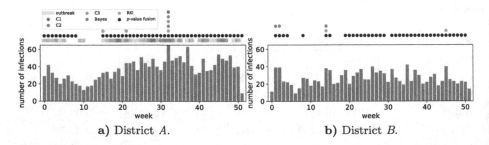

a) District A.   b) District B.

Fig. 4. Exemplary results for two districts associated to two major cities for CAM_{all}. The red lines above the plot indicate ongoing outbreaks at the respective time steps. (Color figure online)

Table 4. Results of the p-value fusion approach for the measure $dAUC_{1\%}$ evaluated on the test data by including or excluding the location.

Approach	Location	SAL_{all}	$SAL_{min=2}$	$SAL_{min=3}$	CAM_{all}	$CAM_{min=2}$	$CAM_{min=3}$
p-value fusion	No	0.1732	0.1901	0.3341	0.1589	0.1516	0.1300
p-value fusion	Yes	0.1867	0.1968	0.3369	0.1689	0.1694	0.1501

districts are incorrect. This assumption is somehow confirmed by our experiments which include features which allow to differentiate between locations. As it can be seen from Table 4, the detection quality benefits from including this additional information.

Coarse Grained Observations: Apart from heterogeneous data, we also face problems due to the aggregation level of the data. In particular for the high populated districts, the reports of multiple different health offices are merged together which makes the identification of small local outbreaks difficult due to the high endemic load. Furthermore, the aggregation over a high population also raises the probability of having many small outbreaks which causes to be in an outbreak for almost all time steps as it can be seen for district A in Fig. 4a. Especially during the learning process, the constant labeling makes it difficult for the ML approaches to identify consistent patterns for outbreaks, since arbitrary increases, as well as decreases of the number infections, are labeled as positive. Our analysis reveals that the p-value approach learns the pattern that there is a high probability of being in an outbreak if the number of total cases is high, which is typically the case for the data of district A, but not for district B.

4 Conclusion

In this work, we introduced an approach for the fusion of outbreak detection methods. We improved over existing approaches, which use the binary output (*alarm* or *no alarm*) of the underlying statistical algorithms as inputs to the learner, by incorporating the p-values instead. For evaluation, we proposed a measure based on partial ROC curves which better adapts to the specific need for a very low false alarm rate in outbreak detection but still considers the trade-off with the detection rate.

Our experimental results on synthetic and real data show that the fusion of p-values improves the performance compared to the underlying statistical algorithms. Contrary to previous work, we could also observe that simple fusion of binary outputs using stacking does not always lead to an improvement. By incorporating additional information to the learning data and by adapting the labeling the supervised machine learning algorithm is able to capture more reliable patterns to detect outbreaks. However, on real data, we face several issues regarding inconsistencies, heterogeneous labeling and the aggregation level of the data which all need to be considered when learning a fusion classifier.

For future work, we plan to analyze more deeply how to pre-process the outbreak annotations in order to optimize the detection of outbreaks. Moreover, stacking allows enriching the detection by additional signals and sources of information in a highly flexible way, such as local weather data or data from neighbor districts, on which further investigations can be spent.

Acknowledgments. This work was supported by the Innovation Committee of the Federal Joint Committee (G-BA) [ESEG project, grant number 01VSF17034]. We thank our project partners the *Health Protection Authority of Frankfurt*, the *Hesse State Health Office and Centre for Health Protection*, the *Hesse Ministry of Social Affairs and Integration*, the *Robert Koch-Institut*, the *Epias GmbH* and the *Sana Klinikum Offenbach GmbH* who provided insight and expertise that greatly assisted the research. Especially, we thank Linus Grabenhenrich, Alexander Ulrich, Theresa Kocher, Madlen Schranz, Sonia Boender and Birte Wagner from the *Robert Koch-Institut* for their valuable feedback, that substantially improved the manuscript, and for providing us the data for the evaluation.

References

1. Jafarpour, N., Precup, D., Izadi, M., Buckeridge, D.: Using hierarchical mixture of experts model for fusion of outbreak detection methods. In: Annual Symposium Proceedings 2013, pp. 663–669, November 2013
2. Jordan, M.I., Jacobs, R.A.: Hierarchical mixtures of experts and the EM algorithm. Neural Comput. **6**(2), 181–214 (1994)
3. Kleinman, K.P., Abrams, A.M.: Assessing surveillance using sensitivity, specificity and timeliness. Stat. Methods Med. Res. **15**(5), 445–464 (2006)
4. Kulessa, M., Loza Mencía, E., Fürnkranz, J.: Improving outbreak detection with stacking of statistical surveillance methods. In: Workshop Proceedings of epi-DAMIK: Epidemiology meets Data Mining and Knowledge Discovery (held in conjunction with ACM SIGKDD 2019) (2019). Also as preprint arXiv:1907.07464
5. Ma, H., Bandos, A.I., Rockette, H.E., Gur, D.: On use of partial area under the ROC curve for evaluation of diagnostic performance. Stat. Med. **32**(20), 3449–3458 (2013)
6. Noufaily, A., Enki, D.G., Farrington, P., Garthwaite, P., Andrews, N., Charlett, A.: An improved algorithm for outbreak detection in multiple surveillance systems. Stat. Med. **32**(7), 1206–1222 (2013)
7. Salmon, M., Schumacher, D., Höhle, M.: Monitoring count time series in R: aberration detection in public health surveillance. J. Stat. Softw. **70**(10), 1–35 (2016)
8. Shmueli, G., Burkom, H.: Statistical challenges facing early outbreak detection in biosurveillance. Technometrics **52**(1), 39–51 (2010)
9. Texier, G., Allodji, R.S., Diop, L., Meynard, J., Pellegrin, L., Chaudet, H.: Using decision fusion methods to improve outbreak detection in disease surveillance. BMC Med. Inform. Decis. Mak. **19**(1), 38 (2019)
10. Ting, K., Witten, I.: Issues in stacked generalization. J. Artif. Intell. Res. **10**, 271–289 (1999)
11. Wolpert, D.H.: Stacked generalization. Neural Netw. **5**(2), 241–259 (1992)
12. Wyner, A.J., Olson, M., Bleich, J., Mease, D.: Explaining the success of AdaBoost and random forests as interpolating classifiers. J. Mach. Learn. Res. **18**(48), 1–33 (2017)

Learning Cancer Drug Sensitivities in Large-Scale Screens from Multi-omics Data with Local Low-Rank Structure

The Tien Mai[1]([⊠])[iD], Leiv Rønneberg[1]([⊠]), Zhi Zhao[1][iD], Manuela Zucknick[1][iD], and Jukka Corander[1,2][iD]

[1] Oslo Centre for Biostatistics and Epidemiology, Department of Biostatistics, University of Oslo, Oslo, Norway
{t.t.mai,leiv.ronneberg}@medisin.uio.no
[2] Department of Mathematics and Statistics, University of Helsinki, Helsinki, Finland

Abstract. The molecular characterization of tumor samples by multiple omics data sets of different types or modalities (e.g. gene expression, mutation, CpG methylation) has become an invaluable source of information for assessing the expected performance of individual drugs and their combinations. Merging relevant information from the omics data modalities provides the statistical basis for determining suitable therapies for specific cancer patients. Different data modalities may each have their own specific structures that need to be taken into account during inference. In this paper, we assume that each omics data modality has a low-rank structure with only a few relevant features that affect the prediction and we propose to use a composite local nuclear norm penalization for learning drug sensitivity. Numerical results show that the composite low-rank structure can improve the prediction performance compared to using a global low-rank approach or elastic net regression.

Keywords: Drug sensitivity · Composite local low-rank · Multi-omics · Nuclear norm penalization · Multi-view · Multi-task · Pharmacogenomic screens · Genomics of Drug Sensitivity in Cancer

1 Introduction

In recent years, large-scale in-vitro pharmacological profiling of cancer drugs on a panel of cancer cell lines, which are well characterised by multiple omics data sets, have been proposed as a promising route to precision medicine [2,4,11, 21]. The omics data can for example consist of genome-wide measurements of mRNA expression, DNA copy numbers, DNA single point and other mutations

The first two authors contributed equally. L.R. is supported by The Norwegian Research Council 237718 through the Big Insight Center for research-driven innovation. The research of T.T.M. and J.C. are supported by the European Research Council (SCARABEE, no. 742158).

or CpG methylation of cell lines, see e.g. [6,13]. These measurements reflect
different molecular profiles of the cancer cell lines, which are heterogeneous with
respect to effect sizes, intra-correlations, measurement scales and signal-to-noise
ratios [14]. The response or sensitivity of a cell line to a drug is characterized
by parameters estimated from the dose-response curve, for example by the half
maximal inhibitory concentration (IC_{50}) [11], or the drug sensitivity score [20].

One of the essential problems in personalized oncology is to predict the sensitivities of cancer cell lines to a panel of drugs [10,16]. There are various computational machine learning methods that have been proposed to tackle the
drug response prediction problem. For example, matrix factorization methods
have been used in [17,19], kernel-based methods are proposed in [8,9], penalized
regression methods have also been studied in [11,12,24], Bayesian approach and
deep learning method have also been applied to this problem [3,5,8]. Discussion as well as comparison of these methods are well documented in some recent
reviews [1,6,13].

Combining different data sources and prior biological knowledge can clearly
help to shed light on the complexities of cancer drug sensitivity prediction. More
specifically, merging relevant information from the omics data modalities provides the statistical basis for determining suitable therapies for specific cancer
patients. Most of the previous approaches using combined multiple omics data
employ a global structure for parameter inference such as low-rank or sparsity.
However, in this application each data source has its own specific structure that
is important for capturing the effects of drug sensitivity.

In this paper, we consider that each omics data modality has a low-rank
structure with only few relevant features that affect the responses (drugs sensitivities). Borrowing motivation from a local low-rank structure explored in a
recent work [15] in another application domain, we propose to use a local low-
rank model for predicting drug sensitivity with multi-omics data. To the best of
our knowledge, it is the first time a composite local low-rank structure is applied
in the context of drug sensitivity prediction with multi-omics data. A composite
nuclear norm penalization is employed to learn the model.

The paper is organized as follow. In Sect. 2, the problem formulation is given
with its local low-rank assumption and the composite nuclear norm inference
method is presented. Section 3 is devoted to extensive numerical studies. Application to a real data set from the Genomics of Drug Sensitivity in Cancer (GDSC)
[21] is presented in Sect. 4. Some discussion and conclusion are in Sect. 5.

2   Materials and Methods

2.1   Model

Let $\mathbf{Y} = \{y_{ij}\} \in \mathbb{R}^{n \times q}$ denote the pharmacological data representing the sensitivities of n samples (e.g. cell lines or patients) to q drugs. We observe K different
(high-dimensional) omics data sets that contain p_k features, $\mathbf{X}_k \in \mathbb{R}^{n \times p_k}$ for each
$k = 1, \ldots, K$. In total, there are $p = \sum_{k=1}^{K} p_k$ features available for n samples
across the K different data modalities.

Let's denote the linear model mapping from high-dimensional covariates data to multivariate responses as

$$\mathbf{Y} = \sum_{k=1}^{K} \mathbf{X}_k \mathbf{B}_k + \mathbf{E} = \mathbf{X}\mathbf{B} + \mathbf{E} \tag{1}$$

where $\mathbf{B} = (\mathbf{B}_1^\top, \ldots, \mathbf{B}_K^\top)^\top \in \mathbb{R}^{p \times q}$ is the unknown regression coefficient matrix partitioned according to the predictor groups $\mathbf{X} = (\mathbf{X}_1, \ldots, \mathbf{X}_K) \in \mathbb{R}^{n \times p}$. The random errors $\mathbf{E} \in \mathbb{R}^{n \times q}$ are assumed to be zero-mean, where specific correlation structures are examined in the simulation study.

We assume, under the model (1), that each omics data set $\mathbf{X}_k$ has its own low-rank coefficient matrix $\mathbf{B}_k$. This means that only a few relevant (latent) features in $\mathbf{X}_k$ affect the drugs sensitivities $\mathbf{Y}$. Note that this local low-rank assumption does not necessarily imply a low-rank structure of the whole coefficient matrix $\mathbf{B}$.

2.2 Learning Method

To estimate a local low-rank structure as assumed in (1), we propose to use a composite nuclear-norm penalization

$$\hat{\mathbf{B}}_{CLR} = \arg \min_{\mathbf{B} \in \mathbb{R}^{p \times q}} \frac{1}{2n} \|\mathbf{Y} - \mathbf{X}\mathbf{B}\|_2^2 + \lambda \sum_{k=1}^{K} w_k \|\mathbf{B}_k\|_*, \tag{2}$$

where $\lambda > 0$ is a tuning parameter and $w_k \geq 0$ are pre-specified weights. Here $\|\mathbf{B}\|_s = (\sum_{ij} |\mathbf{B}_{ij}|^s)^{1/s}$ denotes the matrix ℓ_s-norm and $\|\mathbf{B}\|_* = \sum_{j=1}^{\text{rank}(\mathbf{B})} \sigma_j(\mathbf{B})$ is the nuclear norm with $\sigma_j(\cdot)$ denoting the jth largest singular value of the enclosed matrix.

The weights are used to balance the penalization of different modalities and allows us to use only a single tuning parameter. As studied in [15], the choice

$$w_k = \sigma_1(\mathbf{X}_k)(\sqrt{q} + \sqrt{\text{rank}(\mathbf{X}_k)})/n \tag{3}$$

is used to to adjust for the dimension and scale differences of $\mathbf{X}_k$ n the case of independent and identically distributed (i.i.d.) Gaussian entries $\mathbf{E}$.

Remark 1. Note that problem (2) covers other well-known problems as its special cases such as:

i) global low-rank structure [7], also known as nuclear norm penalized regression: with $p_1 = p, K = 1$ the penalty in (2) becomes the nuclear norm penalty of the whole parameter matrix $\mathbf{B}$.

ii) multi-task learning [23]: with $p_k = 1, p = K$, (2) becomes a special case of multi-task learning that all the tasks share the same set of features and samples.

iii) Lasso and group-Lasso [18,22]: with $q = 1$, (2) becomes a group-Lasso estimator and moreover when $p_k = 1, p = K$ it leads to a Lasso problem.

Remark 2. Some theoretical results of the composite local low-rank estimator (2) have been laid out in [15] for a specific case where $\mathbf{E}$ has i.i.d Gaussian entries. More specifically, non-asymptotic bounds for estimation and prediction are given in the Theorem 1 and Theorem 2 in [15].

2.3 Including Full-Rank Non-omics Data Sources

Often, in real data applications, there are additional data sources available besides omics. These sources are for example clinical variables about patients or cell lines, such as indicators of cancer types, or established predictive markers that we might want to include with their full effect size.

We can extend our model (1) to take into account these additional high-rank data. Formally, let $\mathbf{X}_0 \in \mathbb{R}^{n \times p_0}$ be the additional data. Model (1) becomes

$$\mathbf{Y} = \sum_{k=0}^{K} \mathbf{X}_k \mathbf{B}_k + \mathbf{E},$$

and we can adapt the composite nuclear-norm penalization method (2) by setting the weight $w_0 = 0$. That is

$$\hat{\mathbf{B}}_{CLR} = \arg \min_{\mathbf{B} \in \mathbb{R}^{p \times q}} \frac{1}{2n} \|\mathbf{Y} - \mathbf{X}\mathbf{B}\|_2^2 + \lambda \sum_{k=1}^{K} w_k \|\mathbf{B}_k\|_* + 0 \cdot \|\mathbf{B}_0\|_*.$$

3   Numerical Study

In this section, we conduct simulation studies to examine the performance of the proposed composite local low-rank method. The R code to reproduce the results is available at: https://github.com/ocbe-uio/LocalLowRankLearning.

3.1 Simulation Setups and Details

Using the dimensionalities: $q = 24, n = 200$, $K = 2$, the data are simulated as in linear model (1), where $\mathbf{X} = [\mathbf{X}_1, \mathbf{X}_2]$. Each $\mathbf{X}_k (k = 1, 2)$ is generated from a multivariate normal distribution with mean $\mathbf{0}$ and covariance matrix $\mathbf{\Sigma}_X$. The covariance matrix $\mathbf{\Sigma}_X$ has the diagonal values equal to 1 and all off-diagonal elements equal to $\rho_X \geq 0$. To take into account the correlation between the drugs, we simulate the noise $\mathbf{E}$ from a multivariate normal distribution with mean $\mathbf{0}$ and covariance matrix $\mathbf{\Sigma}_\epsilon$. The covariance matrix $\mathbf{\Sigma}_\epsilon$ has the diagonal values equal to 1 and all off-diagonal elements equal to $\rho_\epsilon \geq 0$.

We vary different correlation setups in omics data $\mathbf{X}$ and the drugs as follows:

- fix $\rho_X = 0$ and vary ρ_ϵ as $0.0; 0.3; 0.6$,
- fix $\rho_\epsilon = 0$ and vary ρ_X as $0.3; 0.6; 0.9$.

Then, for each of the above setups, we consider various settings for the true coefficient matrix $\mathbf{B} = (\mathbf{B}_1^\top, \mathbf{B}_2^\top)^\top$ as:

- S1 *(local low-rank)*: each $\mathbf{B}_k, k = 1, 2$ is a low-rank matrix with $\text{rank}(\mathbf{B}_1) = 5$, $\text{rank}(\mathbf{B}_2) = 20$ which is generated as $\mathbf{B}_k = \mathbf{L}_{\text{rank}(\mathbf{B}_k)} \mathbf{R}_{\text{rank}(\mathbf{B}_k)}^{\top}$ with the entries of $\mathbf{L}_{\text{rank}(\mathbf{B}_k)} \in \mathbb{R}^{p_k \times \text{rank}(\mathbf{B}_k)}$ and $\mathbf{R}_{\text{rank}(\mathbf{B}_k)} \in \mathbb{R}^{q \times \text{rank}(\mathbf{B}_k)}$ both generated from $\mathcal{N}(0, 1)$.
- S2 *(low-rank and sparse)*: $\mathbf{B}_1$ is low-rank as in S1 and $\mathbf{B}_2$ is a sparse matrix where 50% of the elements are non-zero and simulated from $\mathcal{N}(0, 1)$.
- S3 *(global low-rank)*: the whole matrix $\mathbf{B}$ is a rank-6 matrix simulated as in S1.
- S4 *(global sparsity)*: the whole matrix $\mathbf{B}$ is sparse where 20% of the elements are non-zero and simulated from $\mathcal{N}(0, 1)$.

Finally, model performance will depend on the dimensionality of the predictor datasets compared to the number of observations (the ratio p/n), as well as the extent of implicit dimensionality reduction taking place through the low-rank restriction on $\mathbf{B}_k$ (the ratio $\text{rank}(\mathbf{B}_k)/p_k$), in settings S1 and S2. To explore both these situations, we keep the number of observations and rank of $\mathbf{B}_k$s fixed across simulations, but consider two different settings of p_1 and p_2:

- $p_1 = p_2 = 150$,
- $p_1 = p_2 = 500$.

We compare the composite local low-rank estimator (**CLR**) in (2) with a global low-rank estimator (**GLR**) for the reduced-rank regression,

$$\hat{\mathbf{B}}_{GLR} = \arg \min_{\mathbf{B} \in \mathbb{R}^{p \times q}} \frac{1}{2n} \|\mathbf{Y} - \mathbf{XB}\|_2^2, \quad s.t \quad \text{rank}(\mathbf{B}) \leq r,$$

where $r = \min(p, q)$, and the elastic net (sparsity-inducing) estimator (**Enet**)

$$\hat{\mathbf{B}}_{enet} = \arg \min_{\mathbf{B} \in \mathbb{R}^{p \times q}} \frac{1}{2n} \|\mathbf{Y} - \mathbf{XB}\|_2^2 + \lambda(\alpha \|\mathbf{B}\|_1 + (1 - \alpha) \|\mathbf{B}\|_2^2).$$

The algorithm used to estimate the CLR model (2) is based on the original MAT-LAB code from [15]. We use an implementation of the reduced-rank regression from R package 'rrpack'[1] where the rank is chosen by cross-validation. For the elastic net, we use the R package 'glmnet'[2] with λ and α chosen by 5-fold cross-validation.

The evaluations are done by using the mean squared estimation error (**MSEE**) and the mean squared prediction error (**MSPE**)

$$\text{MSEE} := \frac{1}{pq} \|\hat{\mathbf{B}} - \mathbf{B}\|_2^2, \quad \text{MSPE} := \frac{1}{nq} \|\mathbf{Y} - \mathbf{X}\hat{\mathbf{B}}\|_2^2,$$

evaluated on an independent test set. Note that in real-world applications, where the true $\mathbf{B}$ is not known, we can only access the prediction errors. We repeat each experiment setting 100 times and report the mean of the outputs.

---

[1] https://cran.r-project.org/package=rrpack.
[2] https://cran.r-project.org/package=glmnet.

3.2 Numerical Results

Overall, the numerical simulations confirm our intuitions: the proposed CLR method has the smallest prediction error in settings where there is in fact a true, local low rank structure (S1), the GLR when there is a global low rank structure (S3), and elastic net under a global sparsity setting (S4). This can be seen from Table 1 and Table 3. In addition, though the increased dimension of the covariates from $p = 300$ to $p = 1000$ makes the prediction task more difficult, it does not change the overall picture.

In terms of prediction error, all methods seem to do well under increasing correlation between the drugs (Table 1), as the MSPEs does not increase drastically with increasing ρ_ϵ. However, the methods fail to cope with situations where correlations between the covariates are increasing (Table 3). Particularly in the scenario $p = 1000$, where high correlations between high-dimensional predictors really makes the errors blow up. It is also interesting to note that the low-rank methods are often comparable to elastic net in the global sparse setting (S4). This relationship persists across correlation structures and dimensionality, and is mirrored in the estimation errors. Intuitively, there should be some connection between sparsity and low rank, since very sparse matrices can be constructed to also have low rank. It is clear from looking at the MSEE that the low rank estimators are able to get "close" to the elastic net solution in sparse settings. The effect also goes the other way, where elastic net will perform on par with GLR in a global low rank setting (S3), and beat it in settings where more local adaptation is needed (S1 and S2).

The "low-rank and sparse setting" (S2), is interesting. Our CLR model is the top performer, followed by elastic net, and finally GLR. The performance of CLR in this setting could be explained firstly by it's ability to adapt locally to the different components of $\mathbf{B}$, secondly by the ability of low-rank methods to adapt to a sparse setting. Elastic net beating out GLR could be due to it being better at adapting to local differences between the components, which could be the reason why it is beating GLR also in the local low rank setting, S1.

In terms of the estimation errors (Table 2 and Table 4), all models seems relatively robust against increasing correlation, both between drugs, and between covariates. It is surprising how similar GLR and elastic net are in settings with a global structure (S3 and S4), particularly in the high-dimensional setting of $p = 1000$.

3.3 Additional Results Including Nonpenalized Coefficients

We further test the performance of our method in the case with some additional non-penalized coefficients. The simulation is done with $n = 200$, $q = 24$, $K = 2$, p_1, p_2 as in the two previous scenarios, and the clinical variables $p_0 = 15$. Here $\mathbf{X} = [\mathbf{X}_0, \mathbf{X}_1, \mathbf{X}_2]$ where $\mathbf{X}_1, \mathbf{X}_2$ are simulated as above with $\rho_X = \rho_\epsilon = 0$ while the entries of $\mathbf{X}_0$ are binaries and generated from a Bernoulli distribution with

Table 1. MSPE with fixed $\rho_X = 0$ and ρ_ϵ is varied.

$p = 300$	$\rho_\epsilon = 0$			$\rho_\epsilon = 0.3$			$\rho_\epsilon = 0.6$		
	CLR	GLR	Enet	CLR	GLR	Enet	CLR	GLR	Enet
S1 *(local low-rank)*	**433.5**	1254.5	1029.6	**442.3**	1248.4	1030.9	**462.4**	1279.4	1051.4
S2 *(low-rank & sparse)*	**92.1**	277.5	194.0	**93.5**	277.6	194.8	**98.1**	282.2	197.1
S3 *(global low-rank)*	620.8	**613.6**	685.6	613.1	**605.6**	678.6	602.6	**594.7**	664.1
S4 *(global sparsity)*	23.5	23.3	**19.1**	24.2	24.3	**20.1**	26.5	27.2	**22.3**
$p = 1000$									
S1 *(local low-rank)*	**8867.1**	10 014.3	9978.4	**8856.6**	9993.4	9963.6	**8845.9**	9996.9	9973.2
S2 *(low-rank & sparse)*	**1915.2**	2237.5	2229.2	**1867.8**	2181.0	2171.9	**1873.1**	2198.0	2190.6
S3 *(global low-rank)*	4751.6	**4724.5**	4761.1	4750.6	**4727.0**	4768.5	4939.6	**4918.3**	4958.9
S4 *(global sparsity)*	160.9	160.2	**158.3**	161.3	160.6	**158.8**	163.5	162.8	**161.0**

Table 2. MSEE with fixed $\rho_X = 0$ and ρ_ϵ is varied.

$p = 300$	$\rho_\epsilon = 0$			$\rho_\epsilon = 0.3$			$\rho_\epsilon = 0.6$		
	CLR	GLR	Enet	CLR	GLR	Enet	CLR	GLR	Enet
S1 *(local low-rank)*	**1.456**	4.195	3.448	**1.482**	4.185	3.456	**1.542**	4.285	3.524
S2 *(low-rank & sparse)*	**0.308**	0.933	0.653	**0.310**	0.929	0.650	**0.320**	0.938	0.653
S3 *(global low-rank)*	2.065	**2.039**	2.285	2.068	**2.043**	2.279	2.033	**2.006**	2.247
S4 *(global sparsity)*	0.075	0.075	**0.060**	0.077	0.077	**0.063**	0.081	0.083	**0.067**
$p = 1000$									
S1 *(local low-rank)*	**8.919**	10.078	10.048	**8.859**	10.018	9.989	**8.903**	10.067	10.038
S2 *(low-rank & sparse)*	**1.925**	2.250	2.243	**1.861**	2.169	2.162	**1.885**	2.208	2.200
S3 *(global low-rank)*	4.804	**4.779**	4.816	4.772	**4.751**	4.792	4.993	**4.969**	5.009
S4 *(global sparsity)*	0.161	0.160	**0.158**	0.161	0.160	**0.158**	0.162	0.161	**0.159**

Table 3. MSPE with fixed $\rho_\epsilon = 0$ and ρ_X is varied.

$p = 300$	$\rho_X = 0.3$			$\rho_X = 0.6$			$\rho_X = 0.9$		
	CLR	GLR	Enet	CLR	GLR	Enet	CLR	GLR	Enet
S1 *(local low-rank)*	**598.9**	1761.7	1440.3	**1063.2**	3099.5	2555.8	**4228.6**	12 267.0	10 099.0
S2 *(low-rank & sparse)*	**127.8**	396.1	274.0	**218.1**	685.8	476.4	**850.4**	2668.0	1855.9
S3 *(global low-rank)*	843.9	**834.4**	948.0	1483.3	**1469.0**	1712.9	5978.9	**5909.1**	7517.9
S4 *(global sparsity)*	31.4	31.1	**25.3**	53.4	52.8	**43.1**	203.7	200.2	**162.9**
$p = 1000$									
S1 *(local low-rank)*	**12 528.3**	14 152.4	14 123.9	**21 765.6**	24 657.3	24 642.7	**89 166.8**	100 644.8	101 163.5
S2 *(low-rank & sparse)*	**2723.7**	3181.8	3170.6	**4630.4**	5422.5	5401.5	**18 628.9**	21 793.1	21 776.3
S3 *(global low-rank)*	6794.3	**6760.6**	6821.9	12 112.2	**12 029.5**	12 125.8	47 975.2	**47 726.9**	48 091.4
S4 *(global sparsity)*	228.8	228.4	**226.0**	400.4	400.0	**395.3**	1598.7	1594.2	**1577.4**

probability 0.5. Then, for the true coefficient matrix $\mathbf{B} = (\mathbf{B}_0^\top, \mathbf{B}_1^\top, \mathbf{B}_2^\top)^\top$, we consider both the local low-rank setting S1 and the low-rank and sparse setting (S2) for $\mathbf{B}_1^\top, \mathbf{B}_2^\top$ while the entries of $\mathbf{B}_0^\top$ are generated from $\mathcal{N}(0,1)$.

Table 4. MSEE with fixed $\rho_\epsilon = 0$ and ρ_X is varied.

| | $\rho_X = 0.3$ | | | $\rho_X = 0.6$ | | | $\rho_X = 0.9$ | | |
$p = 300$	CLR	GLR	Enet	CLR	GLR	Enet	CLR	GLR	Enet
S1 *(local low-rank)*	**1.463**	4.198	3.448	**1.509**	4.238	3.521	**1.500**	4.206	3.476
S2 *(low-rank & sparse)*	**0.314**	0.943	0.658	**0.311**	0.942	0.659	**0.303**	0.915	0.642
S3 *(global low-rank)*	2.009	**1.988**	2.258	2.011	**1.992**	2.314	2.038	**2.014**	2.555
S4 *(global sparsity)*	0.072	0.072	**0.058**	0.071	0.071	**0.057**	0.069	0.068	**0.055**

$p = 1000$									
S1 *(local low-rank)*	**8.809**	9.949	9.928	**8.806**	9.962	9.957	**8.991**	10.150	10.203
S2 *(low-rank & sparse)*	**1.923**	2.247	2.239	**1.867**	2.187	2.179	**1.875**	2.195	2.193
S3 *(global low-rank)*	4.792	**4.771**	4.815	4.834	**4.804**	4.842	4.831	**4.803**	4.842
S4 *(global sparsity)*	0.161	0.161	**0.159**	0.161	0.161	**0.159**	0.161	0.160	**0.159**

Table 5. Results of CLR method with additional clinical data, in the settings S1 and S2 with $\rho_\epsilon = \rho_X = 0$.

$p = 300$	MSEE	$\frac{1}{p_0 q}\|\hat{\mathbf{B}}_0 - \mathbf{B}_0\|_2^2$	$\frac{1}{p_1 q}\|\hat{\mathbf{B}}_1 - \mathbf{B}_1\|_2^2$	$\frac{1}{p_2 q}\|\hat{\mathbf{B}}_2 - \mathbf{B}_2\|_2^2$	MSPE
S1 *(local low-rank)*	3.116	16.137	1.534	3.396	797.808
S2 *(low-rank & sparse)*	0.527	2.741	0.625	0.207	135.449

$p = 1000$					
S1 *(local low-rank)*	14.182	66.365	10.432	16.367	13 543.627
S2 *(low-rank & sparse)*	2.622	38.170	3.551	0.627	2210.586

We can see from Table 5 that our method still works well with additional clinical data. Particularly in setting S2, which seems to be a slightly easier estimation problem for the CLR model. For the scenario A, where $p = 300$, we can see this by inspecting the estimated ranks for the matrices involved in Fig. 1. In the top row are estimated ranks for setting S1, where the true ranks are 5 and 20, which the model is not able to recover exactly. We see that the model is slightly over- and underestimating the ranks involved, however, it is correctly identifying lower vs. higher rank modalities. In the bottom row, the S2 setting, the true ranks are 5 for the low-rank matrix $\mathbf{B}_1$, and 24 for the full rank, sparse matrix $\mathbf{B}_2$. We see that this is recovered precisely by the model, and yields a better performance in terms of MSPE and MSEE. For the high-dimensional setting where $p = 1000$, which we have not plotted, the model is not able to recover the true ranks precisely in either of the two settings – the higher dimensionality makes for a more difficult estimation problem.

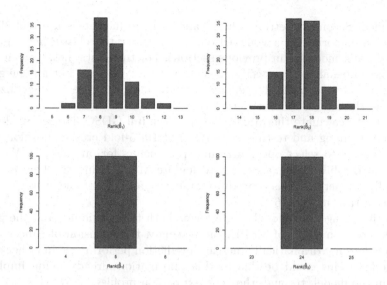

Fig. 1. Histogram of the estimated rank of the coefficient matrices $\hat{\mathbf{B}}_1$ and $\hat{\mathbf{B}}_2$ with $n = 200, q = 24, p_1 = p_2 = 150, p_0 = 15$ over 100 repetitions in setting S1 (top row), and setting S2 (bottom row).

4 Real Data Analysis: GDSC Data

To test our approach on a real dataset, we use data from a large-scale pharmacogenomic study, the Genomics of Drug Sensitivity in Cancer [21], made available online[3] by Garnett *et al.* [11]. The dataset consists of drug sensitivity data (IC$_{50}$) for a large panel of cancer drugs screened on multiple cell lines, in addition to various omics measurements from the cancer cell lines.

We select a subset of the data, consisting of 97 screened drugs and 498 human cancer cell lines from 13 cancer tissue types, such that the response matrix $\mathbf{Y} \in \mathbb{R}^{498 \times 97}$ is fully observed. For each cell line, they measured the mutation status for a panel of known cancer genes, and genome-wide copy number variation and gene expression. For data preprocessing and feature selection, we follow the procedure in [24], which results in

- 2602 continuous gene expression features ($\mathbf{X}_1 \in \mathbb{R}^{498 \times 2602}$),
- 426 integer copy number features ($\mathbf{X}_2 \in \mathbb{R}^{498 \times 426}$),
- 68 binary mutation features ($\mathbf{X}_3 \in \mathbb{R}^{498 \times 68}$),

respectively, and the drug sensitivity being measured as $\log \mathrm{IC}_{50}$. In addition, cancer tissue indicators are available and can be included in high-rank form as in Sect. 2.3. We denote these indicators as $\mathbf{X}_0 \in \mathbb{R}^{498 \times 13}$. Finally, $\mathbf{X}_1$ and $\mathbf{X}_2$ are centered and standardized before weights are calculated.

We evaluate the performance of our model on the GDSC dataset in two different settings. First, a fully penalized setting where only the omics datasets

---

[3] ftp://ftp.sanger.ac.uk/pub4/cancerrxgene/releases/release-5.0/.

are included. Second, a setting where cancer tissue indicators are included in an unpenalized form. As a comparison, we consider the elastic net, and the global low rank model as in previous sections. For the elastic net, we can leave parameters unpenalized by specifying custom penalty factors, but for the global low rank model we cannot. Hence, cancer tissue indicators are also penalized as part of the global low rank restriction.

In order to calculate the MSPE, we split the full GDSC dataset 80/20 randomly into training and testing sets. We perform 5-fold cross validation using the training set to select optimal tuning parameters for our models. We then fit these to the full training set, and calculate MSPE using the held out test set. Finally, we perform this procedure ten times, and report the average MSPE across these runs in Table 6.

In both settings, our model outperforms both the elastic net and the global low rank model in terms of MSPE. Interestingly, the inclusion of cancer tissue indicators does not dramatically increase prediction performance in either of the three models. This could be due to this information already being implicitly present in the model, through the gene expression profiles.

On average across 10 runs, the estimated ranks of each omics data source, $\mathbf{B}_1 \in \mathbb{R}^{2602 \times 97}, \mathbf{B}_2 \in \mathbb{R}^{426 \times 97}, \mathbf{B}_3 \in \mathbb{R}^{68 \times 97}$, from the composite low-rank model are 12.4, 2.2 and 7.0 respectively. Including the non-penalized cancer tissue indicators $\mathbf{X}_0$ reduces the average estimated rank of $\mathbf{B}_1$ to 8.7 (rank of $\mathbf{B}_2$ to 2.1, rank of $\mathbf{B}_3$ to 6.3), which could indicate that part of the information contained in the gene expression profiles is also found in the cancer tissue indicators. For the elastic net, in all runs $\alpha = 0$ was preferred, corresponding to a standard ridge regression, while the global low rank model estimated global ranks between 1 and 21, with an average of 9.3 in both settings. Across CV runs, the estimated error in the global low rank model is relatively similar in the interval between rank 1 and 20. This could also explain why the MSPE is virtually identical between the two settings, with the model essentially fitting an intercept, not really utilizing the omics datasets fully. The same can be said of the elastic net, which seems to prefer a very high level of regularization, averaging at $\lambda = 34.5$, which could mean that very little of the omics information is being utilized in the estimation, and the model is basically fitting only an intercept per tissue.

Table 6. MSPE with real data.

	CLR	GLR	Enet
Real data	**3.0404**	3.4437	3.0796
Real data w/ tissue	**2.9803**	3.4437	3.0299

We can also decompose the MSPE down to individual drugs, and study prediction performance on a per-drug basis. In general, the MSPE varies a lot across different drugs. As an example, the MSPE of the antimetabolite Gemcitabine is 20.0698 for the GLR model, while for Bicalutamide, a testosterone blocker used in treatment of e.g. prostate cancer, it is only 0.3398. While initially surprising, it is a reflection of the heterogeneity in the $\log IC_{50}$ values on the measured cell lines in the GDSC dataset.

Across methods, the general trend of Table 6 is reproduced on a per drug basis, with MSPE of the composite low rank model and elastic net being quite similar across drugs, and GLR performing worse overall. For the twenty first drugs in the GDSC dataset, this is illustrated in Fig. 2 for the models including tissue indicators. Note that the per drug results would be almost identical omitting tissue indicators, just as the overall results in Table 6.

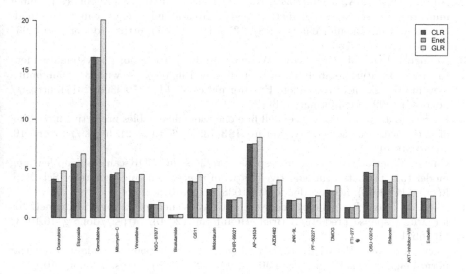

Fig. 2. MSPE on a per-drug basis, across three different methods, for the 20 of 97 drugs in the GDSC dataset, including cancer tissue indicators.

5   Discussion and Conclusion

In this paper, we have studied the problem of drug sensitivity prediction with multi-omics data. Under the assumption that each omics data modality affects the drug sensitivity prediction only through a few latent features (low-rankness), we propose to use a composite local low-rank model that takes into account this local structure. Our numerical results illustrate beneficial performance regarding the prediction errors of the proposed method compared to global methods, such as reduced-rank regression and elastic net. An interesting finding from our simulation study is that low-rank methods perform quite well even when the true generative model is a sparse one.

This paper represents an initial take on the drug prediction based on local low-rank structures. There are some clear limitations in our approach, such as: (i) incorporating correlations between drugs and the heterogeneity of multi-omics data, as in [24], into our model would help to make our method more robust;

(ii) incorporating other local structure rather than low-rankness, could help our method to become more flexible; and (iii) making the model more interpretable. These problems open further venues of research in this area in the future.

References

1. Ali, M., Aittokallio, T.: Machine learning and feature selection for drug response prediction in precision oncology applications. Biophys. Rev. **11**(1), 31–39 (2018). https://doi.org/10.1007/s12551-018-0446-z
2. Ali, M., Khan, S.A., Wennerberg, K., Aittokallio, T.: Global proteomics profiling improves drug sensitivity prediction: results from a multi-omics, pan-cancer modeling approach. Bioinformatics **34**(8), 1353–1362 (2017). https://doi.org/10.1093/bioinformatics/btx766
3. Ammad-Ud-Din, M., Khan, S.A., Wennerberg, K., Aittokallio, T.: Systematic identification of feature combinations for predicting drug response with Bayesian multiview multi-task linear regression. Bioinformatics **33**(14), i359–i368 (2017). https://doi.org/10.1093/bioinformatics/btx266
4. Barretina, J., et al.: The cancer cell line encyclopedia enables predictive modelling of anticancer drug sensitivity. Nature **483**(7391), 603 (2012). https://doi.org/10.1038/nature11003
5. Chang, Y., et al.: Cancer drug response profile scan (CDRscan): a deep learning model that predicts drug effectiveness from cancer genomic signature. Sci. Rep. **8**(1), 1–11 (2018). https://doi.org/10.1038/s41598-018-27214-6
6. Chen, J., Zhang, L.: A survey and systematic assessment of computational methods for drug response prediction. Brief. Bioinform. (2020). https://doi.org/10.1093/bib/bbz164
7. Chen, K., Dong, H., Chan, K.S.: Reduced rank regression via adaptive nuclear norm penalization. Biometrika **100**(4), 901–920 (2013). https://doi.org/10.1093/biomet/ast036
8. Costello, J.C., et al.: A community effort to assess and improve drug sensitivity prediction algorithms. Nat. Biotechnol. **32**(12), 1202 (2014). https://doi.org/10.1038/nbt.2877
9. Ammad-ud din, M., et al.: Drug response prediction by inferring pathway-response associations with kernelized Bayesian matrix factorization. Bioinformatics **32**(17), i455–i463 (2016). https://doi.org/10.1093/bioinformatics/btw433
10. Dugger, S.A., Platt, A., Goldstein, D.B.: Drug development in the era of precision medicine. Nat. Rev. Drug Discov. **17**(3), 183 (2018). https://doi.org/10.1038/nrd.2017.226
11. Garnett, M.J., et al.: Systematic identification of genomic markers of drug sensitivity in cancer cells. Nature **483**(7391), 570 (2012). https://doi.org/10.1038/nature11005
12. Geeleher, P., et al.: Discovering novel pharmacogenomic biomarkers by imputing drug response in cancer patients from large genomics studies. Genome Res. **27**(10), 1743–1751 (2017). https://doi.org/10.1101/gr.221077.117
13. Güvenç Paltun, B., Mamitsuka, H., Kaski, S.: Improving drug response prediction by integrating multiple data sources: matrix factorization, kernel and network-based approaches. Brief. Bioinform. (2019). https://doi.org/10.1093/bib/bbz153
14. Hasin, Y., Seldin, M., Lusis, A.: Multi-omics approaches to disease. Genome Biol. **18**(1), 83 (2017). https://doi.org/10.1186/s13059-017-1215-1

15. Li, G., Liu, X., Chen, K.: Integrative multi-view regression: bridging group-sparse and low-rank models. Biometrics (2018). https://doi.org/10.1111/biom.13006
16. Roses, A.D.: Pharmacogenetics in drug discovery and development: a translational perspective. Nat. Rev. Drug Discov. **7**(10), 807–817 (2008). https://doi.org/10.1038/nrd2593
17. Suphavilai, C., Bertrand, D., Nagarajan, N.: Predicting cancer drug response using a recommender system. Bioinformatics **34**(22), 3907–3914 (2018). https://doi.org/10.1093/bioinformatics/bty452
18. Tibshirani, R.: Regression shrinkage and selection via the lasso. J. R. Stat. Soc. Ser. B (Methodol.) **58**(1), 267–288 (1996)
19. Wang, L., Li, X., Zhang, L., Gao, Q.: Improved anticancer drug response prediction in cell lines using matrix factorization with similarity regularization. BMC Cancer **17**(1), 513 (2017). https://doi.org/10.1186/s12885-017-3500-5
20. Yadav, B., et al.: Quantitative scoring of differential drug sensitivity for individually optimized anticancer therapies. Sci. Rep. **4**(1) (2014).https://doi.org/10.1038/srep05193
21. Yang, W., et al.: Genomics of drug sensitivity in cancer (GDSC): a resource for therapeutic biomarker discovery in cancer cells. Nucl. Acids Res. **41**(D1), D955–D961 (2012). https://doi.org/10.1093/nar/gks1111
22. Yuan, M., Lin, Y.: Model selection and estimation in regression with grouped variables. J. R. Stat. Soc. Ser. B (Stat. Methodol.) **68**(1), 49–67 (2006). https://doi.org/10.1111/j.1467-9868.2005.00532.x
23. Zhang, Y., Yang, Q.: A survey on multi-task learning. CoRR abs/1707.08114 (2017). http://arxiv.org/abs/1707.08114
24. Zhao, Z., Zucknick, M.: Structured penalized regression for drug sensitivity prediction. J. R. Stat. Soc. Ser. C (Appl. Stat.) (2020). https://doi.org/10.1111/rssc.12400

Mass Spectra Interpretation and the Interest of SpecFit for Identifying Uncommon Modifications

Guillaume Fertin[1]([envelope]) [ID], Matthieu David[1,2], Hélène Rogniaux[2,3] [ID], and Dominique Tessier[2,3] [ID]

[1] Université de Nantes, CNRS, LS2N, 44000 Nantes, France
{guillaume.fertin,matthieu.david}@univ-nantes.fr
[2] INRAE, UR BIA, 44316 Nantes, France
{helene.rogniaux,dominique.tessier}@inrae.fr
[3] INRAE, BIBS Facility, 44316 Nantes, France

Abstract. SpecOMS [8] is a software designed to identify peptides from spectra obtained by mass spectrometry experiments. In this paper, we make a specific focus on SpecFit, an optional module of the SpecOMS software. Because SpecOMS is particularly fast, SpecFit can be used within SpecOMS to further investigate spectra whose mass does not necessarily coincide with the mass of its corresponding peptide, and consequently to suggest modifications for these peptides, together with their locations. In this paper, we show that SpecFit is able to identify uncommon peptide modifications that are generally not detected by other software. In that sense, SpecFit is of particular interest since, even today, a large majority of spectra remain uninterpreted.

Keywords: Proteomics · Mass spectrometry · Peptide identification · Open modification search methods · Post-translational modifications

1 Scientific Background

Next generation sequencing is gradually revealing the secrets of genes for all organisms, but the complex world of proteins remains largely unknown despite the critical roles they play in living organisms. As far as we know, diversity of proteins originates either (a) from genetic variations affecting their primary amino acid sequences, due to alternative splicing that generates transcriptional variation or (b) from post-translational modifications (PTMs) that modify the chemistry of the proteins after synthesis. Diversity in PTMs is particularly large, and is thought to explain the gap between the relatively small number of genes and the complex machinery of organisms [15].

Supported by the Conseil Régional Pays de la Loire GRIOTE program (2013–2018) and the French National Research Agency (ANR-18-CE45-004).

P. Cazzaniga et al. (Eds.): CIBB 2019, LNBI 12313, pp. 80–89, 2020.
https://doi.org/10.1007/978-3-030-63061-4_8

Proteomics refers to the scientific discipline that aims at identifying, quantifying and characterizing proteins on a large scale. One common approach in this context is called "bottom-up". In this approach, an unknown protein extract is first hydrolyzed into *peptides*, i.e. (relatively) small amino-acid sequences. Peptides are then separated by liquid chromatography, before they enter a mass spectrometer, in which they are subjected to two mass analysis steps: the first measurement (called MS) provides the masses of the initial peptides, while the second (called MS/MS) generates *experimental spectra*, each such spectrum being obtained from the fragmentation of a set of selected peptides – usually the most intense detected during the MS measurement. Each MS/MS spectrum can thus be seen as a list of peaks, where each peak corresponds to a mass obtained after fragmentation of the corresponding peptide. The goal of a proteomic experiment is to identify each experimental spectrum generated by an MS/MS analysis, i.e., to assign a peptide to each such spectrum. Such interpretation is conducted by pairwise comparisons between each experimental spectrum to what we call *theoretical spectra*, i.e. spectra obtained *in silico* by simulating the measurements of a mass spectrometer on proteins deduced from genomic data (those proteins are available in protein databases). A scoring function measures how closely a given experimental spectrum fits a given theoretical spectrum extrapolated from the protein database. For each experimental spectrum, the top-scoring peptide is assigned to it, resulting in a *peptide-spectrum match* (or PSM). After a PSM is obtained for each experimental spectrum, the complete list of PSMs is sorted by decreasing score. A threshold, based on a measure of statistical significance, determines which PSMs are validated, according to a predetermined false discovery rate (FDR) [6]. The most widely used method for this estimation is based on a decoy database [10]. Then, spectra corresponding to those validated PSMs are considered as *identified*. Finally, identified spectra are used to infer which proteins are most likely to be present in the sample.

Despite the fact that we now have increasingly accurate mass spectrometers, the current rate of identification of spectra is on average only about 25%, leaving a majority of spectra, yet of good quality, uninterpreted. The presence of modifications is considered to be the most likely explanation for this low rate of identification. In addition to improving identification of spectra and of proteins, elucidating these modifications is a major scientific issue in the field of biology and health, since they are assumed to be present in almost all proteins and involved in a large number of biological processes. A complete inventory of the different molecular forms of proteins that originate from the same gene – called proteoforms – is awaited.

Until recently, algorithms were not efficient enough to pairwise compare large volumes of spectra. Consequently, in order to limit the number of comparisons, current standard software compare each experimental spectrum to a *subset* of all the possible theoretical spectra, more precisely those having the same mass as the considered experimental one (plus or minus some device-dependent tolerance). However, this strategy may prevent identification of the correct candidate peptide, if the experimental peptide displays a modification altering its mass. In order to circumvent this limitation, it is possible to incorporate predefined modifications to the modeling of theoretical spectra. This strategy decreases the number of putative missed identifications, but the number of predefined modifications that can be taken into account necessarily remains small (often limited to 3 or 4), due to the induced combinatorial explosion of the search space.

Unlike conventional database methods, open modification search (OMS) methods use very large mass filters or no filter at all, and thus compare each experimental spectrum to all (or almost all) theoretical spectra. Such methods provide for each experimental spectrum S_e one or several triplets of the form $(S_e, P, \Delta m)$, where P is a candidate peptide assigned to S_e, and Δm is defined as the difference between the measured mass of S_e and the calculated mass of P. Thus, Δm represents the mass of potential modifications. Although one may expect OMS methods to be computationally more demanding than conventional ones, some very recent OMS software (see e.g. [2,4,5,8,13,14,16]) manage to achieve very reasonable execution time and memory consumption.

Importantly, OMS methods are now able to increase the percentage of identified spectra compared to those obtained by conventional methods, claiming identification rates up to 70% on certain datasets. However, OMS methods principally highlight the most *frequent* modifications – that are, for a large part, artifacts due to sample preparation. This way of increasing sensitivity can benefit from knowledge described in databases such as Unimod [7], or can be determined by a rapid screening of the frequent modifications present in the sample. But, above all, OMS methods are awaited for identifying *uncommon* modifications – whether they are PTMs or mutations, insertions and/or deletions of amino acids compared to the reference sequence present in the protein data bank – even though such uncommon modifications are much harder to differentiate from random identifications, and very often call for complex experimental validations.

SpecOMS [8] is an OMS method developed by our group, that renews the concept of spectra comparison. SpecOMS is able to compare tens of thousands of experimental spectra from MS/MS analysis to hundreds of thousands of theoretical spectra (e.g., corresponding to the human proteome), in a few minutes on a standard laptop. Because of its fast execution time, SpecOMS is currently the only available OMS method that allows to reevaluate alignments between pairs of spectra by taking into account the mass difference, and subsequently choosing the PSM it considers the most reliable for identification. Such realignment is optional in SpecOMS; if chosen by the user, it is carried out by the SpecFit module. In this paper, we discuss, based on different experimental results, the interest of SpecFit for identifying uncommon modifications in a sample.

2   Materials and Methods

Main Features of SpecOMS and SpecFit. SpecOMS is an open modification search software that works as follows: it first builds an *ad hoc* data structure, called SpecTrees [9], that contains in a condensed form all the necessary information for computing the number of common (or shared) peaks $\mathcal{S}(S_e, S_t)$ (or simply $\mathcal{S}$, if clear from the context) between any two spectra (S_e, S_t), where

S_e (resp. S_t) denotes an experimental (resp. theoretical) spectrum. Then a specific module, called SpecXtract, extracts from SpecTrees all pairs of spectra that share at least T peaks (thus satisfying $S \geq T$, T being a parameter set by the user). This extraction provides an intermediate list of quadruplets of the form $(S_e, S_t, S, \Delta m)$, where every pair (S_e, S_t) such that $S \geq T$ is present in that list. For each such quadruplet, S represents the degree of similarity between the two associated spectra, and is thus considered as the *score* of the PSM. Next, the SpecFit module is (optionally) called. For each PSM previously obtained, SpecFit aims at improving its score, before it computes what it considers to be the best PSM for S_e. For this, it evaluates the increase of the score when a modification, of mass Δm, is applied. This is done by iteratively testing the location of the modification, on each possible amino acid, and at the N-terminal and C-terminal sides of the peptide. Concretely, adding a modification in a spectrum S_t corresponds to shifting a specified subset of its peaks, determined by the location of the modification and by Δm. We thus obtain a modified spectrum S'_t, which in turn modifies the number of peaks that are shared with S_e. SpecFit memorizes the best alignment (of peaks), i.e. the modification location that achieves the best score S. Based on this, SpecOMS chooses the best peptide interpretation for each S_e, and provides for each the following information: S_e, peptide sequence, S before SpecFit, S after SpecFit, best location for Δm, Δm. SpecFit is an important feature in the sense that it allows the identification of PSMs even though they only shared a small number of peaks before SpecFit has been applied.

Datasets. The two MS/MS datasets used in this study were downloaded from the PRIDE proteomics data repository [17]. MS/MS raw datafiles were converted into the MGF format using Raw-Converter version 1.1.0.19 [11].

HEK293 Dataset. This dataset, obtained from human embryonic kidney cells grown in tissue culture, was created to evaluate an open modification search method (see [5]). It consists of 24 raw files (identifier PXD001468) including altogether more than 1 million spectra (more precisely, 1,022,653) charged 2+ and 3+.

Synthetic Dataset. This dataset provides large sets of spectra generated with high accuracy from synthetic peptides representing the human proteome. Experimental details on how the spectra were produced can be found in [18]. The file 01650b BG3-TUM first pool 65 01 01-DDA-1h-R2.raw (identifier PXD004732, pool 65) includes 55,458 spectra charged 2+ and 3+ generated from approximately 1,000 known synthetic peptides.

Protein Databases. Two peptide databases were generated to identify spectra from the HEK293 and the Synthetic datasets. The first database was generated from the human protein database GRCh37 from the Ensemble genome assembly (release 61) [1]. The second database merged synthetic peptides with the Arabidopsis proteome downloaded from the Arabidopsis information portal (Araport11 [3]). An identified peptide from the Synthetic dataset is considered correct if it is one of the 1,000 synthetic peptides, or one derived from common contaminants; it is considered incorrect if it is derived from an Arabidopsis protein. Common contaminants were obtained from the common Repository of Adventitious Proteins (cRAP).

SpecOMS Parameters. The 60 most intense peaks were filtered in the experimental spectra; the ion charge was set between 1 and 3; the peptide length was set between 7 and 30 amino acids; the fragment tolerance was set to 0.02 Da; candidate PSMs were selected for further examination by SpecXtract when the threshold T was greater than or equal to 7; PSMs were immediately validated by SpecFit when $\Delta m = 0$ and $S \geq 9$; the mass of cysteine was modified with the addition of the fixed carbamidomethylation modification; one missed cleavage with trypsin was searched. Concerning mass deltas, only PSMs with a mass delta greater than -500 Da were considered. The above described settings are the default ones, recommended by SpecOMS and are applied on both datasets.

The tests reported in this study were executed on a laptop equipped with an Intel processor (3.3 GHz) and 12 GB of RAM dedicated to the Java Virtual Machine, running under Windows 10. In its V1.2 release, SpecOMS has no multi-threaded implementation.

3   Results

Time and Space Required for SpecFit. SpecFit has been conceived and implemented so as to minimize memory consumption and execution time – see Table 1 for an illustration on the HEK293 dataset. Notice that, even though some realignments operated by SpecFit may involve a large number of spectra, the computational performances of SpecFit remain very good: as shown in Table 1, the average time taken by SpecFit to perform its realignments on the HEK293 dataset is one minute, while its memory consumption is hardly increasing.

Table 1. Computational performances of SpecFit averaged over all 24 HEK293 runs. The number of spectra (theoretical and experimental spectra, together with their decoy version) is around 3 millions for each run.

	SpecFit ON	SpecFit OFF
Execution time (minutes)	35	34
Memory used (Go)	8.5	8.4

Evaluation of SpecFit on a Complex Dataset. Table 2 shows the different results we obtain on the HEK293 dataset, depending on whether SpecFit is applied. Globally, it can be seen that SpecFit has a relatively small impact on the total number of identified spectra (+4.9%). In order to better explain the number of identifications listed in Table 2, it should be noticed that the new score computed by SpecFit may change the choice of the best PSM associated to each of the experimental spectra. Consequently, a certain number of PSMs initially having $\Delta m = 0$ has been replaced, after SpecFit is applied, by PSMs having $\Delta m \neq 0$, but with a better number of shared peaks. However, the decrease in the number of identifications concerning the set of PSMs with $\Delta m = 0$ is accompanied by a decrease in the FDR, which means that the removed PSMs included a large number of errors. As expected, SpecFit identifies many more PSMs associated to positive mass deltas (+31.9%), at the price of a small increase of the FDR. The results we obtain when we compare the number of validated PSMs associated to negative mass deltas (−10.8%), on the other hand, may look surprising. Consequently, in order to understand this phenomenon, we decided to run SpecFit on a better controlled dataset, namely the Synthetic dataset.

Table 2. Influence of SpecFit on the cumulated number of identified spectra from the HEK293 dataset (24 runs). The validation score is the minimum number of shared peaks needed to reach an FDR <1%.

		HEK293	
		SpecFit ON	SpecFit OFF
$\Delta m = 0$	Validation score	7	7
	FDR (%)	0.46	0.91
	# identifications	263,364	277,426
$\Delta m > 0$	Validation score	11	11
	FDR (%)	0.8	0.6
	# identifications	149,794	113,558
$\Delta m < 0$	Validation score	14	11
	FDR (%)	0.96	0.6
	# of identifications	16,545	18,545
Total number of identifications		429,703	409,529

Evaluation of SpecFit on the Synthetic dataset. The Synthetic dataset does not contain any PTMs, but it contains many variations due to the imperfect synthesis of peptides, in addition to modifications generated by sample preparation. The great advantage of this dataset arises from the ability to differentiate between correct and incorrect identifications, since the peptides present in this sample are presumably known. Because we are able to differentiate true positive from false positive identifications in the target database (namely, synthetic or contaminant vs Arabidopsis), we can compare the number of errors due to wrong identifications among the Arabidopsis proteins ("Error rate" in Table 3) to the number of estimated errors from the decoy database ("FDR" in Table 3). As shown in Table 3, no deviation between the estimated FDR using the decoy database and the "reference FDR" using the number of errors is observed. Even though the number of identifications increases when $\Delta m < 0$ in this dataset, the number of shared peaks must be increased by 2 in order to remain at an FDR not exceeding 1% when SpecFit is applied and, even then, the FDR increases from 0.6% to 0.8%. This means that, when $\Delta m < 0$, spectra tend to match more easily to peptides coming from the decoy database. It can be explained as follows: when we identify a peptide p with a negative mass delta, we actually often identify a (potentially small) *subsequence* of p. However, a subsequence that is simultaneously present both in the target and decoy databases alters the FDR evaluation, because the identification in the decoy database does not necessarily measure, as it should, an identification due to chance. This phenomenon is not yet taken into account in the FDR estimation, and likely leads to an overestimation of the FDR when SpecFit is used on spectra with a negative mass delta. This also raises the more general question of determining how to generate a relevant decoy database when using an OMS strategy.

Table 3. Influence of SpecFit on the number of identified peptides on the Synthetic dataset. The validation score is the minimum number of shared peaks needed to reach an FDR <1%. The Error rate computes the percentage of identifications in the Arabidopsis database – which are considered as incorrect.

		Synthetic	
		SpecFit ON	SpecFit OFF
$\Delta m = 0$	Validation score	10	10
	FDR (%)	0.4	0.4
	Error rate (%)	0.4	0.4
	# identifications	3,537	3,537
$\Delta m > 0$	Validation score	13	13
	FDR (%)	0.8	0.5
	Error rate (%)	0.8	0.5
	# identifications	1,390	611
$\Delta m < 0$	Validation score	13	11
	FDR (%)	0.8	0.6
	Error rate (%)	0.8	0.6
	# of identifications	2,937	2,482
Total number of identifications		7,864	6,630

Ability of SpecFit to Identify Uncommon Modifications. If, globally, SpecFit has a relatively small impact on the total number of identified spectra, one can argue that these identifications are more robust. Indeed, the best advantage of SpecFit is certainly that it promotes identifications of "well aligned spectra", i.e. spectra whose mass delta is not only explained, but also (approximately) located. The two tables displayed in Fig. 1 illustrate how the realignment operated by SpecFit can significantly increase the score S, and also show that many different uncommon modifications can be infered and unambiguously interpreted by SpecFit.

It should also be noted that, without SpecFit, a large part of these identifications would simply have been lost.

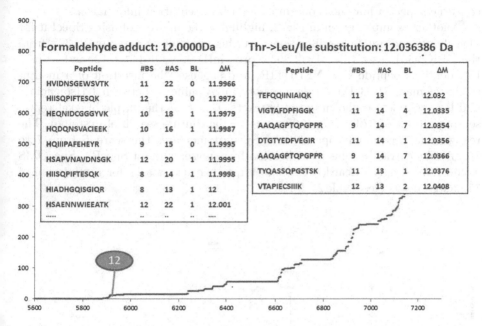

Formaldehyde adduct: 12.0000Da

Peptide	#BS	#AS	BL	ΔM
HVIDNSGEWSVTK	11	22	0	11.9966
HIISQPIFTESQK	12	19	0	11.9972
HEQNIDCGGGYVK	10	18	1	11.9979
HQDQNSVACIEEK	10	16	1	11.9987
HQIIIPAFEHEYR	9	15	0	11.9995
HSAPVNAVDNSGK	12	20	1	11.9995
HIISQPIFTESQK	8	14	1	11.9998
HIADHGQISGIQR	8	13	1	12
HSAENNWIEEATK	12	22	1	12.001
.....	..	..	..	....

Thr->Leu/Ile substitution: 12.036386 Da

Peptide	#BS	#AS	BL	ΔM
TEFQQIINIAIQK	11	13	1	12.032
VIGTAFDPFIGGK	11	14	4	12.0335
AAQAGPTQPGPPR	9	14	7	12.0354
DTGTYEDFVEGIR	11	14	1	12.0356
AAQAGPTQPGPPR	9	14	7	12.0366
TYQASSQPGSTSK	11	13	1	12.0376
VTAPIECSIIIK	12	13	2	12.0408

Fig. 1. Interpretation of mass deltas (Synthetic dataset). PSMs are ordered by increasing mass deltas; their ranks are represented on the x-axis – limited from 5,600 to 7,200 here –, while their mass deltas are on the y-axis. For each PSM, we give the number of shared peaks before SpecFit is applied (column #BS), the number of shared peaks after SpecFit is applied (column #AS), the best location to assign the mass delta (column BL) and the mass delta itself (column Δm).

Once the list of PSMs is validated, the challenging task is then to explain the variety of mass deltas revealed in the dataset [12]. SpecFit also helps the interpretation of the observed mass deltas, because the indication of which amino acids are involved at the sample level can give decisive clues. Concerning the Synthetic dataset, in Fig. 1 (left), several occurrences of a similar mass delta around 12.0000 Da were consistently localized by SpecFit on the Histidine residue located at the N-terminus of the peptide, which suggests a possible formaldehyde adduct on this part of the peptide. In Fig. 1 (right), a mass delta of 12.036386 Da was consistently localized on a Threonine residue, which may arise from a substitution of a Threonine by either a Leucine or an Isoleucine[1]. It should also be noted that the location of this substitution, provided by SpecFit, varies depending on the studied spectrum.

---

[1] Leucine and Isoleucine having the same mass, it is impossible to discriminate one from the other based only on Δm.

Explaining the mass deltas as done in Fig. 1 (i.e. describing a same type of modification occurring in several instances) is sometimes challenging to obtain, because a given mass delta is represented by relatively few spectra and/or the concerned spectra may not contain enough fragmentation information.

Another example, given in Fig. 2, highlights the interest of using SpecFit for identifying uncommon modifications. In this example (Spectrum 43,327 from the HEK293 dataset (file b1922_293T_proteinID_02A_QE3_122212.raw)), Spec-Fit identifies peptide CAGNEDIITIR, and suggests the insertion of a modification of mass 71.03 Da – thus possibly an Alanine –, between I (Isoleucine) and R (Arginine), as shown in red brackets (note that the purple bracket representing a mass of 57.02 Da corresponds to the routine search of cysteine). The realignment accordingly operated by SpecFit increases the score S from 7 to 15, thereby extracting this identification from noise: without SpecFit, SpecOMS would have simply discarded this spectrum, because it asks for a minimum of 11 shared peaks (see Table 2).

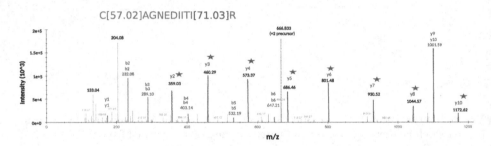

Fig. 2. SpecFit suggests a modification of mass 71.03 Da – which could correspond to the insertion of an Alanine (A) –, between amino acids I and R in peptide CAGNEDI-ITIR, thereby increasing S from 7 (in bold purple) before SpecFit, to 15 after SpecFit. The 8 additionally aligned peaks are the bold red ones topped by a star (Spectrum 43,327 from the HEK293 dataset (file b1922_293T_proteinID_02A_QE3_122212.raw)). (Color figure online)

4   Conclusion

SpecFit is an optional module of SpecOMS that can be used to realign spectra at different possible locations, in order to explain an observed mass delta between a candidate peptide and an experimental spectrum. Its goal is to increase the number of shared peaks between the two, and if validated, the realignment not only provides the suggested modification, but also its location in the peptide. Doing so, SpecFit achieves more identifications (which may have been lost otherwise), while maintaining a reasonable execution time. Since modifications of peptides are considered to be the most likely explanation for the low rate of peptide identification in mass spectrometry experiments, SpecFit can be considered as a promising step towards achieving more and better identifications.

References

1. Aken, B.L., et al.: Ensembl 2017. Nucl. Acids Res. **45**(D1), D635–D642 (2016)
2. Bittremieux, W., Meysman, P., Noble, W., Laukens, K.: Fast open modification spectral library searching through approximate nearest neighbor indexing. J. Proteome Res. **17**(10), 3463–3474 (2018)
3. Cheng, C., Krishnakumar, V., Chan, A.P., Thibaud-Nissen, F., Schobel, S., Town, C.: Araport11: a complete reannotation of the Arabidopsis thaliana reference genome. Plant J. **89**(4), 789–804 (2017)
4. Chi, H., et al.: Comprehensive identification of peptides in tandem mass spectra using an efficient open search engine. Nat. Biotechnol. **36**, 1059–1061 (2018)
5. Chick, J., et al.: A mass-tolerant database search identifies a large proportion of unassigned spectra in shotgun proteomics as modified peptides. Nat. Biotechnol. **33**(7), 743–749 (2015)
6. Choi, H., Nesvizhskii, A.I.: False discovery rates and related statistical concepts in mass spectrometry-based proteomics. J. Proteome Res. **7**(1), 47–50 (2008)
7. Creasy, D.M., Cottrell, J.S.: Unimod: Protein modifications for mass spectrometry. Proteomics **4**(6), 1534–1536 (2004)
8. David, M., Fertin, G., Rogniaux, H., Tessier, D.: SpecOMS: a full open modification search method performing all-to-all spectra comparisons within minutes. J. Proteome Res. **16**(8), 3030–3038 (2017)
9. David, M., Fertin, G., Tessier, D.: SpecTrees: an efficient without a priori data structure for MS/MS spectra identification. In: Frith, M., Storm Pedersen, C.N. (eds.) WABI 2016. LNCS, vol. 9838, pp. 65–76. Springer, Cham (2016). https://doi.org/10.1007/978-3-319-43681-4_6
10. Elias, J., Gygi, S.: Target-decoy search strategy for increased confidence in large-scale protein identifications by mass spectrometry. Nat. Methods **4**, 207–214 (2007)
11. He, L., Diedrich, J., Chu, Y.Y., Yates, J.R.: Extracting accurate precursor information for tandem mass spectra by RawConverter. Anal. Chem. **87**(22), 11361–11367 (2015)
12. Kim, M., Zhong, J., Pandey, A.: Common errors in mass spectrometry-based analysis of post-translational modifications. Proteomics **16**(5), 700–714 (2016)
13. Kong, A., Leprevost, F., Avtonomov, D., Mellacheruvu, D., Nesvizhskii, A.: MSFragger: ultrafast and comprehensive peptide identification in mass spectrometry-based proteomics. Nat. Methods **14**(5), 513–520 (2017)
14. Na, S., Kim, J., Paek, E.: MODplus: robust and unrestrictive identification of post-translational modifications using mass spectrometry. Anal. Chem. **91**(17), 11324–11333 (2019)
15. Prabakaran, S., Lippens, G., Steen, H., Gunawardena, J.: Post-translational modification: nature's escape from genetic imprisonment and the basis for dynamic information encoding. WIREs Syst. Biol. Med. **4**(6), 565–583 (2012)
16. Solntsev, S.K., Shortreed, M.R., Frey, B.L., Smith, L.M.: Enhanced global post-translational modification discovery with MetaMorpheus. J. Proteome Res. **17**(5), 1844–1851 (2018)
17. Vizcaíno, J.A., et al.: 2016 update of the PRIDE database and its related tools. Nucl. Acids Res. **44**(D1), D447–D456 (2015)
18. Zolg, D., et al.: Building ProteomeTools based on a complete synthetic human proteome. Nat. Methods **14**(3), 259–262 (2017)

MSAX: Multivariate Symbolic Aggregate Approximation for Time Series Classification

Manuel Anacleto[1,2], Susana Vinga[2,3] (iD), and Alexandra M. Carvalho[1(✉)] (iD)

[1] Instituto de Telecomunicações, Instituto Superior Técnico, Universidade de Lisboa,
Av. Rovisco Pais 1, 1049-001 Lisboa, Portugal
[2] INESC-ID, Instituto Superior Técnico, Universidade de Lisboa,
R. Alves Redol 9, 1000-029 Lisboa, Portugal
{susanavinga,alexandra.carvalho}@tecnico.ulisboa.pt
[3] IDMEC, Instituto Superior Técnico, Universidade de Lisboa, Lisbon, Portugal

Abstract. Time Series (TS) analysis is a central research topic in areas such as finance, bioinformatics, and weather forecasting, where the goal is to extract knowledge through data mining techniques. Symbolic aggregate approximation (SAX) is a state-of-the-art method that performs discretization and dimensionality reduction for univariate TS, which are key steps for TS representation and analysis. In this work, we propose MSAX, an extension of this algorithm to multivariate TS that takes into account the covariance structure of the data. The method is tested in several datasets, including the Pen Digits, Character Trajectories, and twelve benchmark files. Depending on the experiment, MSAX exhibits comparable performance with state-of-the-art methods in terms of classification accuracy. Although not superior to 1-nearest neighbor (1-NN) and dynamic time warping (DTW), it has interesting characteristics for some classes, and thus enriches the set of methods to analyze multivariate TS.

Keywords: Symbolic aggregate approximation · Time series · Classification · Multivariate analysis

1 Introduction

The vast quantity of available data nowadays is posing new challenges for knowledge discovery, namely, to extract meaningful information such as significant patterns, statistics, and regularities. Temporal data, and in particular time series (TS), are now pervasive in many fields, which fully justifies the development

Supported by national funds through Fundação para a Ciência e a Tecnologia (FCT) through projects UIDB/50021/2020 (INESC-ID), UIDB/50022/2020 (LAETA, IDMEC), UIDB/50008/2020 (IT), PREDICT (PTDC/CCI-CIF/29877/2017), and MATISSE (DSAIPA/DS/0026/2019).

P. Cazzaniga et al. (Eds.): CIBB 2019, LNBI 12313, pp. 90–97, 2020.
https://doi.org/10.1007/978-3-030-63061-4_9

of new methods for their analysis. A discrete TS is a series of n real-valued observations, each one being measured at a discrete time $t \in \{1, \ldots, T\}$, made sequentially and regularly trough T instances of time. In this case, the i-th TS is given by $\{\boldsymbol{x}^i[t]\}_{t \in \{1, \ldots, T\}}$, where $\boldsymbol{x}^i[t] = (x_1^i[t], \ldots, x_n^i[t])$. When $n = 1$, the TS is said to be univariate; otherwise, when $n > 1$, it is multivariate.

Data representation takes a big focus on TS analyses. An abundant wealth of data structures and algorithms for streaming discrete data were developed in recent years, especially by the text processing and bioinformatics communities. To make use of these methods, real-valued TS need symbolic discretizations. Besides this, representation methods also address the TS dimensionality problem arising from the fact that almost all TS datasets are intrinsically of high dimensionality.

In contrast to univariate TS, Multivariate TS (MTS) are characterized not only by serial correlations (auto-correlation) but also by relationships between the attributes measured at the same time point (intra-correlation). Due to considering the attributes individually, their intra-correlations might be poorly captured, as shown in [4,5]. In [6], the necessity of different TS representations for MTS classification was discussed. It was and pointed out as desirable the development of methods that consider all attributes simultaneously, taking into account the relationships between them.

This work proposes a multivariate extension of the well-known TS representation of Symbolic Aggregate Approximation (SAX) [1]. In the SAX method, the TS is normalized to have a temporal mean of zero and a standard deviation of one. A TS normalized in this manner has a Gaussian distribution [2].

If desired, Piecewise Aggregate Approximation (PAA) [3] is then applied, reducing the TS length. This technique divides the TS into w (method parameter) segments of equal length, where each segment is replaced with its average value that is further grouped in a vector representing the TS.

Assuming that the normalized TS has a Gaussian distribution [2], it is possible to divide it into equal size areas under the Gaussian curve trough breakpoints, producing equiprobable symbols. These breakpoints may be determined by a statistical table inspection. After, the discretizing process is done by associating the TS points to a (method parameter that represents the size of the symbolic alphabet) equal area intervals beneath the Gaussian curve associated to the TS to be discretized. An illustrative example of the discretizing process is shown in Fig. 1.

Having a discretized TS, a distance measure between two TS $Q = q_1 q_2 \ldots q_T$ and $C = c_1 c_2 \ldots c_T$, in the new representation space can be defined as:

$$\text{MINDIST}\,(Q, C) = \sqrt{\frac{T}{w}} \sqrt{\sum_{i=1}^{w} dist\,(q[i], c[i])^2}. \tag{1}$$

The function $dist()$ that returns the distance between two symbols is implemented using a lookup table in which the value for entry (r, c) is obtained through the following function, where β represents the breakpoints values:

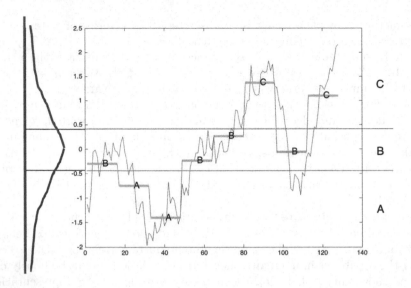

Fig. 1. The TS (in ligth blue) is discretized by first applying the PAA technique and then using predetermined breakpoints to map the PAA coefficients into the symbols. In the example above, with $T = 128$, $w = 8$ and $a = 3$, the TS is mapped to the word BAABBCBC. (Color figure online)

$$cell_{r,c} = \begin{cases} 0, & \text{if} \quad |r - c| \leq 1 \\ \beta_{\max(r,c)-1} - \beta_{\min(r,c)}, & \text{otherwise.} \end{cases} \quad (2)$$

SAX is the only symbolic TS representation, until now, for which the distance measure in the symbolic space lower bounds the distance in the original TS space. This fact is assumed to be one of the reasons for its excellent performance [1,2]. Nevertheless, SAX only works for univariate TS, paving the way for extending its promising results for MTS. In the literature, SAX has been applied to MTS by dealing with each variable independently, disregarding intra-correlations in the discretization process [7,8]. We propose to explore these intra-correlations to understand the benefits of using these dependencies.

2   Materials and Methods

The method proposed in this work, MSAX, expands the SAX algorithm by first performing a multivariate normalization of the MTS. The rationale for this first step is to account for the mean and covariance structure of the data $\boldsymbol{X}[t]$, i.e., $E[\boldsymbol{X}[t]] = \mu$ and $Var[\boldsymbol{X}[t]] = \Sigma_{n \times n}$.

The normalized TS values $\boldsymbol{Z}[t]$ are given by $\boldsymbol{Z}[t] = \Sigma^{-1/2}(\boldsymbol{X}[t] - \mu)$, such that the obtained distribution has zero mean and uncorrelated variables. Assuming a Gaussian distribution, we can identify the cut points and intervals that define equal volumes, a crucial step to identify the areas associated with symbols used in the discretization.

2.1   MSAX Discretization

After the normalization step, and like in the original method, the PAA procedure is applied to each variable individually, to reduce its dimensionality. PAA can be performed individually as the resulting variables are now independent of each other, and so, intra-dependencies do not interfere with temporal ones. First, before the proper discretization of the TS values, the volumes associated with each symbol beneath the multivariate Gaussian curve are defined. With this into consideration, the following reasoning is used to define the volumes and corresponding cut-points.

Due to the normalization step, the new variables of the MTS are now uncorrelated, i.e., the covariance matrix of the TS is the identity matrix. Since the probability density function of the MTS is equal to the product of the probability density function of each variable when no correlation between the variables exist, a Gaussian distribution of $\mathcal{N}(0,1)$ is associated to each series variable of the TS, in the same way as in the original method. Then, each Gaussian curve associated to a variable will be split using breakpoints such that the probability of each space split beneath the Gaussian curve is the same for all divisions. This procedure is done in the same way as the original method following the a parameter (that indicates the alphabet size per variable).

After the split regions under the multivariate Gaussian curve are defined trough the breakpoints intersection for each variable, this results in the variable space to be split in a grid way with each partition of the grid having the same volume under the multivariate Gaussian curve. Finally, the points of the normalized and PAA processed MTS are mapped to the multivariate split space beneath the Gaussian curve associated with multivariate TS. As a result of the entire process, a univariate discrete TS is obtained from the multivariate numerical TS.

An example of the full discretization process is given with bivariate TS normalized points X where x_1 and x_2 represent each dimension. Figure 2 illustrates the Gaussian curve associate to this distribution. If three symbols per variable are used in the discretization process, $a = 3$, the discretization shown in Fig. 2 (right) is obtained, with a total of nine symbols. The final symbol value is obtained by the concatenation of the symbols associated with each variable (these symbols will be designated by variable symbols to distinguish from the final symbols). As an example, consider the purple partition in Fig. 2; its final symbol value is aB, directly obtained by concatenating $x_1 = a$ and $x_2 = B$ (purple partition).

2.2   Dissimilarity Definition

Having introduced this new representation of MTS, a new dissimilarity measure should be defined. Two symbolic univariate TS Q and C of the same length T, obtained from an MTS with n attributes, are considered. The distance measure between two TS using the MSAX representation is given by the sum of the distances between each two-time points, for all the indexes of the TS length, where the distance between two ultimate symbols of the MSAX is obtained by

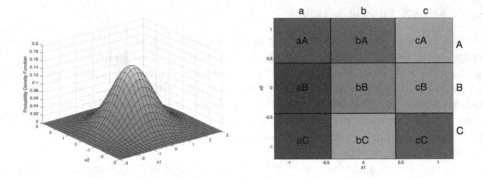

Fig. 2. On the left, plot of the probability density function or Gaussian curve with distribution $\mathcal{N}(0, I)$, for two variables x_1 and x_2. On the right, the areas associated to each symbol on the x_1, x_2 plane, for $a = 3$. Each area with a different color is associated with a symbol. For example, a point situated on the area in orange, the x_1 variable value is associated with b, and the x_2 variable value is associated with A. To this x_1, x_2 example point will be associated final symbol of bA. (Color figure online)

the sum of the difference between the symbol of the variable associated to each variable in this representation:

$$\text{MINDIST\_MSAX}\,(Q, C) = \sqrt{\frac{T}{w}} \sqrt{\sum_{i=0}^{w} \left(\sum_{i=0}^{n} dist\,(q[i], c[i])^2 \right)}. \qquad (3)$$

The distance between two symbols is calculated based on the univariate representations, and by using the corresponding distance defined originally, i.e., obtained through the same table used in the original SAX distance. This result stems from the fact that the breakpoints are the same due to the Gaussian properties.

3   Results

In this section, the MSAX algorithm is evaluated for classification tasks. Tests focused on the comparison between MSAX and the original SAX method applied to each MTS attribute separately, henceforward referred to as SAX_INDP. The behavior of both algorithms is asserted through the use of the first nearest neighbor (1-NN) classifier, varying only the input MTS representation and the respective distance measure.

Benchmark datasets for TS classification tasks included: (i) the *PenDigits* dataset, consisting in multiple labeled samples of pen trajectories [9]; (ii) the *CharacterTrajectories* dataset, representing instead trajectories of characters from the English alphabet; and (iii) 12 datasets with different characteristics from a wide range of areas [10,11].

Firstly, we addressed the comparison between MSAX and SAX_INDP in the PenDigits and CharacterTrajectories datasets varying both the alphabet size and the TS length reduction ratio. Results are depicted in Fig. 3. For both datasets, the accuracy of the SAX_INDP is superior to the MSAX for all parameters configurations on both plots. While fixing the w parameter, both methods on both datasets show a similar behavior by increasing the accuracy as long as the alphabet size increases. When the alphabet size is fixed, and the TS length reduction varies, the behavior differs. In the PenDigits dataset, the accuracy increases as long as the TS length reduction diminishes, whereas, in the CharacterTrajectories dataset, the accuracy remains the same as long the TS length reduction diminishes.

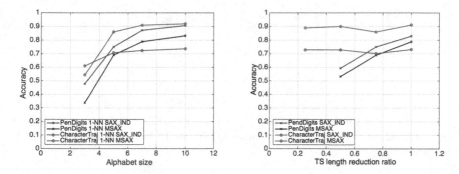

Fig. 3. Comparison between MSAX and SAX_INDP with 1-NN in PenDigits and CharacterTrajectories datasets. On the first plot, the accuracy of the methods is plotted against the parameter a; for these experiments, a fixed value of w was used. On the second, the accuracy is plotted against the TS length reduction ratio (obtained trough the parameter w); for these experiments, a fixed value of a was used.

Additional results comparing both methods were performed by testing 14 datasets with a combination of configurations from an alphabet size varying from 5 to 20 and a TS length reduction ratio from 1/4 to 1. Besides the SAX-based methods, two state-of-the-art classifiers were used: the 1-NN with Euclidean distance and 1-NN with Dynamic Time Warping (DTW). Figure 4 presents the result corresponding to the configuration of parameters that achieved the best accuracy.

On the 14 datasets the accuracy of the SAX_INDP is superior in 12 we compared to the MSAX. In this 12 datasets, the difference is very significant in 6 of them, while in the other 6 the accuracy of both methods is very close. Regarding the comparison of the SAX-based methods with the other two state-of-the-art classifiers, SAX_INDP proves to be very competitive with the Euclidean distance, presenting a small superiority; the results are very similar in 10 datasets, whereas in 4 datasets the SAX_INDP achieves a significantly better result. Concerning the DTW distance, it surpasses, in general, the other algorithms, being

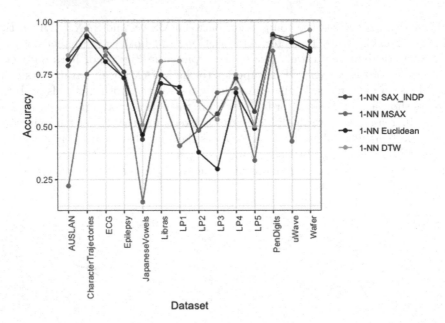

Fig. 4. Accuracy of the four classifiers: 1-NN with SAX_INDP, 1-NN with MSAX, 1-NN with Euclidean distance, and 1-NN with DTW, for 14 benchmarks datasets in TS classification tasks.

the most accurate on most of the datasets. Nonetheless, SAX_INDP achieves very similar and competitive results on a significant number of datasets.

4   Conclusion

In this work, an extension of SAX for multivariate TS, named MSAX, was proposed. Its behavior was assessed in classifications tasks, comparing it with the SAX_INDP and two other state-of-the-art classifiers: 1-NN with the Euclidean distance and 1-NN with DTW. We concluded that the proposed method is overall not competitive with the SAX_INPD, the original SAX algorithm applied independently to each attribute in the MTS. Nonetheless, the obtained results have utility as benchmark values for SAX-based methods in multivariate classifications tasks. It is also noteworthy that for some datasets and specific cases, MSAX surpasses the other techniques. As a future direction, MSAX could be evaluated more deeply in different data mining tasks, such as clustering or forecasting, in which it could be useful and achieve comparable performance with state-of-the-art methods. Possible future applications in bioinformatics include the analysis of patients' data, such as transcriptomics and also time series from electronic health records.

References

1. Lin, J., Keogh, E., Lonardi, S., Chiu, B.: A symbolic representation of time series, with implications for streaming algorithms. In: Proceedings of the 8th ACM SIGMOD Workshop on Research Issues in Data Mining and Knowledge Discovery, pp. 2–11 (2003)
2. Lin, J., Keogh, E., Wei, L., Lonardi, S.: Experiencing SAX: a novel symbolic representation of time series. Data Min. Knowl. Discov. **15**(2), 107–144 (2007)
3. Keogh, E., Chakrabarti, K., Pazzani, M., Mehrotra, S.: Dimensionality reduction for fast similarity search in large time series databases. Knowl. Inf. Syst. **3**(3), 263–286 (2001)
4. Weng, X., Shen, J.: Classification of multivariate time series using locality preserving projections. Knowl. Based Syst. **21**(7), 581–587 (2008)
5. Bankó, Z., Abonyi, J.: Correlation based dynamic time warping of multivariate time series. Expert. Syst. Appl. **39**(17), 12814–12823 (2012)
6. Kadous, M.W., Sammut, C.: Classification of multivariate time series and structured data using constructive induction. Mach. Learn. **58**(2), 179–216 (2005)
7. Esmael, B., Arnaout, A., Fruhwirth, R.K., Thonhauser, G.: Multivariate time series classification by combining trend-based and value-based approximations. In: Murgante, B., et al. (eds.) ICCSA 2012. LNCS, vol. 7336, pp. 392–403. Springer, Heidelberg (2012). https://doi.org/10.1007/978-3-642-31128-4_29
8. Wang, Z., et al.: Representation learning with deconvolution for multivariate time series classification and visualization. CoRR, vol. abs/1610.07258 (2016)
9. Alimoglu, F., Alpaydin E.: Combining multiple representations and classifiers for pen-based handwritten digit recognition. In: Proceedings of the Fourth International Conference on Document Analysis and Recognition, Germany, vol. 2, pp. 637–640 (1997)
10. Bagnall, A.J., et al.: The UEA multivariate time series classification archive, 2018. CoRR, vol. abs/1811.00075 (2018)
11. Bayodan, M.: http://www.mustafabaydogan.com - Mustafa Baydogan website. Accessed 10 Mar 2020

NeoHiC: A Web Application
for the Analysis of Hi-C Data

Daniele D'Agostino[1]([⊠]), Pietro Liò[2], Marco Aldinucci[3], and Ivan Merelli[4]

[1] Institute of Electronics, Computer and Telecommunication Engineering,
National Research Council of Italy, Genoa, Italy
daniele.dagostino@cnr.it
[2] Computer Laboratory, University of Cambridge, Cambridge, UK
Pietro.Lio@cl.cam.ac.uk
[3] Computer Science Department, University of Torino, Turin, Italy
marco.aldinucci@unito.it
[4] Institute for Biomedical Technologies, National Research Council of Italy, Segrate,
MI, Italy
ivan.merelli@itb.cnr.it

Abstract. High-throughput sequencing Chromosome Conformation Capture (Hi-C) allows the study of chromatin interactions and 3D chromosome folding on a larger scale. A graph-based multi-level representation of Hi-C data is essential for proper visualisation of the spatial pattern they represent, in particular for comparing different experiments or for re-mapping omics-data in a space-aware context. The size of the HiC data hampers the straightforward use of currently available graph visualisation tools and libraries. In this paper, we present the first version of NeoHiC, a user-friendly web application for the progressive graph visualisation of Hi-C data based on the use of the Neo4j graph database. The user could select the richness of the environment of the query gene by choosing among a large number of proximity and distance metrics.

Keywords: Hi-C · Graph database · Web application · Graph visualisation

1 Introduction

Modern bioinformatics aims at integrating different omics data to shed light into the mechanisms of gene expression and regulation that give rise to different phenotypes, in order to understand the underlying molecular processes that sustain life and to intervene into these processes by developing new drugs [1,2] when pathological changes occur [3,4]. In this context, the exploration of the 3D organization of chromosomes in the nucleus of cells is of paramount importance for many cellular processes related to gene expression regulation, including DNA accessibility, epigenetic patterns, and chromosome translocations [5,6].

© Springer Nature Switzerland AG 2020
P. Cazzaniga et al. (Eds.): CIBB 2019, LNBI 12313, pp. 98–107, 2020.
https://doi.org/10.1007/978-3-030-63061-4_10

In particular, High-throughput sequencing Chromosome Conformation Capture (Hi-C) allows the study of chromatin interactions and 3D chromosome folding on a larger scale [7,8]. The graph-based representation of Hi-C data produced, for example, by NuChart [9,10] or CytoHic [11], which are software for representing the spatial position of genes in the nucleus, will be essential for creating maps where further omics data can be mapped, in order to characterize different spatially associated domains. This visualisation is an effective complement of the traditional matrix-based representations, for example, produced by Juicer[1] [12] or TADbit[2] [13].

Contact matrices, or better their probabilistic models, allow creating representations that only involve two chromosomes, while graphs can describe the interactions of all the chromosomes using a graph-based approach. This representation highlights the physical proximity of genes in the nucleus in comparison to coordinate-based representations. The very same problem impairs representations based on Circos[3], which can characterize the whole genome in one shot, but fail to describe the physical proximity of genes. In previous works [14,15] we showed some exciting results relying on the possibility of creating metrics for defining how far two genes are one from the other, with possible applications to cytogenetic profiling, to the analysis of the DNA conformation in the proximity of the nucleolus, and for describing the social behavior of genes.

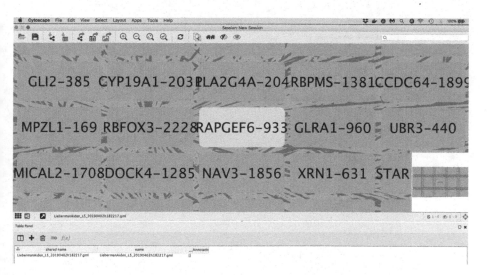

Fig. 1. The visualisation of a Hi-C network using Cytoscape.

However, the typical size of a graph achieved through a Hi-C analysis is in the order of thousands of nodes and hundreds of thousands of edges, which

---

[1] https://github.com/aidenlab/juicer.
[2] https://github.com/3DGenomes/TADbit.
[3] https://circos.ca/.

makes its exploration extremely complex, at least using the tools available. We tested both esyN [16], a tool for the construction and analysis of networks for biological research, and the well-known Cytoscape[4] [17] platform, with a network composed by about 2,400 nodes and 175,000 edges and we found many difficulties in visualizing and analyzing such a massive network with these tools. For example Fig. 1 has been obtained using Cytoscape: it shows all the nodes of the network, but it is not easy to show only a subset of them, i.e. the neighborhood of a selected gene, as discussed later for example in Fig. 2, or also to analyze the edges provided by different experiments.

Such large networks represent an issue also for the effective storage of databases because in Hi-C data, most of the information is represented by the edges connecting genes. Therefore a proper way for their effective management is represented by Graph databases like Neo4j. But this solution has been considered only in these last years, see, for example [18]. At the same time, the most important repositories as STRING [19] or InterMine [20] are still based on relational databases.

For these reasons, we present the first version of NeoHiC, a web application specifically designed to manage and analyze graphs produced by investigating Hi-C data. In this version, we considered Neo4j as graph database management system and NuChart as the tool to compute Hi-C data.

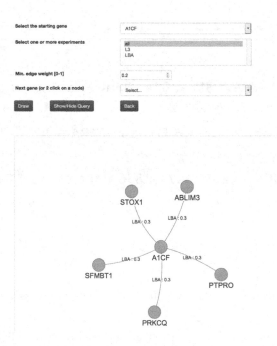

Fig. 2. The basic visualisation of NeoHiC.

---

[4] https://cytoscape.org/.

Select the starting gene

RAPGEF6

Select one or more experiments

all
L3
LBA

Min. edge weight [0-1]

0.0

Next gene (or 2 click on a node)

Select...

Draw   Show/Hide Query   Back

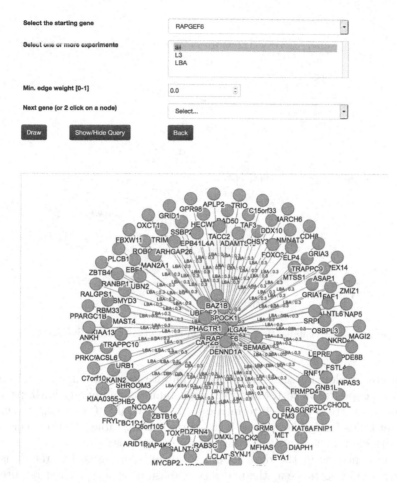

Fig. 3. The initial visualisation of the same network of Fig. 1 after selecting RAPGEF6 as a starting gene.

2   Materials and Methods

NeoHiC relies on the Neo4j graph database and modern web technologies, such as the Node.js JavaScript framework, used in many scientific applications and frameworks [21].

Graph Databases. Graph databases are part of the NoSQL database family created to address the limitations of the existing relational databases. While the graph model explicitly lays out the dependencies between nodes of data, the relational and other NoSQL database models link the data through implicit connections.

In particular, in relational databases, references to other rows and tables are indicated by referring to primary key attributes via foreign key columns. Joins are computed at query time by matching primary and foreign keys of all rows in

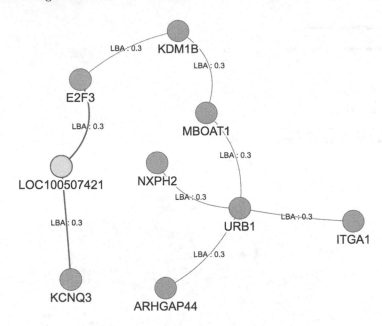

Fig. 4. The visualisation of a path in the network after five steps.

the connected tables. These operations are compute-heavy and memory-intensive and have an exponential cost. Moreover, when many-to-many relationships occur in the model, there is the need to introduce a JOIN table (or associative entity table) that holds foreign keys of both the participating tables, further increasing storage space, and the execution time of join operations.

On the contrary, in the data model of graph databases, the relationships have the same importance as the nodes. Database designers are not required to infer connections among entities using special properties such as foreign keys. For this, graph databases, by design, allow fast and straightforward retrieval of complex hierarchical structures that are difficult to model in relational systems.

Therefore the first step for creating the Web application has been the development of a tool for converting the graph-based representation of Hi-C data in a format that can be directly ingested by a graph database. The tool is a command-line Node.js script responsible for converting the output of NuChart, i.e. a file representing the graph of the chromatin conformation of the analyzed cells, in three files based on the comma-separated value format (CSV). The first file represents just a node describing the experiment, with associated the number of genes and links it produced. The other two files contain the two sets of nodes and edges. It is to note that an experiment usually adds new links between genes already existing in the database.

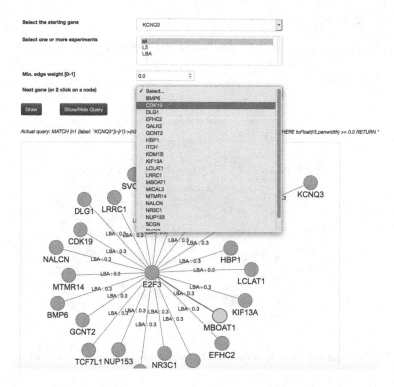

Fig. 5. The selection of the step after E2F3, useful in large networks.

Web Application. The Web application represents the second step. It represents an extension of the Neovis.js[5] visualisation library, which provides general-purpose graph visualisation powered by vis.js with data from Neo4j [22].

NeoHiC interacts with the database through the Javascript driver by performing queries like

```
MATCH (n1 {label: 'A1CF'})-[r1]-(n2) RETURN *
```

whose result is shown in Fig. 2. These queries follow the Neo4j's graph query language Cypher[6]. The above query selects the node of the database labeled with 'A1CF' and retrieves the nodes representing the genes linked with it. This gene is linked with only five other genes that correspond to the nodes matched with $n2$, while the five edges correspond to $r1$. Edges are labeled with the experiments that created them, i.e., the Lieberman-Aiden et al. Hi-C data [8].

---

[5] https://github.com/neo4j-contrib/neovis.js/.
[6] https://neo4j.com/developer/cypher-query-language/.

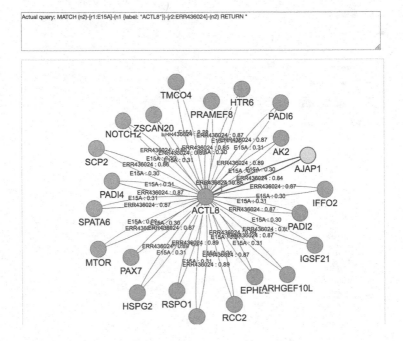

Fig. 6. The selection of a subset of the neighboring nodes of ACTL8.

3   Results

NeoHiC is based on the same approach adopted by STRING, where a protein-protein interaction network is expanded one step at a time by clicking on one of the visible nodes. Examples of visualisation are provided in Fig. 3 and Fig. 4.

The present version of the Web application allows users to select a starting gene and inspecting, step by step, the network described by the Hi-C data. In particular, the result is shown in Fig. 4 corresponds to the query

```
MATCH (n1 {label: "KCNQ3"})-[r1]-
(n2 {label: "LOC100507421"})-[r2]-
(n3 {label: "E2F3"})-[r3]-
(n4 {label: "KDM1B"})-[r4]-
(n5 {label: "MBOAT1"})-[r5]-
(n6 {label: "URB1"})-[r6]-(n7)
WHERE toFloat(r6.penwidth) >= 0.0RETURN *
```

where each of the five gene selection corresponds to add a $n_x - [r_x] - n_{x+1}$ pattern.

In most cases, the selection of a specific gene as the next step is complex, as in Fig. 3, even if it is possible to zoom and move the network. We, therefore, added the possibility to select the next gene as a drop-down menu, as illustrated in Fig. 5.

It is also possible to go back of one step with the *Back* button if the users selected a wrong gene or they want to explore another path. It is also possible to go back to multiple steps by clicking on one edge, for example, on the edge between 'KDM1B' and 'MBOAT1' in Fig. 4.

Furthermore, it is possible to filter the neighboring genes based on the weight associated with the edges and limit the edges to those provided by a subset of the experiments included in the database. Last, the application shows the query corresponding to the shown network configuration, that can be exploited by expert users to interact directly with Neo4j via its command line or Web client to perform specific analysis tasks.

Figure 6 shows the result of a manually-inserted query that filters among the 262 neighboring genes of 'ACTL8', only those having a link for each of the two selected experiments.

4 Conclusion and Future Development

The NeoHiC web application, including the tool for converting the NuChart results, is available via GitHub[7].

It represents a first step in the development of a Web portal [23] for the sharing and analysis of Hi-C data. Currently, we offer a docker container for the execution of NuChart [24] and NeoHiC as two separate tools. Our first future development will be to deploy a cloud service on the HPC4AI platform [25] providing scientists with a portal to explore and publish novel experiments [26].

Besides this, we are adding more filters and options to improve the analysis of the data. At first, we will allow the comparison of the shortest paths linking two genes in two or more experiments, to statistically highlight differences in the chromatin conformation of different cells. Then we will integrate further analyses, such as the significance of the vertex clustering attitude (triangle), as described in [14].

The final goal is represented by the integration of 1D and 2D information on the Hi-C graphs to correlate the 3D conformation of the genome with regulatory and expression patterns and to adopt artificial intelligence to speed up the extraction of relevant results from the data.

Acknowledgments. This work has been funded by the Short-term 2018 Mobility Program (STM) of the National Research Council of Italy (CNR).

References

1. Chiappori, F., Merelli, I., Milanesi, L., Marabotti, A.: Static and dynamic interactions between GALK enzyme and known inhibitors: guidelines to design new drugs for galactosemic patients. Eur. J. Med. Chem. **63**, 423–434 (2013)

---

[7] https://github.com/dddagostino/neohic.

2. Merelli, I., Cozzi, P., D'Agostino, D., Clematis, A., Milanesi, L.: Image-based surface matching algorithm oriented to structural biology. IEEE/ACM Trans. Comput. Biol. Bioinform. **8**(4), 1004–1016 (2010)
3. Viti, F., Merelli, I., Caprera, A., Lazzari, B., Stella, A., Milanesi, L.: Ontology-based, tissue MicroArray oriented, image centered tissue bank. BMC Bioinform. **9**(4), S4 (2008)
4. Banegas-Luna, A.J., et al.: Advances in distributed computing with modern drug discovery. Expert. Opin. Drug Discov. **14**(1), 9–22 (2019)
5. Ling, J.Q., Hoffman, A.R.: Epigenetics of long-range chromatin interactions. Pediatr. Res. **61**, 11R–16R (2007)
6. Phillips-Cremins, J.E., Corces, V.G.: Chromatin insulators: linking genome organization to cellular function. Mol. Cell **50**(4), 461–474 (2013)
7. Duan, Z., Andronescu, M., Schutz, K., Lee, C., Shendure, J., et al.: A genome-wide 3C-method for characterizing the three-dimensional architectures of genomes. Methods **58**(3), 277–288 (2012)
8. Lieberman-Aiden, E., et al.: Comprehensive mapping of long-range interactions reveals folding principles of the human genome. Science **326**, 289–293 (2009). https://doi.org/10.1126/science.1181369. PubMed: 19815776
9. Merelli, I., Lio', P., Milanesi, L.: NuChart: an R package to study gene spatial neighbourhoods with multi-omics annotations. PLoS ONE **8**(9), e75146 (2013)
10. Tordini, F., et al.: NuChart-II: the road to a fast and scalable tool for Hi-C data analysis. Int. J. High Perform. Comput. Appl. **31**(3), 196–211 (2017)
11. Shavit, Y., Lio', P.: CytoHiC: a cytoscape plugin for visual comparison of Hi-C networks. Bioinformatics **29**(9), 1206–1207 (2013)
12. Durand, N.C., et al.: Juicer provides a one-click system for analyzing loop-resolution Hi-C experiments. Cell Syst. **3**(1), 95–98 (2016)
13. Serra, F., Bau, D., Goodstadt, M., Castillo, D., Filion, G., Marti-Renom, M.A.: Automatic analysis and 3D-modelling of Hi-C data using TADbit reveals structural features of the fly chromatin colors. PLOS Comput. Biol. **13**(7), e1005665 (2017)
14. Merelli, I., Tordini, F., Drocco, M., Aldinucci, M., Lio', P., Milanesi, L.: Integrating multi-omic features exploiting chromosome conformation capture data. Front. Genet. **6**, 40 (2015)
15. Tordini, F., Aldinucci, M., Milanesi, L., Lio', P., Merelli, I.: The genome conformation as an integrator of multi-omic data: the example of damage spreading in cancer. Front. Genet. **7**, 194 (2016)
16. Bean, D.M., Heimbach, J., Ficorella, L., Micklem, G., Oliver, S.G., Favrin, G.: esyN: network building, sharing and publishing. PLoS ONE **9**(9), e106035 (2014)
17. Shannon, P., et al.: Cytoscape: a software environment for integrated models of biomolecular interaction networks. Genome Res. **13**(11), 2498–2504 (2003)
18. Have, C.T., Jensen, L.J.: Are graph databases ready for bioinformatics? Bioinformatics **29**(24), 3107 (2013)
19. Szklarczyk, D., et al.: STRING v10: protein-protein interaction networks, integrated over the tree of life. Nucl. Acids Res. **43**(D1), D447–D452 (2014)
20. Smith, R.N., et al.: InterMine: a flexible data warehouse system for the integration and analysis of heterogeneous biological data. Bioinformatics **28**(23), 3163–3165 (2012)
21. Galizia, A., Roverelli, L., Zereik, G., Danovaro, E., Clematis, A., D'Agostino, D.: Using Apache Airavata and EasyGateway for the creation of complex science gateway front-end. Future Gener. Comput. Syst. **94**, 910–919 (2019)
22. Lyon, W.: Graph Visualization With Neo4j Using Neovis.js. (2018). https://bit.ly/2vOmPkj

23. D'Agostino, D., et al.: A science gateway for exploring the X-ray transient and variable sky using EGI federated cloud. Future Gener. Comput. Syst. **94**, 868–878 (2019)
24. Merelli, I., Fornari, F., Tordini, F., D'Agostino, D., Aldinucci, M., Cesini, D.: Exploiting Docker containers over Grid computing for a comprehensive study of chromatin conformation in different cell types. J. Parallel Distrib. Comput. **134**, 116–127 (2019)
25. Aldinucci, M., et al.: HPC4AI, an AI-on-demand federated platform endeavour. In: ACM Computing Frontiers, Ischia, Italy (2018). https://doi.org/10.1145/3203217. 3205340
26. Aldinucci, M., et al.: Parallel stochastic systems biology in the cloud. Brief. Bioinform. **15**(5), 798–813 (2014)

Random Sample Consensus
for the Robust Identification of Outliers
in Cancer Data

André Veríssimo[1,2], Marta B. Lopes[3], Eunice Carrasquinha[4],
and Susana Vinga[1,2(✉)]

[1] INESC-ID, Instituto Superior Técnico, Universidade de Lisboa, R. Alves Redol 9,
1000-029 Lisbon, Portugal
susanavinga@tecnico.ulisboa.pt
[2] IDMEC, Instituto Superior Técnico, Universidade de Lisboa, Lisbon, Portugal
[3] Instituto de Telecomunicações, Instituto Superior Técnico, Universidade de Lisboa,
Av. Rovisco Pais 1, 1049-001 Lisbon, Portugal
[4] Center for the Unknown, Champalimaud Foundation, Av. Brasília,
1400-038 Lisbon, Portugal

Abstract. Random sample consensus (RANSAC) is a technique that
has been widely used for modeling data with a large amount of noise.
Although successfully employed in areas such as computer vision, extensive testing and applications to clinical data, particularly in oncology,
are still lacking. We applied this technique to synthetic and biomedical datasets, publicly available at The Cancer Genome Atlas (TCGA)
and the UC Irvine Machine Learning Repository, to identify outliers in
the classification of tumor samples. The results obtained by combining
RANSAC with logistic regression were compared against a baseline classical logistic model. To evaluate the robustness of this method, the original
datasets were then perturbed by generating noisy data and by artificially
switching the labels. The flagged outlier observations were compared
against the misclassifications of the baseline logistic model, along with
the evaluation of the overall accuracy of both strategies. RANSAC has
shown high precision in classifying a subset of core (inlier) observations
in the datasets evaluated, while simultaneously identifying the outlier
observations, as well as robustness to increasingly perturbed data.

Keywords: RANSAC · Machine learning · Outlier detection ·
Oncological data

A. Veríssimo and M. B. Lopes—joint first author.
Supported by national funds through Fundação para a Ciência e a Tecnologia
(FCT) through projects UIDB/50021/2020 (INESC-ID), UIDB/50022/2020 (LAETA,
IDMEC), UID/EEA/50008/2019, SFRH/BD/97415/2013, PREDICT (PTDC/CCI-
CIF/29877/2017), MATISSE (DSAIPA/DS/0026/2019) and BINDER (PTDC/CCI-
INF/29168/2017).

1   Introduction

Obtaining a noise-free dataset that adequately represents a complex biological system is a utopic goal, regardless of how well the experiment was designed and conducted. Many factors contribute to the loss of quality in the data, translated into errors and noise in the variables measured and the outcome, which may hamper parameter estimation and the correct identification of the underlying model. Such factors are related to i) the data acquisition process, e.g., measurement errors or corrupted/missing data; ii) data processing, such as normalization, computation of new variables or batch-correcting, which can add uncertainty or create unwanted artifacts/loss of information; and iii) the presence of outlying observations, with responses that deviate from the population behavior. This problem has lead to the emergence of robust regression methods, which allow building models with high prediction accuracy even in the presence of outliers.

RANdom SAmple Consensus (RANSAC) [1] was proposed as an algorithm that introduces sampling to build smaller models from a minimal subset of the data and then proceeds to expand it to find a better representation of the underlying distribution. It simultaneously identifies inlier and outlier observations in the dataset. RANSAC has shown to be a useful tool in computer vision applications, e.g., [2–4], with its strength showing in its ability to incrementally detect different modes in the observations and effectively separating the data.

The capabilities of RANSAC have been seldom explored in biomedical applications, and were mostly to analyse gas chromatography/mass spectrometry proteomic and metabolomic data [5,6]. In the context of precision medicine, RANSAC is especially promising, as non-identified misdiagnosed patients compromise the development of accurate models, and severely impact therapy decision and biomarker discovery. Additionally, the detection of individual abnormal molecular values for correctly classified patients might provide valuable hints on the molecular understanding of the disease, and the identification of putative biomarkers and therapeutic targets.

We evaluated RANSAC in the classification of synthetic and breast cancer data using the logistic model. Our results show that this strategy is advantageous for the analysis of biomedical data and the concomitant tasks of outlier detection and robust regression.

The outline of this work is as follows: in Sect. 2, RANSAC and the logistic regression model are explained in detail. The results concerning application examples and parameter estimation are presented in Sect. 3. Finally, conclusions are addressed in Sect. 4.

2   Materials and Methods

2.1   Random Sample Consensus

RANdom SAmple Consensus (RANSAC) was first introduced by [1], as a new paradigm for robust parameter estimation. The success of RANSAC came from its efficiency in minimizing an objective function, $\mathcal{C}$, while identifying a set of potential outliers. This efficiency is achieved by solving $\mathcal{C}$ in a smaller subset of randomly chosen data points, before expanding to the whole dataset.

The RANSAC algorithm is described as follows (Algorithm 1). First, a subset S of *inliers* (size n) of the original data D (size N) is sampled and used to fit a model M. Given this model, all data points in $D \setminus S$ are tested against model M to determine whether they are within some pre-defined *error tolerance*, ϵ; if so, they are included in a consensus set S^*, expanding the set of *inliers*. In this work, we use Pearson's residuals (for logistic regression) as the error function to measure the discrepancy between the observed and predicted response under model M. If S^* has more than t elements, i.e., the consensus set has cardinality larger than t, the model is considered valid, and a refitted model M^* is then estimated based on all these (*inliers*) points. The best model is chosen using an error function, which can be, e.g., the cardinality of the S^* set, a residual, the log-likelihood, or AUC (Area Under the Curve) ROC (Re-ceiver Operating Characteristics).

Input : Original data D
　　　　n, the minimum number of data points to fit the model
　　　　ϵ, the inlier threshold for distance
　　　　t, the minimum number of inliers to consider the model valid
　　　　B, the number of iterations of the algorithm
Output: $\hat{M}$ best model
for 1 **to** B **do**
　　$S \leftarrow$ Randomly choose n data points from the original data D;
　　$M \leftarrow$ Model fitted with S set;
　　for *inlierTest* **In** $D \setminus S$ **do**
　　　　if $distance(inlierTest, M) < \epsilon$ **then**
　　　　　　$S^* \leftarrow Union(S^*, inlierTest)$

　　end
　　if $size(S^*) \geq t$ **then**
　　　　$M^* \leftarrow$ Refit model with S^* set
　　else
　　　　skip and go to the next iteration
　　end
　　if M^* *is better than* $\hat{M}$ **then**
　　　　$\hat{M} \leftarrow M^*$

end

Algorithm 1: The RANSAC algorithm.

The iterative procedure of RANSAC is based on the repetition of the previous procedure B times, which eventually leads to an accurate consensus model for a large subset of the original data. The advantage is to have a final model that fits well a subset of the data while simultaneously identifying the outlying elements.

This strategy is somehow the opposite of other smoothing techniques since it starts from a subset of the data and expands this set with observations that are consistent with the estimated model.

In this work, we studied data with a categorical response, i.e., cancer/non-cancer tissue and malign/benign tumor. Logistic regression is a classical approach to model this type of clinical binary response. We used the logistic model both coupled with RANSAC and as the baseline model in our comparison framework. The RANSAC algorithm was implemented in a prototype R package[1] to support logistic regression using either the `glmnet` or the `stats` R packages, although in the results we only use the `stats` package.

2.2 Logistic Regression

In this work we studied data with a categorical response that represents for each instance a binary outcome, cancer/non-cancer tissue or malign/benign tumor. The generalized linear model (GLM) [7] is a flexible and widely used tool that generalizes linear regression by allowing the mean of a population to depend on a linear predictor through a nonlinear link function. In the particular case of categorical responses, the *logit* function used in logistic regression can be used to classify this type of binary clinical outcome. The logistic probability function is given by

$$p_i = \frac{\exp(\mathbf{x}_i^T \boldsymbol{\beta})}{1 + \exp(\mathbf{x}_i^T \boldsymbol{\beta})},\tag{1}$$

where $\mathbf{x}$ is the $n \times p$ design matrix (n is the number of observations and p is the number of covariates or features), p_i is the probability of success (*i.e.* $Y_i = 1$) for observation i and $\boldsymbol{\beta} = (\beta_1, \beta_2, \ldots, \beta_p)$ are the regression coefficients associated to the p independent variables. The parameters of the model are estimated by maximizing the log-likelihood function given by

$$l(\boldsymbol{\beta}) = \sum_{i=1}^{n} \left\{ y_i \mathbf{x}_i^T \boldsymbol{\beta} - \log \left(1 + e^{\mathbf{x}_i^T \boldsymbol{\beta}} \right) \right\}.\tag{2}$$

We used the logistic model both coupled with RANSAC and as the baseline model, as a means for a comparison framework where the model itself is not a variable.

In a high-dimensional context, i.e., in case the number of features p largely exceeds the sample size n, logistic regression is unlikely to be used without any modification. Different approaches are available in the literature to tackle this problem. One of the most common approaches is the application of sparsity-inducing regularizers [10], which mathematically corresponds to adding a penalization to the cost function. Least absolute shrinkage and selection operator (lasso) regression shrinks feature's coefficient to zero, which works well for feature selection in case of high-dimensional data. The elastic net [11] is the result of combining the ℓ_1 (lasso) and ℓ_2 (ridge) norms. The most common regularizers

---

[1] https://github.com/sysbiomed/ransac.

are based on the ridge, lasso, and elastic net penalties. The difference between them lies in the combinations of the L_p norms used.

Synthetic and biomedical datasets were used to assess the performance of RANSAC, regarding parameter sensitivity and model accuracy across different data perturbation scenarios, in classification problems, as described next.

2.3 Datasets

RANSAC was tested with the logistic model in synthetic data and biomedical datasets from cancer clinical trials.

Synthetic Data. The synthetic data was generated from 4 normal distributions in a 2-dimensional space, with 225 sample data points per center (Fig. 1). The first distribution generates data with class 0 in $(1, 1)$ and class 1 in $(1, -1)$, while the second is centered in $(-1, -1)$ and $(-1, 1)$ for classes 0 and 1, respectively. This is a hard problem to tackle with linear classifiers, as the set is not linearly separable. Only two dimensions were used to allow a visual interpretation of the results, as the goal was to test the robustness of RANSAC to outliers and noise, and a higher variables' dimensional space is out of the scope of this work.

Breast Invasive Carcinoma Data. The Breast Invasive Carcinoma (BRCA) datasets were retrieved from the University of California Irvine (UCI) Machine Learning Repository[2] and The Cancer Genome Atlas (TCGA)[3]. These data comprise different types of variables, such as cytological features and gene expression in breast tissue.

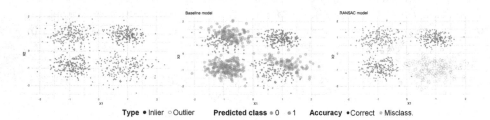

Fig. 1. Misclassifications by RANSAC and the baseline model over a 2-dimensional synthetic dataset (blue dots represent class 1 and red dots stand for class 0). (Color figure online)

The Wisconsin diagnostic BRCA dataset from the UCI repository contains 569 instances over 30 real-valued features computed for each cell nucleus. The response binary variable indicates whether the samples were taken from the tumor are *malignant* or *benign*, accounting for 357 and 212 cases, respectively.

---

[2] http://archive.ics.uci.edu/ml.
[3] https://cancergenome.nih.gov/.

The BRCA transcriptomic dataset from TCGA was imported using the R prototype package brca.data[4]. The dataset is composed of 54, 820 variables for a total of 1222 samples (1102 with primary solid tumor, 7 metastatic, and 113 with normal tissue) from 1097 individuals. The BRCA subset used in this study is composed of paired samples, i.e., one tumor sample and a normal tissue sample per BRCA patient, corresponding to a total of 113 individuals and 226 samples.

Given the high dimensionality of the BRCA (TCGA) dataset, dimensionality reduction was performed in 2 steps. First, by identifying a set of relevant variables through the classification of *tumor* and *normal* tissue samples. This reduction was achieved through sparse logistic regression, using the glmnet R package [8], via a modified DEGREECOX [9] regularization using LASSO [10] and elastic net [11] ($\alpha = 0.5, 0.7$). The optimum λ value for each run was obtained by 10-fold cross-validation. A total of 42 genes were selected for further analysis.

In the second step, a subset of the original dataset was created considering only the genes identified in the previous step and further reducing the number of variables by stepwise logistic regression using the stats R package. This procedure resulted in the selection of two genes, $ENSG00000072778$ and $ENSG00000235505$.

It should be noted that while in the first step the resulting model yielded perfect separation with a minimal subset of the genes used, i.e., 42 out of the 54, 820, in the second, the two genes selected could not correctly classify all individuals.

3   Results

The results obtained by RANSAC were evaluated regarding parameter sensitivity and accuracy as noise was introduced. In both cases, the RANSAC's results were compared against a baseline model where logistic regression was applied to the full dataset.

To assess how the different parameters in RANSAC affect the results, we performed a series of experiments on the synthetic and Wisconsin datasets. These experiments were twofold: first, we tested on both the synthetic and Wisconsin datasets by fixing all but one of RANSAC parameters and observing how the results changed in terms of inliers detected and AUC across different noise scenarios generated by a uniform distribution; secondly, we tested how increasing the number of noisy observations on the datasets impacted the misclassification rate.

The parameter t was excluded from testing as it only affects the performance of the algorithm by being more restrictive on the iterations that are accepted to be refitted. The parameter n was also fixed to $n = 16$, as it dependents on the type of model used by RANSAC: the glm function, used in this study, requires a minimum of 8 observations from each class. By sampling n data points we can infer the residuals' distribution from the fitted model using the initial data

---

[4] https://github.com/sysbiomed/data-archives/releases/download/ransac/brca.tar.gz.

points (Fig. 2), following a Gaussian or Log-Normal distribution. When changing n between $n \in \{16, 32\}$ for a set of 100 different initial models, little changes to the distribution of residuals could be observed (results not shown), meaning that RANSAC is not sensitive to this parameter.

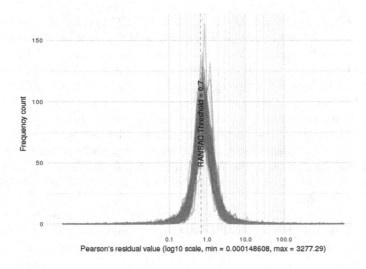

Fig. 2. Distribution of residuals for a set of 100 different initial models, considering $\epsilon = 0.7$.

The crucial parameter for RANSAC is the ϵ threshold that determines whether a given data point is an inlier or not. The choice of parameter ϵ is dependent on the data, as the distribution center varied widely when testing different datasets, as well as the number of outliers. The majority of the data points in this distribution seem to follow a Gaussian or Log-Normal distribution (Fig. 2). As the optimal ϵ value is dependent on the dataset, we used a different range of values per dataset. For the synthetic dataset we used $\epsilon \in \{0.1, 0.2, \ldots, 0.9, 1\}$, while for the Wisconsin dataset $\epsilon \in \{10^{-7}, 10^{-6}, \ldots, 10^{-1}, 0.5, 1\}$ was tested, since the optimal ϵ value is data dependent.

The model accuracy, given by the AUC for the inlier observations and the misclassification rate, was used for evaluating the parameter sensitivity across real (Wisconsin) and synthetic datasets. The data are perturbed by adding noisy observations that follow a uniform distribution. Each experiment was performed five times with different sampling of the noise data. We can immediately observe from the sensitivity results, partially shown in Fig. 3 and Fig. 4, that RANSAC is not sensitive to the ϵ parameter, as the indicators are stable in all noise scenarios until they reach $\epsilon = 1$, where the models start to produce worse results. This outcome is expected, as $\epsilon > 1$ starts to include misclassified observations, which justifies the decline in the model performance. The number of inliers also increases with ϵ, as the threshold to include them becomes more relaxed.

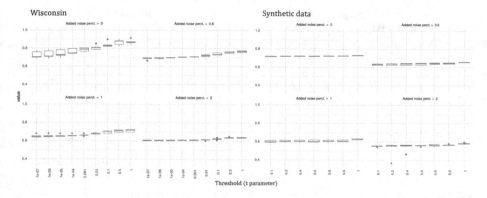

Fig. 3. Inliers included as the threshold parameter changes (with increased noise in the dataset), for the Wisconsin BRCA and synthetic datasets.

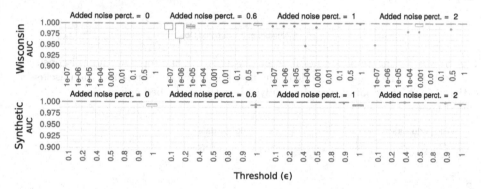

Fig. 4. AUC variation calculated using the inlier observations as the threshold parameter changes (with increased noise in the dataset), for the Wisconsin BRCA and synthetic datasets.

When comparing RANSAC and the baseline model regarding misclassifications, a more evident superiority is observed for RANSAC (Figs. 1 and 5), especially in the synthetic case, designed to be a hard classification problem. The most interesting finding when looking at the predictive power of RANSAC is its confidence to classify the inliers, as revealed by the AUC results (Fig. 4) showing that for $\epsilon < 1$ the RANSAC model correctly classifies all inliers. This result is a major strength of RANSAC, as it not only creates a good classifier but also provides tentative sets of inlier and outlier observations.

Data contamination was also evaluated in the transcriptomic BRCA dataset from TCGA to investigate the nature and precision of the outliers identified by RANSAC compared to the baseline model. Data perturbation, seen in Fig. 6, was performed by randomly switching 20 labels of raw observations (circled in orange) and running RANSAC B times, ensuring that the models converge to a solution. In the original data, RANSAC identified four outliers with high precision (circled in red), while with the logistic baseline model, 8 observations were

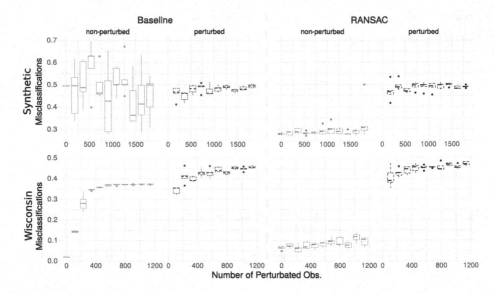

Fig. 5. Misclassification rate per dataset (Synthetic in the top row and Wisconsin on the bottom) as more noise is added. The non-perturbed columns show the misclassifications for the original observations only, while the perturbed columns show only the added noise.

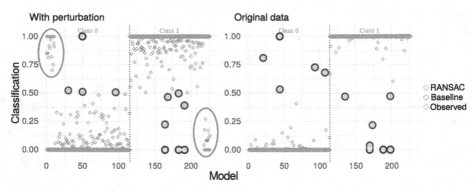

Fig. 6. Outliers identified by RANSAC and the baseline model in the original (right panel) and perturbed (left panel) BRCA (TCGA) datasets. RANSAC is more robust to perturbations of the original labels, resulting in higher stability on the identified outliers. (Color figure online)

flagged as outliers (circled in green). In the contaminated dataset, RANSAC accurately identified the same 4 outlier observations plus the 20 synthetic outliers generated. In turn, the baseline model flagged 4 out of the previous 8 outlier observations identified, plus 4 new observations and the 20 synthetic outliers, though with considerably less precision compared to RANSAC. This small and interpretable example clearly shows the suitability of RANSAC to biomedical settings, where mislabeling often takes place and further compromises data modeling and clinical decision making.

4   Conclusion

Modern biotechnologies provide a range of contaminated data originated from, e.g., sequencing errors, artifacts or sample mislabeling that, if not detected, deteriorate the models and compromise the extraction of reliable disease knowledge. We have shown that RANSAC can be successfully used in a clinical setting as a means to provide high-accurate classification of inliers and identify outlier observations in gene expression data, major strengths that made this algorithm widely used in computer vision applications. It is a model-based outlier detection method, enabling the estimation of both parameters and outliers simultaneously. Outlier identification is becoming a key aspect in modeling clinical data, in particular for improved diagnostics and prognostics assessment. Classification is playing a major rule towards personalised medicine and the automatic and accurate patient stratification constitutes a major challenge that can be tackled using robust sparse logistic regression coupled with RANSAC. The proposed approach has shown to be robust to increased random noise and mislabeling, which makes it particularly promising in the biomedical field, in particular for classification problems in clinical decision support systems in oncology research.

References

1. Fischler, M.A., Bolles, R.C.: Random sample consensus: a paradigm for model fitting with applications to image analysis and automated cartography. Commun. ACM **24**(6), 381–395 (1981)
2. Zhou, F., Cui, Y., Wang, Y., Liu, L., Gao, H.: Accurate and robust estimation of camera parameters using RANSAC. Opt. Lasers Eng. **51**(3), 197–212 (2013)
3. Nurunnabi, A., West, G., Belton, D.: Outlier detection and robust normal-curvature estimation in mobile laser scanning 3D point cloud data. Pattern Recognit. **48**, 1404–1419 (2015)
4. Stewart, C.: Robust parameter estimation in computer vision. SIAM Rev. **41**(3), 513–537 (1999)
5. Teoh, S.T., Kitamura, M., Nakayama, Y., Putri, S., Mukai, Y., Fukusaki, E.: Random sample consensus combined with partial least squares regression (RANSAC-PLS) for microbial metabolomics data mining and phenotype improvement. J. Biosci. Bioeng. **122**(2), 168–175 (2016)
6. Pluskal, T., Castillo, S., Villar-Briones, A., Orešič, M.: MZmine 2: modular framework for processing, visualizing, and analyzing mass spectrometry-based molecular profile data. BMC Bioinform. **11**(1), 395 (2010)
7. Nelder, J.A., Wedderburn, R.W.M.: Generalized linear models. J. R. Stat. Society. Ser. A (Gen.) **135**(3), 370–384 (1972)
8. Friedman, J., Hastie, T., Tibshirani, R.: Regularization paths for generalized linear models via coordinate descent. J. Stat. Softw. **33**(1), 1–22 (2010)
9. Veríssimo, A., Oliveira, A.L., Sagot, M.-F., Vinga, S.: DegreeCox - a network-based regularization method for survival analysis. BMC Bioinform. **17**, 109–121 (2016)

10. Tibshirani, R.: Regression shrinkage and selection via the lasso. J. R. Stat. Society. Ser. B (Methodol.) **58**(1), 267–288 (1996)
11. Zou, H., Hastie, T.: Regularization and variable selection via the elastic net. J. R. Stat. Soc. Ser. B (Stat. Methodol.) **67**(2), 301–320 (2005)

Solving Equations on Discrete Dynamical Systems

Alberto Dennunzio[1] , Enrico Formenti[2(✉)] , Luciano Margara[3] ,
Valentin Montmirail[4] , and Sara Riva[2]

[1] DISCo, Università degli Studi di Milano-Bicocca, Milan, Italy
dennunzio@disco.unimib.it
[2] Université Côte d'Azur, CNRS, I3S, Nice, France
{enrico.formenti,sara.riva}@univ-cotedazur.fr
[3] Università degli Studi di Bologna, Campus di Cesena, Cesena, Italy
margara@cs.unibo.it
[4] Avisto Telecom, Vallauris, France
valentin.montmirail@avisto.com

Abstract. Discrete dynamical systems (DDS) are a useful tool for modelling the dynamical behavior of many phenomena occurring in a huge
variety of scientific domains. Boolean automata networks, genetic regulation networks, and metabolic networks are just a few examples of DDS
used in Bioinformatics. Equations over DDS have been introduced as a
formal tool to check the model against experimental data. Solving generic
equations over DDS has been proved undecidable. In this paper we propose to solve a decidable abstraction which consists in equations having
a constant part. The abstraction we focus on consists in restricting the
solutions to equations involving only the periodic behavior of DDS. We
provide a fast and scalable method to solve such abstractions.

Keywords: Discrete dynamical systems · Decidability · Boolean
automata networks

1 Introduction

A dynamical system is a formal tool to study phenomena evolving along time.
They are used in a huge variety of scientific fields ranging from physics to biology and social sciences. Examples of DDS used in bioinformatics are Boolean
automata networks [6], genetic regulation networks [3], cellular automata [2],
and many others. In the very broad view, a dynamical system is a structure
$\langle \chi, f \rangle$ where χ is the *set of states* of the system and f is a function from χ to
itself called *next state function*.

When studying a phenomenon of interest which evolves along time, one can
be interested in reconstructing the dynamical behavior of the phenomenon from
experimental data. Both experimental data and partial knowledge or hypotheses
on the true ongoing system can be rephrased into an equation (or a series of

© Springer Nature Switzerland AG 2020
P. Cazzaniga et al. (Eds.): CIBB 2019, LNBI 12313, pp. 119–132, 2020.
https://doi.org/10.1007/978-3-030-63061-4_12

equations) over DDS. The solution to these equations will be a finer description of the dynamical systems which explain the experimental data.

In [4], a general algebraic framework to write equations of discrete dynamical systems has been provided. Unfortunately, in the same paper, it has also been proved that finding if generic equations on DDS have solutions is undecidable. This paper proposes a strategy to solve equations over DDS in practical situations. In the hypotheses validation case, the equations have a constant part and a polynomial part. The overall idea is to take a sequence of decidable abstractions of an equation which model an hypothesis, efficiently find solutions to these abstractions and use them to provide approximate solutions to the initial equation to validate or not the hypothesis modelled.

Indeed, the evolution of a finite DDS can be divided into two parts, namely, the transient and the periodic part. In this paper, we focus on the periodic part since it characterizes the long-term behavior of the system. We provide new algebraic results and a general method to solve equations restricted to the periodic behavior. The first step is to provide some basic algebraic results and notation which allow to express the problem in a more convenient manner. Then, we reduce the overall problem to the solution of (a series of) polynomial equations of degree one. Finally, we propose the *colored-tree* method to solve those equations. The idea of the method is to build a tree containing partial solutions and then build up solutions of the abstraction from them, starting from the leaves and moving towards the root. Branches of the tree may be *colored* in different colors according to the algebraic operations one should perform on them to propagate the solution towards the root. We prove that the method is sound, complete and terminating (Proposition 7 to 9) and, finally, we provide some experimental results to illustrate the scaling properties and to compare with brute force approach to show that the method is worth the effort (Sect. 6).

2   Background

A discrete dynamical system (DDS) is a structure $\langle \chi, f \rangle$ where χ is a finite **set of states** and $f : \chi \to \chi$ is a function called the **next state map**. Any DDS $\langle \chi, f \rangle$ can be identified with its **dynamics graph** $G \equiv \langle V, E \rangle$ where $V = \chi$ and $E = \{(a, b) \in V \times V, f(a) = b\}$.

In [4], an abstract algebraic setting for representing the dynamical evolution of finite DDS has been proposed. The authors proposed to define the following operations on DDS.

Definition 1 (Sum of DDS). *Given two discrete dynamical systems $\langle X, f \rangle$ and $\langle Y, g \rangle$, their sum, denoted $\langle X, f \rangle + \langle Y, g \rangle$, is the discrete dynamical system $\langle X \sqcup Y, f \sqcup g \rangle$ where the function $f \sqcup g : X \sqcup Y \to X \sqcup Y$ is defined as:*

$$\forall (v, i) \in X \sqcup Y \quad (f \sqcup g)(v, i) = \begin{cases} (f(v), i) & \text{if } v \in X \land i = 0 \\ (g(v), i) & \text{if } v \in Y \land i = 1 \end{cases}$$

where $\sqcup$ is the disjoint union. The disjoint union $A \sqcup B$, between two sets A and B, is defined as $A \sqcup B = (A \times \{0\}) \cup (B \times \{1\})$.

Definition 2 (Product of DDS). *Given two discrete dynamical systems* $\langle X, f \rangle$ *and* $\langle Y, g \rangle$, *their product* $\langle X, f \rangle \cdot \langle Y, g \rangle$, *is the discrete dynamical system* $\langle X \times Y, f \times y \rangle$ *where* $\forall (x, y) \in X \times Y$, $(f \times g)(x, y) = (f(x), g(y))$.

The two operations on DDS given above can also be restated in terms of dynamics graphs. Indeed, given two dynamics graphs $G_1 = \langle V_1, E_1 \rangle$ and $G_2 = \langle V_2, E_2 \rangle$, their **sum** $G_1 + G_2$ is defined as $\langle V_1 \sqcup V_2, E_1 \sqcup E_2 \rangle$. Whilst the **product** $G_1 \cdot G_2$ is the structure $\langle V', E' \rangle$ with $V' = V_1 \times V_2$ and $E' = \{((a, x), (b, y)) \in V' \times V', (a, b) \in E_1 \text{ and } (x, y) \in E_2\}$.

A dynamical system $\langle X, f \rangle$ is isomorphic to $\langle Y, g \rangle$ if and only if there exists a bijection γ such that $g \circ \gamma = \gamma \circ f$. According to the isomorphic relation, it is possible to define R, the set of equivalence classes of dynamical systems. It is easy to see that $\langle R, +, \cdot \rangle$ is a commutative semiring in which $\langle \emptyset, \emptyset \rangle$ is the neutral element *w.r.t.* $+$ and $\langle \{a\}, \{(a, a)\} \rangle$ is the neutral element *w.r.t.* multiplication. We stress that, from now on, when speaking of a DDS, we will always refer to its dynamics graph.

Now, consider the semi-ring $R[x_1, x_2, \ldots, x_n]$ of polynomials over R in the variables x_i, naturally induced by R. Polynomial equations of the form (1) model hypotheses about a certain dynamics deduced from experimental data (each X_i represents a variable $x_i^{w_i}$ for $w_i \in \mathbb{N} \setminus \{0\}$).

$$a_1 \cdot X_1 + a_2 \cdot X_2 + \ldots + a_k \cdot X_k = C \tag{1}$$

The known term C is the dynamical system deduced from experimental data. The coefficients a_i are hypothetical sub-dynamical systems that should cooperate to produce the observed dynamics C. Finding valid values for the unknown terms in (1) provides a finer structure for C which can bring further knowledge about the observed phenomenon.

In literature, some complexity results of solving some types of equations are already known.

Theorem 1 [4]. *Given two polynomials* $P(x_1, \ldots, x_n)$ *and* $Q(x_1, \ldots, x_n)$ *belonging to* $R[x_1, \ldots, x_n]$, *consider the following equation*

$$P(x_1, \ldots, x_n) = Q(x_1, \ldots, x_n). \tag{2}$$

The problem of finding a solution to Eq. 2 is undecidable. Moreover, if Eq. 2 is linear or quadratic, then finding a solution is in NP. *Finally, when* $P(x) = const$, *where the polynomial is in a single variable and all its coefficients are systems consisting of self-loops only, the equation is solvable in polynomial time.*

In order to limit the complexity, one can follow at least two strategies: either further constrain the polynomials or solve approximated equations which can provide information on the real solutions.

In this paper, we follow the second option. First of all, we consider equations of type (1) since neither subtraction nor division are definable over R. Then, we restrict the DDS involved in the equation to their periodic part. Finally, we propose an effective method to solve these new equations. Technically, this last operation is an abstraction, hence, on one hand, solutions found by our method are just candidate solutions for the original equation and on the other hand, solutions to the original equation must satisfy our modified equations.

3   More Background and a Useful Notation

Starting from an equation over DDS, we are interested to create a method that models and solves a decidable abstraction about the cyclic behavior. In other words, we need a mathematical characterization and an algorithmic approach to modelled and validate hypotheses about the asymptotic behavior of a phenomenon model through a DDS.

For any function f, let f^n denote the n-fold composition of f with itself, and let f^0 be the identity map. Given a DDS $\langle \chi, f \rangle$, a point $x \in \chi$ is **periodic** if there exists a positive number $p \in \mathbb{N}$ such that $f^p(x) = x$. The smallest p with the previous property is the **period** of the cycle. When $f(x) = x$, i.e. a cycle which consists of a single state, x is a **fixed point**. A **cycle** is a set $\{x, f(x), ..., f^{p-1}(x)\}$ for some periodic point $x \in \chi$ of period p. Denote Π the set of periodic points of $\langle \chi, f \rangle$. It is clear that $f(\Pi) = \Pi$ and hence $\langle \Pi, f|_\Pi \rangle$ is a sub-dynamical system of $\langle \chi, f \rangle$ (here $f|_A$ means the restriction of f to A) called the **sub-dynamical system induced by** Π. It is also clear that the dynamics graph G' of $\langle \Pi, f|_\Pi \rangle$ is the subgraph of the dynamics graph G of $\langle \chi, f \rangle$ made of the strongly connected components of G. Two DDS $A \equiv \langle \chi, f \rangle$ and $B \equiv \langle \chi', g \rangle$ are topologically conjugate if there exists a homeomorphism h from χ to χ' such that $h \circ f = g \circ h$. It is not difficult to see that the relation of being topologically conjugate is an equivalence relation between DDS. In particular, if $x \in \chi$ is a periodic point of f with period p, then $h(x)$ is periodic for g with period p. Moreover, if A and B are topologically conjugate then the dynamics graphs induced by the sub-dynamical system induced by Π are isomorphic. For these reasons, when studying the asymptotic behavior of a DDS one only needs to know the number of cycles and their periods but not what they stand for. Hence, we introduce the following notation which will be very useful in the sequel.

Definition 3. *Consider a DDS $A \equiv \langle \chi, f \rangle$ and let Π be the set of its periodic points. A cycle $\{x, f(x), ..., f^{p-1}(x)\}$ of **length** p of A is denoted C_p^1. Then,*
$$\bigoplus_{i=1}^{k} C_{p_i}^{n_i}$$
denotes the sub-dynamical system induced by Π saying that there are a total of $\sum_{i=1}^{k} n_i$ cycles of which n_i have period p_i for $i \in \{1, ..., k\}$.

Figure 1 illustrates the new notation just introduced.

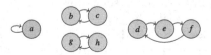

Fig. 1. A DDS with four cycles: $(C_1^1 \oplus C_2^2 \oplus C_3^1)$ in our notation.

4 Contributions

From now on, given $A \equiv \langle \chi, f \rangle$ and Π its set of periodic points, we denote $\bar{A}$ the DDS induced by Π.

After describing a new way to represent a discrete dynamical system, it is important to understand how this notation changes the previous reasoning about the operations of sum and product.

Definition 4. *Given* $A \equiv \langle \chi, f \rangle$ *(resp.,* $B \equiv \langle \chi', g \rangle$*), let* $\bar{A} \equiv \bigoplus_{i=1}^{k} C_{p_i}^{n_i}$ *(resp.,*

$\bar{B} \equiv \bigoplus_{j=1}^{w} C_{q_j}^{m_j}$*).*

Then $\bar{A} \oplus \bar{B}$ *is* $\bigoplus_{i=1}^{k} C_{p_i}^{n_i} \oplus \bigoplus_{j=1}^{w} C_{q_j}^{m_j}$.

The following result is immediate from the definitions.

Proposition 1. *Given* $A \equiv \langle \chi, f \rangle$ *(resp.,* $B \equiv \langle \chi', g \rangle$*), let* $\bar{A} \equiv \bigoplus_{i=1}^{k} C_{p_i}^{n_i}$ *(resp.,*

$\bar{B} \equiv \bigoplus_{j=1}^{w} C_{q_j}^{m_j}$*). Then* $\bar{A} \oplus \bar{B}$ *respects the following equivalence:*

$$\bigoplus_{i=1}^{k} C_{p_i}^{n_i} \oplus \bigoplus_{j=1}^{w} C_{q_j}^{m_j} = C_{p_1}^{n_1} \oplus C_{p_2}^{n_2} \oplus ... \oplus C_{p_k}^{n_k} \oplus C_{q_1}^{m_1} \oplus C_{q_2}^{m_2} \oplus ... \oplus C_{q_w}^{m_w}.$$

Definition 5. *Given* $A \equiv \langle \chi, f \rangle$ *(resp.,* $B \equiv \langle \chi', g \rangle$*), let* $\bar{A} \equiv \bigoplus_{i=1}^{k} C_{p_i}^{n_i}$ *(resp.,*

$\bar{B} \equiv \bigoplus_{j=1}^{w} C_{q_j}^{m_j}$*). Then,* $\bar{A} \odot \bar{B}$ *is* $\bigoplus_{i=1}^{k} C_{p_i}^{n_i} \odot \bigoplus_{j=1}^{w} C_{q_j}^{m_j} = \bigoplus_{i=1}^{k}\bigoplus_{j=1}^{w} C_{p_i}^{n_i} \odot C_{q_j}^{m_j}$.

The following proposition provides the precise value of $C_{p_i}^{n_i} \odot C_{q_j}^{m_j}$ given in the previous definition.

Proposition 2. *Assume* $\bar{A} \equiv C_p^m$ *and* $\bar{B} \equiv C_q^n$*. Then,*

$$\bar{A} \odot \bar{B} \equiv C_p^m \odot C_q^n = C_{\text{lcm}(p,q)}^{mn \cdot \gcd(p,q)} \ . \tag{3}$$

Proof. Assume $m = n = 1$. It is easy to see that $\bar{A}$ (resp., $\bar{B}$) can be identified with the cyclic group $\mathbb{Z}_p$ (resp., $\mathbb{Z}_q$). Similarly, $\bar{A} \odot \bar{B}$ can be identified with $\mathbb{Z}_p \times \mathbb{Z}_q$. For $(x, y) \in \mathbb{Z}_p \times \mathbb{Z}_q$, let $H_{x,y} = \{(w, z) \in \mathbb{Z}_p \times \mathbb{Z}_q \,|\, \exists k \in \mathbb{N} \text{ s.t. } w = x + k \text{ and } z = y + k\}$. Clearly, $H_{x,y}$ is a cyclic subgroup of $\mathbb{Z}_p \times \mathbb{Z}_q$ and its period is $\text{lcm}(p, q)$. Since

$$\mathbb{Z}_p \times \mathbb{Z}_q = \bigcup_{(x,y) \in \mathbb{Z}_p \times \mathbb{Z}_q} H_{x,y} \ ,$$

the number of disjoint sets of type $H_{x,y}$ is $\frac{|\mathbb{Z}_p \times \mathbb{Z}_q|}{\text{lcm}(p,q)} = \frac{pq}{\text{lcm}(p,q)} = \gcd(p,q)$. In the case of m or n different from 1, this means that each product operation is done for each of these components, so in general the result is duplicated $m \cdot n$ times. $\square$

We would like to prove the distribution properties of the new notation but before we need some propositions which have an immediate proof.

Proposition 3. *Assume* $\bar{A} \equiv C_p^n \oplus C_q^m$ *and* $\bar{B} \equiv C_r^a$. *Then,*

$$\bar{A} \odot \bar{B} \equiv (C_p^n \oplus C_q^m) \odot C_r^a = (C_p^n \odot C_r^a) \oplus (C_q^m \odot C_r^a). \tag{4}$$

The previous proposition can be generalized as follows.

Proposition 4. *Assume* $\bar{A} \equiv \bigoplus_{i=1}^{k} C_{p_i}^{n_i}$ *and* $\bar{B} \equiv C_r^a$, *then*

$$\bar{A} \odot \bar{B} \equiv \bigoplus_{i=1}^{k} C_{p_i}^{n_i} \odot C_r^a = (C_{p_1}^{n_1} \odot C_r^a) \oplus (C_{p_2}^{n_2} \odot C_r^a) \oplus \ldots \oplus (C_{p_k}^{n_k} \odot C_r^a). \tag{5}$$

Proposition 5. *Assume* $\bar{A} \equiv \bigoplus_{i=1}^{k} C_{p_i}^{n_i}$ *and* $\bar{B} \equiv \bigoplus_{j=1}^{w} C_{q_j}^{m_j}$, *then* $\bar{A} \odot \bar{B}$ *is given by*

$$\bigoplus_{i=1}^{k} C_{p_i}^{n_i} \odot \bigoplus_{j=1}^{w} C_{q_j}^{m_j} = (\bigoplus_{i=1}^{k} C_{p_i}^{n_i} \odot C_{q_1}^{m_1}) \oplus (\bigoplus_{i=1}^{k} C_{p_i}^{n_i} \odot C_{q_2}^{m_2}) \oplus \ldots \oplus (\bigoplus_{i=1}^{k} C_{p_i}^{n_i} \odot C_{q_w}^{m_w}). \tag{6}$$

Let us remind that each X_i represents a variable $x_i^{w_i}$. Therefore, it is necessary to know how to retrieve the solutions for the original x_i.

Proposition 6. *Given* $2m$ *integers* $p_i, k_i \in \mathbb{N}$, *with* $p_i > 0$ *for all* $i \in \{1, \ldots, m\}$, *let* $g(p_1, p_2, \ldots, p_m, k_1, k_2, \ldots, k_m)$ *be the gcd between the* p_i *for which* $k_i \neq 0$ *and let* $l(p_1, p_2, \ldots, p_m, k_1, k_2, \ldots, k_m)$ *be the lcm between the* p_i *for which* $k_i \neq 0$. *Assume* $\bar{A} \equiv C_{p_1}^1 \oplus C_{p_2}^1 \oplus \ldots \oplus C_{p_m}^1$. *Then,*

$$(\bar{A})^n \equiv \bigoplus_{i=1}^{m} C_{p_i}^{p_i^{n-1}} \oplus \bigoplus_{\substack{k_1+k_2+\ldots+k_m=n \\ 0 \leq k_1, k_2, \ldots, k_m < n}} \binom{n}{k_1, k_2, \ldots, k_m} C_l^{g \cdot \prod_{\substack{t=1 \\ k_t \neq 0}}^{m} p_t^{k_t - 1}}.$$

Proof. Using the multinomial theorem one finds

$$(\bar{A})^n \equiv (C_{p_1}^1 \oplus C_{p_2}^1 \oplus \ldots \oplus C_{p_m}^1)^n = \bigoplus_{k_1+k_2+\ldots+k_m=n} \binom{n}{k_1, k_2, \ldots, k_m} \bigodot_{t=1}^{m} (C_{p_t}^1)^{k_t}$$

$$= \bigoplus_{i=1}^{m} (C_{p_i}^1)^n \oplus \bigoplus_{\substack{k_1+k_2+\ldots+k_m=n \\ 0 \leq k_1, k_2, \ldots, k_m < n}} \binom{n}{k_1, k_2, \ldots, k_m} \bigodot_{t=1}^{m} (C_{p_t}^1)^{k_t}. \tag{7}$$

The resulting Formula (7) is obtained by extrapolating the cases in which a $k_i = n$. Another transformation is possible according to Proposition 2.

$$\bigoplus_{i=1}^{m} (C_{p_i}^1)^n \oplus \bigoplus_{\substack{k_1+k_2+\ldots+k_m=n \\ 0 \leq k_1,k_2,\ldots,k_m < n}} \binom{n}{k_1, k_2, \ldots, k_m} \bigodot_{t=1}^{m} (C_{p_t}^1)^{k_t}$$

$$= \bigoplus_{i=1}^{m} (C_{p_i}^1)^n \oplus \bigoplus_{\substack{k_1+k_2+\ldots+k_m=n \\ 0 \leq k_1,k_2,\ldots,k_m < n}} \binom{n}{k_1, k_2, \ldots, k_m} C_{l(p_1,p_2,\ldots,p_m,k_1,k_2,\ldots,k_m)}^{g(p_1,p_2,\ldots,p_m,k_1,k_2,\ldots,k_m) \cdot \prod_{t=1}^{m} p_t^{k_t-1}}$$

$$= \bigoplus_{i=1}^{m} C_{p_i}^{p_i^{n-1}} \oplus \bigoplus_{\substack{k_1+k_2+\ldots+k_m=n \\ 0 \leq k_1,k_2,\ldots,k_m < n}} \binom{n}{k_1, k_2, \ldots, k_m} C_{l(p_1,p_2,\ldots,p_m,k_1,k_2,\ldots,k_m)}^{g(p_1,p_2,\ldots,p_m,k_1,k_2,\ldots,k_m) \cdot \prod_{t=1}^{m} p_t^{k_t-1}} \,.$$

$\square$

For k equal to 0 we assume that $(\bar{A})^0$ is equal to C_1^1, the neutral element of the product operation. Let us go back to Eq. (1) which is the problem that we want to solve. It can be rewritten as follows:

$$(\bigoplus_{j=1}^{s_1} C_{p_{1j}}^1 \odot X_1) \oplus (\bigoplus_{j=1}^{s_2} C_{p_{2j}}^1 \odot X_2) \oplus \ldots \oplus (\bigoplus_{j=1}^{s_k} C_{p_{kj}}^1 \odot X_k) = \bigoplus_{j=1}^{m} C_{q_j}^{n_j} \qquad (8)$$

with s_i, the number of different cycles in the system i, p_{ij} is the period of the jth cycle in the system i. In the right term, there are m different periods, where for the jth different period, n_j is the number of cycles, and q_j the value of the period. However, Eq. (8) is still hard to solve in this form. We can simplify it further by performing a **contraction step** which consists in cutting Eq. (8) into two simpler equations

$$\begin{cases} C_{p_{11}}^1 \odot X_1 = \bigoplus_{i=1}^{m} C_{q_i}^{u_i} & (9a) \\ \\ C_1^1 \odot Y = \bigoplus_{i=1}^{m} C_{q_i}^{v_i} & (9b) \end{cases}$$

with $Y = (\bigoplus_{i=2}^{s_1} C_{p_{1i}}^1 \odot X_1) \oplus (\bigoplus_{j=1}^{s_2} C_{p_{2j}}^1 \odot X_2) \oplus \ldots \oplus (\bigoplus_{j=1}^{s_k} C_{p_{kj}}^1 \odot X_k)$ and $n_i = u_i + v_i$ for $i \in \{1, \ldots, n\}$.

By recursively applying contraction steps, and for all possible values u_i, v_i in Equations (9a) and (9b), solving Eq. (8) boils down to solve multiple times the following type of equation:

$$C_p^1 \odot X = C_q^n \,. \qquad (10)$$

If the variable X in Eq. (10) has a power different from one, then Lemma 6 is used to find the final solution. The above method allows to find solutions to equations of type (8) at the price of solving a huge number of equations of type (10). Indeed, in the solving process one might have to compute the intersection between the set of solutions, so we need to enumerate the solutions set of each equation of type (10). Therefore, we need an efficient method for finding their solution. This is the matter of the next section.

5   The Colored-Tree Method

In this section, we formally introduce the problem and its complexity. The colored-tree method is presented and some examples are shown.

Definition 6 (SOBFID). *The* SOBFID *(SOlve equation on BIjective Finite DDS) problem is a decision problem which takes in input $p, n, q \in \mathbb{N} \setminus \{0\}$ and returns true iff Eq. (10) admits a solution.* EnumSOBFID *is the problem which takes the same input as* SOBFID *and outputs the list of all solutions to Eq. (10).*

Lemma 1. SOBFID *is in* NP.

Proof. A problem is in NP iff the verification of a solution takes polynomial time (w.r.t. the size of the input). Given $p, n, q \in \mathbb{N} \setminus \{0\}$ and a possible solution $\bigoplus_{i=1}^{m} C_{q_i}^{n_i}$, one need to check that: $n = \sum_{i=1}^{m} \gcd(p, n_i)$ and for all $i \in \{1, \ldots, m\}$, $\mathrm{lcm}(p, q_i) = q$. All these computations can be made in polynomial time. □

Solving SOBFID is hard but still tractable. Indeed, the following lemma classifies our enumeration problem in EnumP. Recall that EnumP is the complexity class of enumeration problems for which a solution can be verified in polynomial time. It can be seen as the enumeration counterpart of the NP complexity class. For a full comprehension on the complexity classes of enumeration problems, we redirect the reader to [5].

Lemma 2. EnumSOBFID *is in* EnumP.

Proof. One just needs to be able to check if a given value is a solution in polynomial time. This is the case according to Lemma 1. □

The colored-tree method provides a solution to EnumSOBFID by establishing a connection between the enumeration of the solutions to Eq. (10) and the well-known Change-Making problem [7].

The Change-Making problem, in its enumeration version, consists to find all the possible ways to change a total amount τ using only coins of some specific units. In the optimization version, one must find the best way (the smaller number of coins) to express τ using the available coin units. The n cycles of right hand of Eq. (10) must be generated using the factor C_p^1 and some other value $C_{r_k}^k$. There are several possibilities for choosing $C_{r_k}^k$. Moreover, $C_p^1 \odot C_{r_k}^k$ might generate only a subset of the C_q^n expected cycles.

Hence, one can consider that the total amount τ is the set of n cycles of size q, and the possible coins are the divisors of q.

The `colored-tree` technique uses this connection coupled with a completeness-check to explore the feasible solutions space. The method is composed of two main phases: the tree building and the solutions aggregation. This algorithmic approach decomposes n in every possible way and represents these possibilities as a tree (memorized in tabular form to improve memory usage). Nodes of the tree are colored. The colors represent different independent subspaces of the solutions space of the change-making problem that are necessary to explore for enumerating the solutions. In fact, the connection with the change-making is iterated over each node (or subset of cycles) to study all the solutions of the equation. The cycles are then divided into child subsets. Iterating this technique on each subset produced, the method arrives to enumerate all the possible ways to generate the components involved in the right part of (10). During the second phase, the method computes the real solutions of the equation represented in the tree.

The `colored-tree` method is pretty involved, we prefer to illustrate it by two examples.

Example 1. Consider the equation $C_6^1 \odot X = C_6^6$. The algorithm consists of two distinct phases: tree building and solutions aggregation. In the first phase, the algorithm enumerates all the divisors $\mathcal{D}$ of $q = 6$ *i.e.* $\{6, 3, 2, 1\}$. Then, it applies a making-change decomposition algorithm (MCDA) [1] in which the total sum is 6 and the allowed set of coins is $\mathcal{D}' = \mathcal{D} \backslash \{6\}$. MCDA decomposes 6 as $3+3$. MCDA is then applied recursively (always using $\mathcal{D} \backslash \{i\}$ as the set of coins to decompose an amount i). We obtain $(6 = 3 + 3)$, $(3 = 2 + 1)$ and $(2 = 1 + 1)$ as reported in Table 1. At this point, a check is performed to ensure that all possible ways of decomposing 6 using $\mathcal{D}'$ are present in the tree (this is necessary to explore the whole solutions space). In our case, we already have $[3, 3]$ found by the first run of MCDA. We also found: $[3, 2, 1]$, $[2, 2, 1, 1]$, $[1, 1, 2, 1, 1]$, $[1, 1, 1, 1, 1, 1]$ by the recursive application of MCDA. By performing the check, we discover that the decomposition of 6 as $[2, 2, 2]$ is not in the current tree. For this reason, $[2, 2, 2]$ is added to the set of decompositions of 6. As illustrated in Fig. 2, it is assigned a new color and a recursive application of MCDA is started on the newly added nodes, if it is necessary (in other words if they are not already represented in the table). A new check ensures that all decompositions are present. This ends the building phase. The resulting tree is reported in Fig. 2. After this first phase of construction of the tree, the aggregation of solutions starts. Remark that each node m represents the equation $C_p^1 \odot X = C_q^m$ that we call the **node equation**. The single cycle solution is called the **node solution** and it is obtained thanks to Lemma 2, $C_{\frac{q}{p} \times m}^1$ whenever a **feasible solution** exists *i.e.* if $\gcd(p, \frac{q}{p} \times m) = m$ and $\mathrm{lcm}(p, \frac{q}{p} \times m) = q$. For example, for $m = 3$ one finds $x = C_3^1$. Moreover, for each node equation, one needs to compute the solutions represented in the subtree that has root m. To find all the solutions for the current node, it is necessary to take the Cartesian product of the solutions sets in the subtrees of the same color and then the union of the solution sets of nodes of different colors (different splits). All the solutions can be found in Table 1.

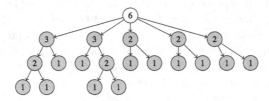

Fig. 2. The colored tree for the equation $C_6^1 \odot X = C_6^6$ after the completeness check.

Example 2. Consider the equation $C_6^1 \odot X = C_{12}^3$. In the first phase, the algorithm takes all the divisors $\mathcal{D}$ of $q = 12$ *i.e.* $\{12, 6, 4, 3, 2, 1\}$. Then, it applies a making-change decomposition algorithm where for each node m computes an optimal decomposition based on $\mathcal{D} \setminus \{m\}$ as the set of coins to decompose m. We obtain $(3 = 2 + 1)$ and $(2 = 1 + 1)$ as reported in Table 2. At this point, a check is performed to ensure that all possible ways of decomposing 3 using $\mathcal{D} \setminus \{3\}$ are represented in the tree. In our case, we already have $[2, 1]$ found by the first run of MCDA. We also found $[1, 1, 1]$ by the recursive application of MCDA. The check returns that all the possible decompositions of 3 are represented in the current tree. This ends the building phase. The resulting tree is reported in Fig. 3. After this first phase, the aggregation of solutions starts. In this case, the tree has only one color. Remark that if in the Cartesian product an empty set is involved, the result of the operation is then the empty set. For example, for $m = 2$, from the subtrees of the node one finds empty sets, but with the union of the solution of the node, the subtree solutions set for $m = 2$ is $\{C_4^1\}$. Moreover, the final solution set for the node 3 is the empty set, in fact in the Cartesian product $m = 1$ is involved (empty set). In this case, the method returns an empty set of solutions *i.e.* the equation has no solutions.

Table 1. Final data-structure storing all the decompositions, each solution for each value and at each step, the set of all solutions for a given value.

Node	Splits	Node solution	Subtree solutions set
6	[3,3][2,2,2]	C_6^1	$\{C_6^1, C_3^2, C_1^1 \oplus C_2^1 \oplus C_3^1, C_3^1 \oplus C_1^3,$ $C_2^1 \oplus C_1^4, C_1^6, C_2^3, C_1^2 \oplus C_2^2\}$
3	[2,1]	C_3^1	$\{C_3^1, C_1^1 \oplus C_2^1, C_1^3\}$
2	[1,1]	C_2^1	$\{C_1^2, C_2^1\}$
1	$\emptyset$	C_1^1	$\{C_1^1\}$

Table 2. Final data-structure storing all the decomposition, each solution for each value and at each step, the set of all solutions for a given value.

Node	Splits	Node solution	Subtree solutions set
3	[2,1]	{}	{}
2	[1,1]	C_4^1	$\{C_4^1\}$
1	$\emptyset$	{}	{}

Fig. 3. The tree represented in the table for $C_6^1 \odot X = C_{12}^3$, after the check of completeness.

The following propositions prove that the `colored-tree` method is sound, complete, and always terminating. For any $n, p, q \in \mathbb{N} \setminus \{0\}$, let $T_{p,q}^n$ denote the set of solutions of Eq. (10) and $S_{p,q}^n$ the set of solutions returned by the `colored-tree` method.

Proposition 7 (Soundness). *For all* $n, p, q \in \mathbb{N} \setminus \{0\}$, $S_{p,q}^n \subseteq T_{p,q}^n$.

Proof. Let us prove the soundness by induction on the depth of the tree from leaves to root. *Induction base*: if there is only one step, by Lemma 2 there is a solution iff $gcd(p, \frac{q}{p} \times m) = m$ and $lcm(p, \frac{q}{p} \times m) = q$, and because there is only one leaf in the tree, we have all the solutions. *Induction hypothesis*: assume that we have all the solutions at depth n and let us show that we can obtain all the solutions at depth $n + 1$. *Induction step*: it is easy to see that a solution exists if and only if it comes from a decomposition. Thus, by performing a Cartesian product between the set of solutions at depth n (which is true by IH) and the node solution (which is true by Induction base, since the node can be seen as a leaf), we know that we will obtain all the solution coming from the possible decomposition in the sub-tree. If a solution is coming from another sub-tree, since we perform an exhaustive check where we assign a different color to the other sub-tree, we know again, by IH and because we are taking the union of all the possible solutions, that we have all the possible solutions at a depth $n + 1$. □

Proposition 8. (Completeness). *For all* $n, p, q \in \mathbb{N} \setminus \{0\}$, $T_{p,q}^n \subseteq S_{p,q}^n$.

Proof. By contradiction, assume that there is a solution $r \in T_{p,q}^n$ such that $r \notin S_{p,q}^n$. This means that the colored-tree method does not return it. Hence, there exists a decomposition of n, which leads to r, which is not in the tree. This is impossible since an exhaustive check is performed to ensure that all the decompositions are there.   □

Proposition 9. (Termination). *The colored-tree method always terminates.*

Proof. The building phase always terminates since the colored-tree has maximal depth $|\mathcal{D}'| = \lfloor div(q, n) \rfloor$ and the number of different possible colors is bounded by 2^k where k is the size of the multi-set containing n/p_i copies of the divisor p_i per each divisor in $\mathcal{D}'$. The aggregation phase always terminates since it performs a finite number of operations per each node of the colored tree.   □

After the presentation of the `colored-tree` method, let us recall the general point of view of this work. We introduced a general abstraction about the periodic part of the dynamics of DDSs. The `colored-tree` method is the algorithmic technique solving the basic equations of our abstraction. We present a simple example which illustrates how to solve the general abstraction.

Example 3. Given the periodic behavior $C_6^6 \oplus C_{12}^3$, let us introduce the hypothesis that this behavior can be expressed like a factor of the cyclic behavior C_6^1. The corresponding equation in the form (8) is $C_6^1 \odot X = C_6^6 \oplus C_{12}^3$. To validate this hypothesis, we need to solve two basic equations, namely, $C_6^1 \odot X = C_6^6$ and $C_6^1 \odot X = C_{12}^3$. In order to find the solutions, we need to compute the Cartesian product between the two sets of solutions. From the previous examples, we know that the second equation is impossible, consequently, we cannot validate the initial hypothesis. If we consider the equation $C_3^1 \odot X = C_6^6 \oplus C_{12}^3$, and the two corresponding basic equation $C_3^1 \odot X = C_6^6$ and $C_3^1 \odot X = C_{12}^3$. In this case, the following solutions set $\{C_6^2 \oplus C_{12}^1, C_6^2 \oplus C_4^3, C_6^1 \oplus C_2^3 \oplus C_{12}^1, C_6^1 \oplus C_2^3 \oplus C_4^3, C_2^6 \oplus C_{12}^1, C_2^6 \oplus C_4^3\}$ validates the hypothesis.   □

6   Experimental Evaluation

The `colored-tree` method provides a complete set of solutions for equations of type 10. An important parameter that affects the computational time of the algorithm is the number of nodes in the colored tree. This number gives an idea of the number of operations (Cartesian products and unions) that are necessary to find the solutions set.

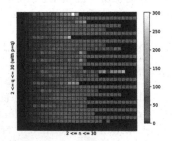

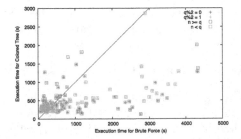

Fig. 4. Number of nodes in the colored tree as a function of n and q. Black parts correspond to out of memory cases with 30 GB of RAM.

Fig. 5. The brute force approach vs. `colored-tree` method *w.r.t.* execution time (in seconds).

Figure 4 shows how the dimension of the tree grows as a function of n and q. For this case, we set $p = q$ to ensure that we always have at least one solution and therefore a tree-decomposition. Notice that, in some cases, the complexity is particularly high due to specific analytical relations between the input parameters; we will address this problem in the near future. Notice also that our method seems to have a weakness when q is an even number. This is easily explained: in many cases, all the divisors can be expressed by the other ones. Therefore the check that ensures that all the decompositions are present is particularly time- and memory-consuming.

Since there is no other competitor algorithm at the best of our knowledge, we compared the `colored-tree` method to a brute force algorithm. We test our algorithm for $n, p \in \{1, \ldots, 20\}$ with $p = q$. Results are reported in Fig. 5. As expected, the `colored-tree` method outperforms the brute force solution, sometimes with many orders of magnitude faster. However, when the input equation has small coefficients, the `colored-tree` method performs worse. This can be explained considering that building the needed data structures requires a longer time than the execution of the brute force algorithm.

7   Conclusion

In [4], Dennunzio *et al.* argued that many questions about Boolean automata networks, used in biological modelling for genetic regulatory networks and metabolic networks, can be rewritten as equations over DDS. One of the core routines of the algorithm uses a time/memory expensive check for the change-making problem which clearly affects the overall performances. Therefore, a natural research direction consists in finding a better performing routine. One possibility would consider a combination of parallelism and logic reformulation of the problem. Another interesting research direction consists in better understanding the computational complexity of SOBFID. We are still working to improve the model and the performances of the algorithm in order to provide a handy tool that can be exploited by bioinformaticians.

References

1. Adamaszek, A., Adamaszek, M.: Combinatorics of the change-making problem. Eur. J. Comb. **31**(1), 47–63 (2010)
2. Alonso-Sanz, R.: Cellular automata and other discrete dynamical systems with memory. In: Smari, W.W., Zeljkovic, V. (eds.) Proceedings of HPCS, p. 215. IEEE (2012)
3. Bower, J.M., Bolouri, H.: Computational Modeling of Genetic and Biochemical Networks. MIT Press, Cambridge (2004)

4. Dennunzio, A., Dorigatti, V., Formenti, E., Manzoni, L., Porreca, A.E.: Polynomial equations over finite, discrete-time dynamical systems. In: Proceedings of ACRI 2018, pp. 298–306 (2018)
5. Mary, A., Strozecki, Y.: Efficient enumeration of solutions produced by closure operations. Discrete Math. Theoret. Comput. Sci. **21**(3) (2019). https://doi.org/10.23638/DMTCS-21-3-22. https://dmtcs.episciences.org/5549
6. Sené, S.: On the bioinformatics of automata networks. HDR, University of Évry Val d'Essonne, France (2012). https://tel.archives-ouvertes.fr/tel-00759287
7. Wright, J.: The change-making problem. J. ACM (JACM) **22**(1), 125–128 (1975)

SW+: On Accelerating Smith-Waterman Execution of GATK HaplotypeCaller

Meysam Roodi[1,2](✉) and Andreas Moshovos[1]

[1] Electrical and Computer Engineering Department, University of Toronto, Toronto, Canada
meysam.roodi@mail.utoronto.ca, moshovos@ece.toronto.edu
[2] Huawei Technologies, Canada Research Center, Toronto, Canada

Abstract. Next Generation sequencing is widely used today in several applications such as in studying hereditary diseases or in prenatal genetic testing. Genome Analysis ToolKit (GATK) workflow is currently the best practice flow in use in industry and academia. Variant Calling, the last step in the GATK pipeline, is performed by GATK HaplotypeCaller. It is one of the most time consuming steps in the whole pipeline. In this paper, we investigated the Smith-Waterman implementation of HaplotyeCaller and achieved an up to 40% reduction in Smith-Waterman execution of the HaplotypeCaller by proposing a new optimization where when possible we conclude the Smith-Waterman results by running a simpler linear comparison function. The optimization reduces the HaplotypeCaller run time by up to 10%.

Keywords: Gene sequencing · Variant Calling · HaplotypeCaller · Smith-Waterman

1 Introduction

The decoding of human genome by the "Human Genome Project" in 2003 started a new era in biology and bio-informatics. The human genome is used as reference in gene sequencing, a process that has been improving and finding an increasing number of applications. These include applications in forensic sciences, studying hereditary diseases, cancer studies, and prenatal monitoring to name a few. The amount of data produced in genetics is exponentially increasing while the cost of gene sequencing dropped dramatically in the past decade.

Currently, the GATK [1–3] workflow is accepted as the best practice gene sequencing pipeline in industry and academia. The pipeline is composed of five major computation steps, illustrated in Fig. 1, where each step processes the output of the preceding one producing the input to the next stage. In *Alignment*, the first stage of the pipeline, the DNA samples, in the form of short reads, are first aligned to the human reference genome. The short reads are then sorted based on their alignment chromosome and position in the second stage. Subsequently, duplicated short reads are marked. These are short read pairs which

© Springer Nature Switzerland AG 2020
P. Cazzaniga et al. (Eds.): CIBB 2019, LNBI 12313, pp. 133–141, 2020.
https://doi.org/10.1007/978-3-030-63061-4_13

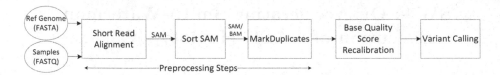

Fig. 1. Gene sequencing pipeline

have common aligned positions. The fourth step is *Base Quality Score Recalibration* which adjusts base pair (bp) quality scores and removes any potential systematic sequencer error/bias in generating bp quality scores.

The last step in this flow is *Variant Calling* which is performed by the GATK HaplotypeCaller tool. The HaplotypeCaller differentiates real variation between input data and the reference genome from variations that are introduced by either the sequencer at input data generation phase or by any of the tools or algorithms along the previous steps of the pipeline. GATK HaplotyeCaller finishes the *Variant Calling* step by performing Genotyping via a statistical approach through employing Pair Hidden Markov Models (PHMM) algorithm. Another set of haplotyping tools try to tackle this problem by dividing the reads into two main categories associating each of them to one copy of chromosome strands through sovling the NP-hard Minimum Error Correction (MEC) problem. These approaches are reviewed in [10]. Authors in [11] and [9] take a further step to accelerate MEC based methods by exploiting distributed systems.

The Smith-Waterman (SW) [4] algorithm finds the best alignment match between two sequences. SW is incorporated in both the read alignment "*bwa mem*" [5,6] tool and the variant calling HaplotypeCaller tool. The Haplotype-Caller employs SW to carefully align candidate haplotypes to the reference genome. SW finds the best alignment through the calculation of a score matrix, a computationally demanding task; it registers as taking a significant portion of the overall execution time for both the read alignment and the haplotype caller tool [7].

SW is a general and effective alignment algorithm, however, this generality comes at a price as its computational complexity is $O(n^2)$. In this work we take a *data aware* approach and study the properties of the data often being used in practice. In particular, we analyze the use of SW in the GATK HaplotypeCaller and show that most invocations of SW could be safely replaced with simple, linear sequence comparisons without any loss of accuracy. These SW calls include pattern comparisons which return either full match or a result with limited number of mismatches. Linear sequence comparisons are much less computationally demanding than SW as their computational complexity is $O(n)$. By avoiding a large fraction of the SW calls, our optimization improves the performance of the HaplotypeCaller. We show that for the three input datasets we used, NA18507, NA19240 and NA12878, our optimization replaces more than 50% of the SW calls of the HaplotypeCaller. This reduces total SW elapsed time by up to 40% improving overall HaplotypeCaller run time by 10% essentially for free.

The rest of this document is organized as follows: Smith-Waterman is described in Sect. 2. Section 3 details our proposed optimization and Sect. 4 reports experimental measurements demonstrating the utility of our optimization. Finally the last section concludes the paper.

2   Smith-Waterman Algorithm

Let $A = a_1a_2...a_n$ and $B = b_1b_2...b_r$ be two input sequences, where for our purposes, A is a reference and B is a pattern we want to find in the reference. In some cases, B will appear exactly somewhere in A, but in general B may not necessarily appear as-is within A. For example, some of b_i maybe missing in A or some of a_i may be missing in B. In this case, we wish to find a suitable subsequence within A that best matches with B. The Smith-Waterman [4] algorithm assigns scores to different ways of matching B with A with the goal of finding the "best" match between these two input sequences. SW assigns scores to each of these possibilities by constructing a $(n + 1) \times (r + 1)$ scoring matrix. To do so, the algorithm uses four pre-specified score/penalty weights to populate score values in the matrix. These score/penalty weights are:

- *Match Score*: A positive score applied where sequence elements match
- *Mismatch Score*: A negative penalty value applied where sequence elements do not match
- *Insertion penalty*: A negative penalty value applied where SW introduces an insertion; an element which exists in the pattern but does not exist in the reference
- *Deletion penalty*: A negative penalty value applied where SW introduces a deletion; an element which exists in the reference but does not exist in the pattern

$$s(a_i, b_j) = \begin{cases} match\ score\ \text{if } a_i == b_j \\ mismatch\ penalty\ \ \text{if } a_i \neq b_j \end{cases} \tag{1}$$

$$H_{i,j} = max \begin{cases} H_{i-1,j-1} + s(a_i, b_j) \\ H_{i-1,j} - P_{ins} \\ H_{i,j-1} - P_{del} \\ 0 \end{cases} \tag{2}$$

As the example of Fig. 2b shows, the columns of the scoring table correspond to the reference sequence whereas the rows correspond to the pattern string. The algorithm first initializes the first row and column scores to zero. Then, starting from the upper left corner, and as Fig. 2a illustrates, the score value of each matrix element is calculated based on the value of its three predecessor score values according to the Formula 2. $H(i, j)$ represents the score value of matrix's (i, j) cell in Formula 2, and $s(a_i, b_j)$ is set to *match score* if a_i matches b_j. It is assigned with *mismatch penalty* if a_i mismatches b_j.

This is shown in Formula 1. P_{ins} and P_{del} are the insertion and deletion penalties respectively. Figure 2b depicts the SW score matrix of $ATCGATTAGGTC$ and $CGATAGCT$ sequences. In this example, *match score* is $+3$ and *mismatch, insertion* and *deletion* penalties are all -1.

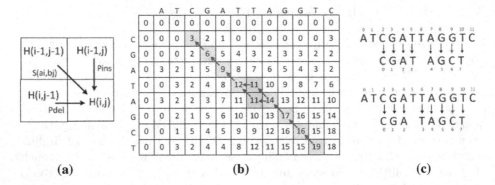

(a)   **(b)**   **(c)**

Fig. 2. (a) Smith-Waterman matrix cell score calculation (b) Best alignment between ATCGATTAGGTC and CGATAGCT. Match score is $+3$ and all penalties including mismatch, insertion and deletion penalties are -1 (c) The two possible best alignments between ATCGATTAGGTC as reference and CGATAGCT as pattern (Color figure online)

Once the matrix is fully populated, SW finds the cell with the maximum score and assembles the best match by tracing back the maximum score path from the maximum score element. The directions of the maximum score path determines the alignment result at each point; a diagonal move is a *match/mismatch*, a left move is a *deletion* and an up move is an *insertion*. On a diagonal move, moving to a lower score cell infers a *match* whereas moving to a higher score cell infers a *mismatch*. The maximum score cell is highlighted in pink in Fig. 2b and the trace back paths are highlighted in yellow. The two resulting best alignments are shown in Fig. 2c.

Like most other software implementations, the HaplotypeCaller implements SW in the *calculateMatrix* function using two nested loops each traversing the characters of one of the input sequences. The SW implementation in the HaplotypeCaller returns the Compact Idiosyncratic Gapped Alignment Report (CIGAR) and the alignment position of the pattern in the reference. CIGAR is a string notation which summarizes the alignment result between two sequences.

3   Our Optimization

In this section we explain our solution to optimize the SW implementation in the HaplotypeCaller. We observed, through conducting a data analysis, that most SW calls return full alignment match in the HaplotypeCaller. Our solution targets those SW calls; we try to filter SW calls and run a linear comparison between the two sequences prior to running the actual SW algorithm. If the result of SW calls could be determined with the linear comparison, we return the results and bypass the SW call. Otherwise, we let the SW call to proceed. This approach imposes an overhead to those SW calls for which we need to run the actual algorithm. On the other hand, it alleviates the computation burden of running all SW calls. In the following we detail our solution.

The heart of our solution is a *for loop* which compares the bps of the two sequences from left. In order to be consistent with SW, we want to ensure that our optimization finds the best score that SW is searching for. Therefore, we bypass SW calls only in two cases. The first case is where the pattern completely matches with the reference without any mismatches which could be safely bypassed because there is no better imaginable alignment than that which produces the maximum score. The second case is where we have non consecutive mismatches between the pattern and the reference separated by a matching sequence containing all possible bp types for cases where the length of the reference and pattern are the same. In the following we explain why the best score of the aforementioned case could be determined by our proposed optimization.

The HaplotypeCaller uses a $+30$ score for bp matches, a -10 for bp mismatches, a -10 for opening a gap whether an insertion or a deletion and a -2 penalty for extending a gap. For our second bypass case, we set the condition that the length of the pattern and the reference should be the same. Let's assume that n is the length of our pattern and reference. The maximum possible score is then $30 \times n$. Now, let's assume that we encounter k mismatches between the pattern and the reference. Using HaplotypeCaller's *match/mismatch* numbers, the score will be $30 \times (n - k) - 10 \times k$. However, we have to differentiate real mismatches from the mismatches that are produced by insertions or deletions and ensure that no better score could be derived in the k mismatch case.

Equation 3 illustrates the case of a bp mismatch between $a_1 a_2 a_3 ... a_n$ and $b_1 b_2 b_3 ... b_n$ at the mth position. Equation 4 on the other hand depicts the case of insertion and deletion. Certainly, our incorporated *for loop* finds the first mismatch at the mth base pair while traversing characters from left. To ensure that the detected bp mismatch is a real mismatch and it is not produced by an insertion or deletion, we inspect the next bp comparisons till the next possible mismatch at l^{th} position: $a_{m+1} a_{m+2} a_{m+3} ... a_l$ versus $b_{m+1} b_{m+2} b_{m+3} ... b_l$; $a_{m+1} a_{m+2} a_{m+3} ... a_l$ matches with $b_{m+1} b_{m+2} b_{m+3} ... b_l$ in a real mismatch case at m^{th} bp position whereas it matches with $b_m b_{m+1} b_{m+2} ... b_{l-1}$ in case of a deletion and $b_{m+2} b_{m+3} b_{m+4} ... b_{l+1}$ in case of an insertion.

$$a_1 a_2 ... a_{m-1} = b_1 b_2 ... b_{m-1}$$
$$a_m \neq b_m \tag{3}$$
$$a_{m+1} a_{m+2} ... a_n = b_{m+1} b_{m+2} ... b_n$$

$$a_1 a_2 ... a_{m-1} = b_1 b_2 ... b_{m-1}$$
$$a_m a_{m+1} ... a_{n-1} = b_{m+1} b_{m+2} ... b_n (Insertion) \tag{4}$$
$$a_{m+1} a_{m+2} ... a_n = b_m b_{m+1} ... b_{n-1} (Deletion)$$

This condition cannot capture repetitive patterns however. Based on Eqs. 3 and 4, the optimization condition confuses a real deletion case with a mismatch case if $b_{m+1}b_{m+2}b_{m+3}...b_l$ equals $b_m b_{m+1}b_{m+2}...b_{l-1}$ which essentially means that all bp in the two sequences are the same. Now, let's consider a 2 bp deletion case which is illustrated in Eq. 5. The optimization condition is not able to detect a 2 bp deletion case if $b_m b_{m+1}b_{m+2}...b_{l-1}$ matches $b_{m+2}b_{m+3}b_{m+4}...b_{l+1}$. This case actually exhibits a 2 bp repetitive pattern. Likewise, a 3 bp deletion could not be detected if there is a 3 bp repetitive pattern in the reference/pattern. These cases are illustrated in Fig. 3. This figure shows that a true 1, 2 or 3 bp deletion might be confused with a single mismatch if we have repetitive patterns in the suffix of the mismatch. Repetitive pattern of size 1, 2 and 3 could happen with only using 1, 2 and 3 bp types respectively and therefore we set an additional condition that all bp types should occur in the matching sequence between two non-consecutive mismatches to capture repetitive patterns which enables us to differentiate true mismatches from mismatches that are produced by insertions/deletions of up to 3 bp.

Fig. 3. 1, 2 and 3 bp deletion cases for repetitive patterns

Although our optimization is not capable of separating real mismatches from deletions/insertions larger than 4 bps, we set yet another condition that the maximum number of mismatches in our optimization is 8 and here we show that insertions/deletions larger than 4 bps produce a lower score. We saw that a k mismatch case produces a score of $30 \times (n - k) - 10 \times k$. Now, let's assume that the mismatches are produced by at least one deletion of size p larger than 4 bp (that we cannot detect in our optimization). This produces a score of $30 \times (n - p \times 2) - 10 - 2 \times (p - 1)$. We should subtract $2 \times p$ from match score because the bps that are consumed for deletion from reference will leave the same number of bp at the end of pattern unmatched. This value is less than the score for the case of mismatches and therefore it is guaranteed that we found the best score.

$$a_1 a_2 ... a_{m-1} = b_1 b_2 ... b_{m-1}$$
$$a_{m+2} a_{m+3} ... a_n = b_m b_{m+1} ... b_{n-2} (Deletion) \tag{5}$$

Our conditions might be conservative and we might be able to relax some of them to bypass more SW calls. However, we want to ensure that insertion and deletion cases are not handled by the linear comparison.

Finally, the optimization does not include cases where the pattern length is longer than the reference, because the result will likely include an insertion which is not supported by the linear search.

Table 1. The total number of HaplotypeCaller SW calls and the number of bypassed SW calls

Input sample	Local Assembly SW calls			Realign Reads SW calls		
	Total SW calls	Bypassed SW calls	% of total	Total SW calls	Bypassed SW calls	% of total
NA12878	555,524	261,886	46.9	6,439,520	3,476,694	53.9
NA18507	650,411	331,908	51.0	5,919,079	3,110,175	52.5
NA19240	630,286	333,836	52.9	6,606,748	3,447,696	52.1

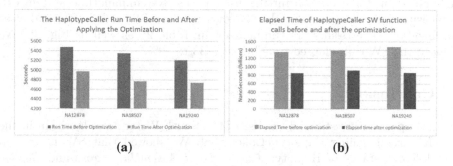

(a)   (b)

Fig. 4. (a) The HaplotypeCaller run time before and after applying the optimization (b) The elapsed time of HaplotypeCaller SW calls measured with Java nanotime system call before and after applying the optimization

4   Measurements

In this section we report experimental measurements demonstrating the utility of our optimization. We ran our experiments on a 2-way multiprocessor system with Intel Xeon E3 2640 v2 chips and with 48 GB of main memory. We ran the GATK release 3.8.1 using one thread and modified the GATK java code and specifically the SWPairwiseAlignment.java source file which contains the SW calculate matrix implementation to incorporate the optimization solution. We ran the HaplotypeCaller step on the three input sample files, NA12878, NA19240 and NA18507, with and without the optimization and successfully verified that the HaplotypeCaller produces the same VCF file after applying the optimization. The data set is freely available at [8]. We also compared the results of SW calls and verified that all linear comparisons return the same results as real SW calls.

The measurements presented in Table 1 shows the percentage of bypassed SW calls from the total SW calls in the two stages of HaplotypeCaller. The Table 1 demonstrates that except for one case, more than 50% of SW calls is bypassed after applying our optimization.

Figure 4a shows the HaplotypeCaller run time before and after applying the optimization in seconds. The optimization solution provides a minimum of 9% run time performance improvement for the NA19240 input sample. NA18507 experiences the best performance improvement of 10.8%. The numbers are

Table 2. (a) The HaplotypeCaller run time before and after applying the optimization (b) Elapsed time of the HaplotypeCaller SW calls measured with Java nanotime system call

	before (sec)	after (sec)	Improvement %		before (nanosec)	after (nanosec)	Reduction %
NA12878	5477	4972	9.2	NA12878	1334B	848B	36.4
NA18507	5351	4768	10.8	NA18507	1384B	916B	33.8
NA19240	5210	4738	9.0	NA19240	1465B	865B	40.9
(a)				(b)			

highlighted in Table 2a. We further used java's System.nanotime system call to measure the elapsed time of the HaplotypeCaller's SW calls before and after optimization. The results are summarized in Table 2b as well as Fig. 4b. We reduced the SW calls elapsed time between 33 to 40%.

5 Conclusion

In this paper, we investigated the use of the Smith-Waterman algorithm in the GATK variant calling HaplotypeCaller tool. We showed that SW is used in two different stages of HaplotypeCaller; *Local Assembly Engine* and *Realign Reads Engine*. We demonstrated that most SW calls of the HaplotypeCaller return full alignment through running experiments on three different input samples. We proposed an optimization whereby we bypass most fully aligned SW calls by a much simpler linear comparison *for loop*. We almost doubled the performance of SW calls in the HaplotypeCaller and achieved an up to 10% run time speed up for the HaplotypeCaller. This approach can technically be deployed in any step of gene sequencing which itself relies heavily on incorporating Smith-Waterman algorithm. This solution does not depend on the length of its inputted sequences and thus can be effectively employed with long-reads as well. As Smith-Waterman implementation has been adopted in many hardware implementations, this proposed optimization could also be exploited to further accelerate hardware implementations.

Acknowledgments. The authors would like to thank Huawei Canada Research Center for providing the opportunity and facilities for conducting this research. The rights of the technology developed in this work belongs to Huawei Technologies. Patent filing procedure of this research is in progress.

References

1. DePristo, M.A., et al.: A framework for variation discovery and genotyping using next-generation DNA sequencing data. Nat. Genet. **43**, 491–498 (2011)
2. GATK. https://software.broadinstitute.org/gatk/
3. McKenna, A., et al.: The Genome Analysis Toolkit: a MapReduce framework for analyzing next-generation DNA sequencing data. Genome Res. **20**, 1297–1303 (2010)

4. Smith, T.F., Waterman, M.S.: Identification of common molecular subsequences. Mol. Biol. **147**, 195–195 (1981)
5. Li, H., Durbin, R.: Fast and accurate long-read alignment with Burrows-Wheeler transform. Bioinformatics **26**, 589–595 (2010)
6. Li, H., Durbin, R.: Fast and accurate short read alignment with Burrows-Wheeler transform. Bioinformatics **5**, 1754–1760 (2009)
7. Roodi, M., Moshovos, A.: Gene sequencing: where time goes. In: 2018 IEEE International Symposium on Workload Characterization (IISWC) (2018)
8. SMASH Dataset. http://smash.cs.berkeley.edu/datasets.html
9. Tangherloni, A., Spolaor, S., Rundo, L., et al.: GenHap: a novel computational method based on genetic algorithms for haplotype assembly. BMC Bioinformatics **20**, 172 (2019). https://doi.org/10.1186/s12859-019-2691-y
10. Tangherloni, A., et al.: High performance computing for haplotyping: models and platforms. In: Mencagli, G., et al. (eds.) Euro-Par 2018. LNCS, vol. 11339, pp. 650–661. Springer, Cham (2019). https://doi.org/10.1007/978-3-030-10549-5_51
11. Bracciali, A., Aldinucci, M., Patterson, M., et al.: PWHATSHAP: efficient haplotyping for future generation sequencing. BMC Bioinformatics **17**, 342 (2016). https://doi.org/10.1186/s12859-016-1170-y

Algebraic and Computational Methods
for the Study of RNA Behaviour

Algebraic Characterisation
of Non-coding RNA

Stefano Maestri[1,2]([✉]) [iD] and Emanuela Merelli[1] [iD]

[1] School of Science and Technology, University of Camerino, Camerino, Italy
{stefano.maestri,emanuela.merelli}@unicam.it
[2] CPT - Centre de Physique Théorique, Aix-Marseille University, Marseille, France

Abstract. Process calculi have been proved to be a powerful tool for describing biological processes. They allowed us to study the folding process of RNAs and proteins and identify an abstraction level at which these two classes of molecules are behaviourally equivalent and perform functions of the same complexity. In this work, we go one step further, by exploring the expressiveness of process algebras in modelling the functions representing the behaviour of non-coding RNA molecules; the characterisation of the RNA catalytic activity defines a congruence class. Basing on these results, we propose a methodology suitable to generate an algebraic specification of a multiagent simulation. This approach is designed not only for theoretical purposes, but mostly to support the study of cellular processes and pathologies involving non-coding RNAs, by constructing agent-based models and validating hypotheses through model simulation. It might equally promote the development of future applications of non-coding RNA mediated inhibition of influenza infections.

Keywords: Ribozyme functions · Multiagent simulation · Formal methods · CCS

1 Introduction

The relation between structures and functions is a relevant topic in biology, whose investigation received a significant contribution by different computational approaches, from process calculi to topological data analysis [3,5,15,16,20].

In particular, formal languages and graph grammars have been successfully applied in modelling the properties that correlate the functions expressible by RNA molecules and specific substructures involved in their folding - the process that allows a linear biopolymer to reach a three-dimensional conformation by forming hydrogen bonds between non-consecutive monomers [14,22].

In a previously published work, we pushed forward this approach and proved that the complexity of RNA functions can be traced back to the inner potentiality of each nucleotide to interact with the others in the same sequence. This result has been obtained by comparing the RNA folding with that performed

© Springer Nature Switzerland AG 2020
P. Cazzaniga et al. (Eds.): CIBB 2019, LNBI 12313, pp. 145–158, 2020.
https://doi.org/10.1007/978-3-030-63061-4_14

by proteins, in order to identify an abstraction level at which these two classes of molecules show the same structural and functional complexity [13]. We refer to this level as *congruence level*. Reaching such a goal was possible thanks to the expressiveness of process algebras [1], through which we modelled both RNA and protein folding.

In the present article, we want to hypothesise the functions that characterise the *congruence level* and further explore the applicability of process algebras in modelling the related biological processes.

The resulting models will form the basis of a multiagent simulation [9]. In an agent-based simulation, agents are discrete software elements whose interactions correspond to those performed by the components of the modelled system, quite faithfully to the actual behaviour of a biological process [18]. In process algebras, processes are concurrent, autonomous and reactive; all these properties are also shared by agents populating a multiagent environment, making process algebras suitable specification languages for multiagent systems.

However, biological processes are complex systems whose emerging behaviour is not always possible to predict, due to the incompleteness of observed data. To incorporate this property in an agent-based model of a biological system, agents' interactions should have an aleatory nature or the simulation environment should be non-predictable (this implies that each run of the simulation is affected by statistical uncertainty). For that reason, a further step is needed to provide an effective specification of the environment, hopefully by referring to interactive computation modelling [17].

The multiagent simulator referred in this work, was developed to study the molecular interactions characterising metabolic pathways, and analyse the emergence of global properties from local interactions [4,21]. As depicted in Fig. 1, we simulated a complete enzymatic reaction by modelling the molecules involved (enzymes, metabolites and complexes) as autonomous and interactive agents.

The RNA models we propose in this article are algebraic specifications of new functionalities that will enrich the simulator. We expect that, similarly to the results we obtained regarding metabolic reactions [21], analysing the behaviour emerging from agents' interactions will yield additional information on the biological properties of RNAs.

2   Materials and Methods

The models we provide in this article are specified with Milner's CCS process algebra (more details about its application to reactive systems can be found in the book of Aceto et al. [1]). It consists of a collection of constructors for building a new process description from existing ones, by representing them as systems that exhibit a behaviour and interact via synchronised communication. A process can be viewed as a black box with a name and a set of communication channels. An output or input action on the channel a is indicated using the labels $\bar{a}$ or a respectively.

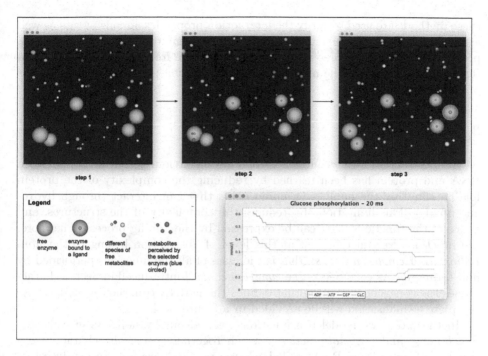

Fig. 1. Agent-based simulation of the molecular interactions involved in an enzymatic reaction. In the upper part of the figure, we show the steps needed for an enzyme to bind to a cognate ligand in a simplified run of the simulation: in the *first step*, all the enzymes (yellow spheres) and metabolites (small spheres of various colours) are freely immerse in the three-dimensional environment; the *second step* shows how a selected enzyme (highlighted by a blue circle) perceives all the affine metabolites it can reach (coloured in blue); in the *third step*, the enzyme binds one of the identified metabolites. The plot shown on the lower right is the output of an actual short simulation (20 ms) of the reaction catalysed by hexokinase. (Color figure online)

Let $\mathcal{P}$, $\mathcal{Q}$ be processes, the main process constructors are:

- **action prefixing:** if a is an action, $a.\mathcal{P}$ is a process that begins by performing the action a and behaves like $\mathcal{P}$ thereafter;
- **choice operator:** $\mathcal{P} + \mathcal{Q}$ is a process that may behave like $\mathcal{P}$ or $\mathcal{Q}$;
- **parallel composition:** $\mathcal{P}|\mathcal{Q}$ are processes that run in parallel, proceeding independently or communicating via complementary channels;
- **restriction:** if $\mathcal{L}$ is a set of channel names, then $\mathcal{P} \setminus \mathcal{L}$ is a process in which the scope of the channel names in $\mathcal{L}$ is restricted to $\mathcal{P}$; this means that those channel names can only be used for communication within $\mathcal{P}$.

CCS is a process-based specification language while a multiagent system is an agent-based model suitable for computer simulation. Being able to express agents as processes, in general, allows one to verify if the behaviour of a specified system conforms the simulated model. Moreover, in the specific case of the RNA domain, these methods are used to verify the interaction properties, among agents as well as between agents and the environment (when the environment is modelled as a process). A schematic representation of the transition from the biological domain (experimental data) to the multiagent simulation, via process-based models, is provided in Fig. 2.

3   Results

At the abstraction level we are exploring, the behavioural equivalence between RNA and protein has been reached by reducing the complexity of the protein folding (limiting the number of amino acids that can interact through hydrogen bonds). This limitation also reduces the complexity of the structures, and hence of the functions, that can be expressed by the folding process. Therefore, the functions we can represent at this level of abstraction belong to the *non-coding RNA congruence class*, that is the class of all the functions performed by non-coding RNAs (ncRNAs). The *congruence level* introduced in Sect. 1 characterises the congruence relation that defines the ncRNA congruence class, whose complete formalisation will be provided in a future work.

In this work, we model two functions carried out by ncRNAs in cells, *ligand binding* and *enzymatic activity*, which together specifically characterise a subclass of non-coding RNAs called *ribozymes*. They are able to catalyse biochemical reactions similarly to protein enzymes, carrying out fundamental roles in cellular processes [10,25].

3.1   Ligand Binding Function

Ribozymes can bind, through specific binding-sites, small molecules necessary to carry out their enzymatic functions. As an example, the binding of GlcN6P to the glmS ribozyme is fundamental for enabling the glmS catalytic activity [7,27].

In our models, the ligand binding function consists in gaining a ligand, through a binding site of the RNA molecule, in order to:

- store the ligand;
- trigger or interrupt another function of the same molecule.

A ligand can bind to a free binding site only if it shows steric and electrostatic complementarity to this site (two properties labelled sc and ec respectively). If a steric hindrance (sn) or an electrostatic non-complementarity (en) is present, the binding of the ligand is not possible.

The model of this functional role is provided by the *Ligand Binding* process ($\mathcal{B}$), which takes a free RNA binding site (bs) and a ligand (l) as input and checks the sc and ec constraints.

If both these conditions are satisfied, it produces an occupied binding site (bs_*) as output; otherwise the binding site remains free and the RNA molecule is ready to check the compatibility of another ligand.

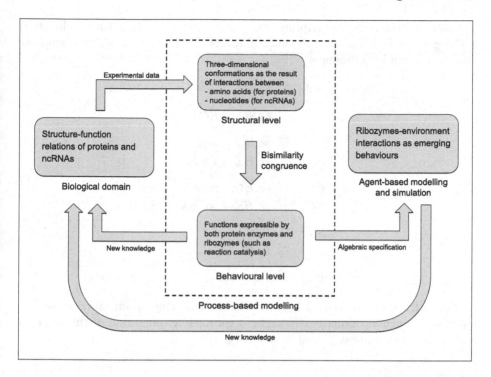

Fig. 2. Schematic representation of the modelling approach proposed in our work. *Experimental data* retrieved from *in vivo* and *in vitro* studies on proteins and RNAs provide the fundamental information and knowledge upon which we constructed the CCS models of their respective folding processes. At the *structural level*, these models correlate the interactions between the elementary units of proteins and RNAs (amino acids and nucletides, respectively) to their three-dimensional conformations. Discovering an abstraction level in which the two kinds of folding processes are *bisimilar*, gave us the perspective needed to identify a class of functions of the same complexity, which can be equally performed by proteins and RNAs; it also yielded *new knowledge* on the *biological domain* [13]. In the present paper, we outline an *algebraic specification* of this class of functions, which will be at the basis of an *agent-based model*, eventually resulting in the related computer simulation.

To remain as faithful as possible to the biological process and avoid the common problem of state explosion during the simulation, we abstract the parallel verification of the steric and electrostatic constraints as a non-deterministic choice.

When the binding site is occupied, three events can be triggered:

1. the binding site is maintained occupied in order to store the bound ligand;
2. the ligand is released;
3. a second function is activated or interrupted.

Basing on the above description, we provide the following CCS specification of the process which allows checking if a ligand can be stored, producing as output an occupied binding site (bs_*):

$$
\begin{aligned}
RNA &\stackrel{\text{def}}{=} bs.\mathcal{B}; \\
\mathcal{B} &\stackrel{\text{def}}{=} l.(SC_v + EC_v); \\
SC_v &\stackrel{\text{def}}{=} sc.SC + sn.SN; \\
SC &\stackrel{\text{def}}{=} ec.BS_* + en.EN; \\
EC_v &\stackrel{\text{def}}{=} ec.EC + en.EN; \\
EC &\stackrel{\text{def}}{=} sc.BS_* + sn.SN; \\
SN &\stackrel{\text{def}}{=} \overline{bs}.\mathcal{B}; \\
EN &\stackrel{\text{def}}{=} \overline{bs}.\mathcal{B}; \\
BS_* &\stackrel{\text{def}}{=} \overline{bs_*}.0
\end{aligned}
\qquad (1)
$$

The ncRNA is represented here and in the following specifications and formulas as the general process RNA. For a complete explanation of the symbols used in our models, refer to Table 1.

3.2   Enzymatic Function

Ribozymes perform a variety of enzymatic activities in cells, for which several analogies have been found with those carried out by proteins [6].

Since the present work is intended to outline a model of the functions characterising the *congruence level* that relates RNAs and proteins [13], we can generalise the enzymatic activity of ribozymes as the catalysis of a reaction.

Formalising this process requires first to provide a basic model of a chemical reaction.

A reaction, such as $S \rightleftharpoons P$, can be modelled in its key properties with two complementary *reaction directions*, represented by the following processes:

- *Forward Reaction Direction* ($\mathcal{R}_{fd}$): starting from a substrate, generates one or more products;
- *Backward Reaction Direction* ($\mathcal{R}_{bd}$): starting from the products, generate the original substrate.

The choice between $\mathcal{R}_{fd}$ and $\mathcal{R}_{bd}$ is determined by the value of the respective *free energy change* (ΔG): only the reaction direction with a negative ΔG can occur. This property has been modelled by placing both $\mathcal{R}_{fd}$ and $\mathcal{R}_{bd}$ in parallel composition with the ΔG process; it produces the three possible outputs representing the types of values that the free energy variation can assume: negative, positive or zero (dgn, dgp and dgz respectively).

Table 1. ncRNA processes, states and action labels

Process	Description
$\mathcal{B}$	Ligand binding
$\mathcal{C}$	Catalysis
ΔG	Free energy variation
ΔG_{fd}	Free energy variation in the forward reaction direction
ΔG_{bd}	Free energy variation in the backward reaction direction
$\mathcal{E}$	Enzymatic activity
$\mathcal{R}$	Reaction
$\mathcal{R}_{fd}$	Forward reaction direction (from substrate to product)
$\mathcal{R}_{bd}$	Backward reaction direction (from product to substrate)

State	Description
BS_*	Binding site occupied
EC	Electrostatic complementarity
EC_v	Electrostatic complementarity check
EN	Electrostatic non-complementarity
ES	Enzyme-substrate complex
P_{fd}	Product in the forward reaction direction
P_{bd}	Product in the backward reaction direction
S_{fd}	Substrate in the forward reaction direction
S_{bd}	Substrate in the backward reaction direction
SC	Steric complementarity
SC_v	Steric complementarity check
SN	Steric non-complementarity
TS_{fd}	Transition state of the forward reaction direction
TS_{bd}	Transition state of the backward reaction direction

Label	Description
aer	Activation energy reduction
as	Active site free
bs	Binding site free
bs_*	Binding site occupied
dgn	Negative free energy variation
dgp	Positive free energy variation
dgz	Null free energy variation
ec	Electrostatic complementarity
en	Electrostatic non-complementarity
es	Enzyme-substrate complex
l	Ligand
p	Product
s	Substrate
sc	Steric complementarity
sn	Steric non-complementarity
ts	Transition state

A *Reaction* process $(\mathcal{R})$ can be specified in CCS as follows:

$$\mathcal{R} \quad \overset{\text{def}}{=} \quad (\mathcal{R}_{fd}|\Delta G)\backslash\{dgn, dgp, dgz\}$$
$$+(\mathcal{R}_{bd}|\Delta G)\backslash\{dgn, dgp, dgz\};$$

$$\Delta G \quad \overset{\text{def}}{=} \quad \overline{dgn}.\Delta G + \overline{dgp}.\Delta G + \overline{dgz}.\Delta G;$$

$$\mathcal{R}_{fd} \quad \overset{\text{def}}{=} \quad s.S_{fd};$$
$$S_{fd} \quad \overset{\text{def}}{=} \quad p.\Delta G_{fd};$$
$$\Delta G_{fd} \quad \overset{\text{def}}{=} \quad dgn.P_{fd};$$
$$P_{fd} \quad \overset{\text{def}}{=} \quad \overline{ts}.TS_{fd}; \tag{2}$$
$$TS_{fd} \quad \overset{\text{def}}{=} \quad \overline{p}.\mathcal{R};$$

$$\mathcal{R}_{bd} \quad \overset{\text{def}}{=} \quad p.P_{bd};$$
$$P_{bd} \quad \overset{\text{def}}{=} \quad s.\Delta G_{bd};$$
$$\Delta G_{bd} \quad \overset{\text{def}}{=} \quad dgn.S_{bd};$$
$$S_{bd} \quad \overset{\text{def}}{=} \quad \overline{ts}.TS_{bd};$$
$$TS_{bd} \quad \overset{\text{def}}{=} \quad \overline{s}.\mathcal{R};$$

We want to point out that the modelled reaction (and eventually the corresponding multiagent simulation) is driven by the free energy reduction. The ΔG_{fd} and ΔG_{bd} processes check if the ΔG of the related reaction direction is negative.

Before producing its final output (p for $\mathcal{R}_{fd}$ and s for $\mathcal{R}_{bd}$), each reaction direction has an intermediate output, the transition state (ts).

The *Enzymatic Activty* process $(\mathcal{E})$ takes this transition state as input to catalyse the reaction, along with an active site (as). The latter is a catalytic binding site, therefore, similarly to what described for the ligand binding function, it must show steric and electrostatic complementarity with the transition state, in order for the $\mathcal{E}$ process to proceed.

If these constraints are satisfied, the $\mathcal{E}$ process makes a transition to the ES state, representing the formation of the enzyme-substrate (ES) complex. Otherwise, if there is steric non-complementarity (sn) or electrostatic noncomplementarity (en), the active site remains free and the ribozyme can check another transition state. As in the case of the $\mathcal{B}$ process, this verification as been modelled as a non-deterministic choice.

On the ES complex acts the binding energy of the enzyme to perform the *catalysis*, modelled with the process $\mathcal{C}$, which causes the reduction of the activation energy of the reaction (aer), in order to obtain the ouput of one of the two reaction directions.

Here we propose a simplified specification for the model of the $\mathcal{E}$ process:

$$RNA \stackrel{\text{def}}{=} as.\mathcal{E};$$

$$\mathcal{E} \stackrel{\text{def}}{=} ts.(SC_v + EC_v);$$

$$SC_v \stackrel{\text{def}}{=} sc.SC + sn.SN;$$
$$SC \stackrel{\text{def}}{=} ec.ES + en.EN;$$

$$EC_v \stackrel{\text{def}}{=} ec.EC + en.EN; \tag{3}$$
$$EC \stackrel{\text{def}}{=} sc.ES + sn.SN;$$

$$SN \stackrel{\text{def}}{=} \overline{as}.\mathcal{E};$$
$$EN \stackrel{\text{def}}{=} \overline{as}.\mathcal{E};$$

$$ES \stackrel{\text{def}}{=} es.\mathcal{C};$$
$$\mathcal{C} \stackrel{\text{def}}{=} \overline{aer}.(TS_{fd} + TS_{bd});$$

To further clarify how this process works, Fig. 3 shows its Labelled Transition System (LTS) specification, automatically generated with the aid of the web-based tool CAAL [2].

The models of *ligand binding* and *enzymatic activity* are part of the engineering life cycle for the simulation of ribozyme functions, where they outline the *process modelling*; as depicted in Fig. 4, the subsequent step is represented by the *model verification*. We will discuss this step in the next section, so that this article can cover the whole first phase of the engineering life cycle. In future works, we will provide the *modelling, simulation* and *validation* of the system in which ribozymes and metabolites will be represented as concurrent agents.

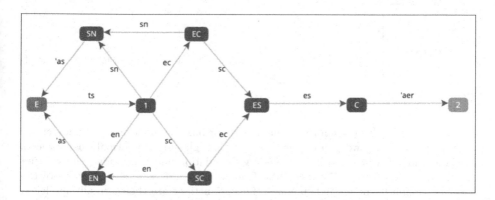

Fig. 3. Labelled Transition System (LTS) of the $\mathcal{E}$ process. In an LTS, each transition $P \stackrel{a}{\rightarrow} P'$ means that the process P can become the process P' by performing the action a. Each state has been transliterated from the CCS model, while action labels are left unchanged; output actions are indicated with a quotation mark. The state 1 represents the $SC_v + EC_v$ choice, while the state 2 corresponds to $TS_{fw} + TS_{bw}$.

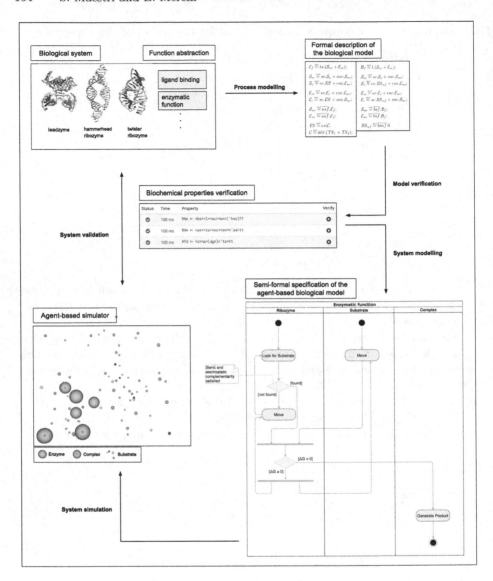

Fig. 4. Engineering life cycle for the simulation of ribozyme functions. We can identify five steps enclosed in two phases (represented through different formalisms): process modelling and verification; system modelling, simulation and validation. The starting point is the actual biological system [12], from which we derive an abstraction of the functions we aim to model and simulate. These functions are then formally modelled using process algebras (CCS in our case), and the properties of the models obtained verified through the best fitting method for model checking (for our models, we chose the Hennessy-Milner Logic). This phase is the one explored in the present article; the second phase will be defined upon the multiagent simulator described in Sect. 1 and in Fig. 1. It will involve the definition of a low-level specification, the generation of the actual agent-based simulation and the validation of the results obtained, intended to make the MAS model more faithful to biological system. For the first step of this phase, we provide in this figure a semi-formal example using a UML activity diagram.

3.3   Model Checking

To show the validity of the models described in the previous section, we provide the verification of two biochemical properties of ribozyme functions; we also verify that all the reactions are driven by the free energy reduction. Such biochemical properties are expressed as Hennessy-Milner Logic (HML) formulas so that we can ensure, via model checking, that they are satisfied [11].

- If a free binding site and a ligand have steric complementarity but they do not also show electrostatic complementarity, the binding site cannot be occupied:

$$RNA \vDash \langle bs \rangle \langle l \rangle \langle sc \rangle \langle en \rangle [\overline{bs_*}] ff \tag{4}$$

- If the free active site of a ncRNA has electrostatic complementarity with a transition state but, at the same time, a steric hindrance is present, the active site cannot be occupied (i.e., it remains free - as):

$$RNA \vDash \langle as \rangle \langle ts \rangle \langle ec \rangle \langle sn \rangle \langle \overline{as} \rangle tt \tag{5}$$

- In order for a substrate and a product to form a transition state, the ΔG of the reaction must be negative:

$$\mathcal{R}_{fd} \vDash \langle s \rangle \langle p \rangle [dgn] \langle \overline{ts} \rangle tt \tag{6}$$

The verification that these formulas has been made with the aid of the model checking function of the web-based tool CAAL [2]. The results are shown in Fig. 5.

Status	Time	Property	Verify
✓	100 ms	RNA ⊨ <bs><l><sc><en>['bso]ff	▶
✓	100 ms	RNA ⊨ <as><ts><ec><sn><'as>tt	▶
✓	100 ms	Rfd ⊨ <s><p>[dgn]<'ts>tt	▶

Fig. 5. Verification of some biochemical properties of the ribozyme functions, expressed as HML formulas. It has been performed through the CAAL web-based tool [2]; the checkmarks on the "Status" column indicate that all the formulas are satisfied. Output actions are represented with a quotation mark; the "bs_*" action label has been transliterated as "bso".

4   Conclusions

In this article, we provide a formal description of the functions that can be performed by RNA molecules at the abstraction level where thy have the same complexity of proteins [13]. We show how CCS, thanks to its expressiveness, can handle the complexity of modelling non-coding RNA functions, and specifically those performed by ribozymes. These functions characterise the congruence classes defined by the RNA catalytic activity. The validity of these models has been tested using the Hennessy-Milner Logic, to perform the model checking, and confirmed through an automated tool.

These results are solid basis upon which a multiagent simulator of molecular interactions can be enriched by implementing the functions of non-coding RNAs [21]. The models we provide in this work should be intended as the first phase of the engineering life cycle for the simulation of ribozyme functions (see Fig. 4). Considering the results we obtained on metabolic reactions, we are optimistic that the analysis of the behaviour emerging from agents' interactions will bring new knowledge on the properties of ribozymes.

This molecules, beyond their biological function, have been applied in the treatment of respiratory viral infections; it was possible due to their ability to cleave specific RNA segments of influenza viruses, like the influenza A virus or the SARS-coronavirus [8,19,26]. The simulations based on the models we propose in this article, might allow providing *in silico* support to further applications of ribozyme mediated inhibition of influenza infections.

Moreover, we are taking just the first steps towards a broader modelling and simulation approach, intended to study the behaviour of the more complex class of long non-coding RNAs (lncRNAs). In recent years, it is being increasingly acknowledged the relevance of these molecules in fundamental cellular processes, as well as their involvement in several diseases, such as in tumour progressions, where they carry out either the oncogenic or the tumour-suppressive role [23,24]. We think that applying formal models in the study of non-coding RNA functions can provide the perspective needed to fully understand the behaviour of this class of molecules and therefore contribute with a concrete support to handle the pathologies in which they are involved.

Acknowledgments. This article is the result of the research project developed at the Bioshape and Data Science Lab of the Camerino University and funded by the Future and Emerging Technologies (FET) programme within the Seventh Framework Programme (FP7) for Research of the European Commission, under the FET-Proactive grant agreement TOPDRIM (www.topdrim.eu), number FP7-ICT- 318121.

Competing interests. The authors declare no competing interests.

References

1. Aceto, L.: Reactive Systems: Modelling. Specification and Verification. Cambridge University Press (2007). https://doi.org/10.1017/CBO9780511814105
2. Andersen, J.R., et al.: CAAL: Concurrency Workbench, Aalborg edition. In: Leucker, M., Rueda, C., Valencia, F.D. (eds.) ICTAC 2015. LNCS, vol. 9399, pp. 573–582. Springer, Cham (2015). https://doi.org/10.1007/978-3-319-25150-9_33

3. Bernini, A., Brodo, L., Degano, P., Falaschi, M., Hermith, D.: Process calculi for biological processes. Natural Comput. **17**(2), 345–373 (2018). https://doi.org/10.1007/s11047-018-9673-2
4. Cannata, N., Corradini, F., Merelli, E.: Multiagent modelling and simulation of carbohydrate oxidation in cell. Int. J. Model. Ident. Control **3** (2008). https://doi.org/10.1504/IJMIC.2008.018191
5. Danos, V., Laneve, C.: Formal molecular biology. Theoret. Comput. Sci. **325**(1), 69–110 (2004). https://doi.org/10.1016/j.tcs.2004.03.065. Computational Systems Biology
6. Doudna, J., Lorsch, J.: Ribozyme catalysis: not different, just worse. Nat. Struct. Mol. Biol. **12**, 395–402 (2005). https://doi.org/10.1038/nsmb932
7. Ferré-D'Amaré, A.: The glmS ribozyme: use of a small molecule coenzyme by a gene-regulatory RNA. Q. Rev. Biophys. **43**, 423–47 (2010). https://doi.org/10.1017/S0033583510000144
8. Fukushima, A., et al.: Development of a chimeric DNA-RNA hammerhead ribozyme targeting SARS virus. Intervirology **52**, 92–9 (2009). https://doi.org/10.1159/000215946
9. Jennings, N.R.: An agent-based approach for building complex software systems. Commun. ACM **44**(4), 35–41 (2001). https://doi.org/10.1145/367211.367250
10. Jimenez, R., Polanco, J., Lupták, A.: Chemistry and biology of self-cleaving ribozymes. Trends Biochem. Sci. **40** (2015). https://doi.org/10.1016/j.tibs.2015.09.001
11. Larsen, K.: Proof systems for satisfiability in Hennessy-Milner logic with recursion. Theoret. Comput. Sci. **72**, 265–288 (1990). https://doi.org/10.1016/0304-3975(90)90038-J
12. Lucasharr: Ribozyme structure picutres, CC BY-SA 4.0 (2014). https://commons.wikimedia.org/wiki/File:Ribozyme_structure_picutres.png. Accessed 06 June 2020
13. Maestri, S., Merelli, E.: Process calculi may reveal the equivalence lying at the heart of RNA and proteins. Sci. Rep. **9**(1), 559 (2019). https://doi.org/10.1038/s41598-018-36965-1
14. Mamuye, A., Merelli, E., Tesei, L.: A graph grammar for modelling RNA folding. Electron. Proc. Theoret. Comput. Sci. **231**, 31–41 (2016). https://doi.org/10.4204/EPTCS.231.3
15. Mamuye, A., Rucco, M., Tesei, L., Merelli, E.: Persistent homology analysis of RNA. Mol. Based Math. Biol. **4** (2016). https://doi.org/10.1515/mlbmb-2016-0002
16. Merelli, E., Pettini, M., Rasetti, M.: Topology driven modeling: the IS metaphor. Natural Comput. **14**(3), 421–430 (2014). https://doi.org/10.1007/s11047-014-9436-7
17. Merelli, E., Wasilewska, A.: Topological interpretation of interactive computation. In: Bartocci, E., Cleaveland, R., Grosu, R., Sokolsky, O. (eds.) From Reactive Systems to Cyber-Physical Systems. LNCS, vol. 11500, pp. 205–224. Springer, Cham (2019). https://doi.org/10.1007/978-3-030-31514-6_12
18. Merelli, E., Young, M.: Validating mas simulation models with mutation. Multiagent Grid Syst. **3**, 225–243 (2007). https://doi.org/10.3233/MGS-2007-3206
19. Pandey, A., Kumar, P., Sanicas, M., Meseko, C., Khanna, M., Kumar, B.: Advancements in nucleic acid based therapeutics against respiratory viral infections. J. Clin. Med. **8**, 6 (2018). https://doi.org/10.3390/jcm8010006
20. Phillips, A., Cardelli, L., Castagna, G.: A graphical representation for biological processes in the stochastic pi-calculus. Trans. Comput. Syst. Biol. TCSB **4230**, 123–152 (2006). https://doi.org/10.1007/11905455_7

21. Piangerelli, M., Maestri, S., Merelli, E.: Visualising 2-simplex formation in metabolic reactions. J. Mol. Graph. Model. **97**, 107576 (2020). https://doi.org/10.1016/j.jmgm.2020.107576
22. Quadrini, M., Tesei, L., Merelli, E.: An algebraic language for RNA pseudoknots comparison. BMC Bioinformatics **20** (2019). https://doi.org/10.1186/s12859-019-2689-5
23. Rahmani, Z., Mojarrad, M., Moghbeli, M.: Long non-coding RNAs as the critical factors during tumor progressions among Iranian population: an overview. Cell Biosci. **10** (2020). https://doi.org/10.1186/s13578-020-0373-0
24. Renganathan, A., Felley-Bosco, E.: Long noncoding RNAs in cancer and therapeutic potential. Adv. Exp. Med. Biol. **1008**, 199–222 (2017). https://doi.org/10.1007/978-981-10-5203-3_7
25. Serganov, A., Patel, D.: Ribozymes, riboswitches and beyond: regulation of gene expression without proteins. Nat. Rev. Genet. **8**, 776–790 (2007). https://doi.org/10.1038/nrg2172
26. Tang, X., Hobom, G., Luo, D.: Ribozyme mediated destruction of influenza a virus. J. Med. Virol. **42**, 385–95 (1994). https://doi.org/10.1002/jmv.1890420411
27. Zhang, J., Lau, M., Ferré-D'Amaré, A.: Ribozymes and riboswitches: modulation of RNA function by small molecules. Biochemistry **49**, 9123–31 (2010). https://doi.org/10.1021/bi1012645

Bi-alignments as Models of Incongruent Evolution of RNA Sequence and Secondary Structure

Maria Waldl[1] ⓘ, Sebastian Will[1] ⓘ, Michael T. Wolfinger[1,2] ⓘ,
Ivo L. Hofacker[1,2] ⓘ, and Peter F. Stadler[1,3,4,5,6(✉)] ⓘ

[1] Faculty of Chemistry, Department of Theoretical Chemistry, University of Vienna,
Währingerstraße 17, 1090 Vienna, Austria
{maria,will,mtw,ivo,studla}@tbi.univie.ac.at
[2] Faculty of Computer Science, Research Group Bioinformatics and Computational
Biology, University of Vienna, Währingerstraße 29, 1090 Vienna, Austria
[3] Department of Computer Science and Interdisciplinary Center for Bioinformatics,
Leipzig University, Härtelstraße 16-18, 04109 Leipzig, Germany
[4] Max Planck Institute for Mathematics in the Sciences,
Inselstraße 22, 04103 Leipzig, Germany
[5] Facultad de Ciencias, Universidad National de Colombia, Bogotá, Colombia
[6] Santa Fe Institute, 1399 Hyde Park Road, Santa Fe, NM 87501, USA

Abstract. RNA molecules may be subject to independent selection pressures on sequence and structure. This can, in principle, lead to the preservation of structural features without maintaining the exact position on the conserved sequence. Consequently, structurally analogous base pairs are no longer formed by homologous bases, and homologous nucleotides do not preserve their structural context. In other words, the evolution of sequence and structure is incongruent. We model this phenomenon by introducing bi-alignments, defined as a pair of alignments, one modeling sequence homology; the other, structural homology, together with an alignment of the two alignments that models the relative shifts between conserved sequence and conserved structure. Bi-alignments therefore form a special class of four-way alignments. A preliminary survey of the Rfam database suggests that incongruent evolution is not a very rare phenomenon among structured ncRNAs and RNA elements.

Keywords: RNA secondary structure · RNA alignment · Incongruent evolution · 4-way alignment

1 Introduction

The secondary structure of many functional RNAs is well conserved over long evolutionary timescales. Paradigmatic examples include rRNAs, tRNAs,

---

A preliminary version of this contribution was presented at the 16th International Conference on Computational Intelligence methods for Bioinformatics and Biostatistics (CIBB 2019) [13].

© Springer Nature Switzerland AG 2020
P. Cazzaniga et al. (Eds.): CIBB 2019, LNBI 12313, pp. 159–170, 2020.
https://doi.org/10.1007/978-3-030-63061-4_15

spliceosomal RNAs, small nucleolar RNAs, the precursors of miRNAs, many families of regulatory RNAs in bacteria, as well as some regulatory features in mRNAs, such as iron-responsive (IRE) or selenocystein insertion (SECIS) elements. The Rfam database [7] collects these RNAs and presents them as an alignment of sequences from different species annotated by a consensus secondary structure. In such families, the variation of the secondary structure is limited to small deviations from the consensus (additional or omitted base pairs). Even more stringently, the notion of a consensus structure implies that conserved base pairs are formed by pairs of homologous nucleotides.

If selection acts to preserve base pairs, then base pairs provide additional information on the homology of nucleotides. As a consequence, Sankoff's algorithm [10] to simultaneously compute an alignment of the sequences and a consensus structure results in an improvement of both the alignment—over structure-unaware sequence alignment—and the predicted secondary structure—over 'homology-unaware' prediction from the single RNAs. Although this assumption of *congruent evolution* of sequence and structure is appealing and has been very fruitful for modeling RNA families in the Rfam database, our partial survey of Rfam reveals several families that do not follow congruent evolution.

Figure 1 shows two alignments of the sequence and secondary structure of two paralogous subfamilies of mir-30 precursors. The two families presumably are a product of the vertebrate-specific (2R) genome duplication and have evolved independently for the last 600 Myr. While the two alignments agree in the outer part of the stem loop structure, we observe that the structure alignment (bottom) slightly misaligns well matching sub-sequences, in order to properly align corresponding structure. These substructures are aligned nicely by the sequence alignment (top), which shows a much weaker consensus structure. As key observation, sequence and structure cannot be reconciled in this case. Insisting on matching common sequence patterns necessarily disrupts base pairs, while matching up the base pairs implies that the corresponding sequences appear "shifted" relative to each other.

In this contribution, we introduce a very simple mechanism to go beyond the typical assumption of strong negative selection on both sequence and structure. To bring about incongruencies between sequence and structure, we moreover assume (1) that the selective pressures on sequence and structure are mechanistically independent, and (2) the exact position of the individual base pairs are less important than the overall 'shape' (e.g. the cloverleaf of a tRNA) of the secondary structure. For example, in such a model, a stem may "move" by losing a base pair on one end and introducing a new base pair at the other end. While this kind of stem moving would still be consistent with a consensus structure, in which all inner base pairs are conserved, this consensus structure would not capture the actual analogy of the structure. Even more remarkably, our model allows for more unusual evolutionary transitions.

In the simple example of evolutionary stem sliding (Fig. 2) the sequences of the two sides of a stem or entire stem-loop structure allow two different pairings with disjoint sets of base pairings but comparable energy. Single substitutions

```
mmu-mir-30a ss    (((((((((((..(((((((((((((((....))).....)))))))))-)))))))))))))
mmu-mir-30a       UGUAAACAUCCUCGACUGGAAGCUGUGAAGCCACAAAUGGGCUUUCA-GUCGGAUGUUUGCA
consensus         UGUAAACAUCC..ACU...AGCUGU.A...CA....U.GGCU...A.GU.GGAUGUUU.C.
mmu-mir-30b       UGUAAACAUCCUACACUC--AGCUGUCAUA-CAUGCGUUGGCUGGGAUGU-GGAUGUUUACG
mmu-mir-30b ss    (((((((((((.(((((--(((((.....-.......))))))))).)))-)))))))))))
consensus ss      [[[[[[[[[[[.........[[[[...............]]]].......-]]]]]]]]]]]

mmu-mir-30a       UGUAAACAUCCUCGACUGGAAGCUGUGAAGCCACAAAUGGGCUUUC-AGUCGGAUGUUUGCA
mmu-mir-30a ss    (((((((((((..(((((((((((((((....))).....)))))))-)))))))))))))
consensus ss      [[[[[[[[[[[...[[[[[[[[[...............]]]]]]].]]].]]]]]]]]]]]
mmu-mir-30b ss    (((((((((((.--(((((((((((.....-.......))))))))).)))-)))))))))))
mmu-mir-30b       UGUAAACAUCCU--ACACUCAGCUGUCAUA-CAUGCGUUGGCUGGGAUGU-GGAUGUUUACG
consensus         UGUAAACAUCCU..AC....AGCUGU.A...CA....U.GGCU.....GU.GGAUGUUU.C.
```

Fig. 1. Two alignments of the mouse miRNA precursors mir-30a and mir-30b, comparing once their sequences (top) and once their mfe structures (bottom). The alignments are annotated by consensus sequences and consensus secondary structures. The two paralogous miRNAs have diverged since the genome duplications in the ancestor of the jawed vertebrates. Aligning the miRNAs based on their sequence (top) suggests that parts of the stem structure have shifted relative to the sequence (shown in red). The identically matched nucleotides in these regions are highlighted in the same way. Due to this shift the highlighted structure parts are misaligned and, therefore, do not appear in the consensus structure. The structure-based alignment (bottom) manages to match the structure almost perfectly, while in turn the underlying sequences are partially misaligned. To visually emphasize this difference between the two alignments, we color the same base pairs and nucleotides as in the upper alignment; moreover, the subsequences that are shifted between the two alignments are marked by gray boxes. The sequence and structure alignment of the two miRNAs cannot be reconciled in a single alignment, but differ by slight shifts. This suggests incongruent evolution of sequence and structure. (Color figure online)

or indels may stabilize either one or the other structural alternative, leading to very similar sequences that also have very similar structures, while base pairs are no longer conserved for homologous nucleotides. As a consequence, the sequence alignment (describing *homologous* nucleotides) and the alignment of secondary structures (describing *analogous* base pairs) are incongruent. Stem sliding may explain the evolution of the mir-30 paralogs in Fig. 1: selective pressures at sequence level are dominated by stabilizing selection on the mature miRNA product, while pressures on the structure require only a sufficiently stable stem-loop structure to maintain Dicer processing, independent of the exact position of the mature product in the precursor hairpin.

The incongruence between sequence and structure alignment violates the assumptions underlying the consensus structure model: sequence-based alignments cannot capture the incongruent structure; consequently tools such as RNAalifold [2] that determine consensus structures are likely to fail. Conversely, tools like RNAforester [6] can produce nearly perfect alignments of the analogous structure. However, due to incongruent evolution, the resulting alignments match non-homologous bases, such that the turnover of base pairs is grossly overpredicted. Usually, combined sequence/structure alignments based on the Sankoff algorithm [10], such as LocARNA [14] or alternatives such as cmfinder [15] would be superior to either sequence-based alignment of structure-based

```
.(((((....))))).        .(((((....-))))).
ACCCCCUCCGGGGGGA        ACCCCCUCCG-GGGGGA
CCCCCCUCCGGGGGGA        CCCCCCUCCG-GGGGGA
CCCCCCUCCCGGGGGA        -CCCCCUCCCGGGGGA
 (((((.....))))).       -(((((.....))))).
```

Fig. 2. Evolutionary stem sliding. The two hairpins shown in "dot-parenthesis" notation have no base pair in common. The middle structure folds into both structures with similar energy, the mutants fix different alternatives.

alignment. However, here they do not provide remedy, since these methods rely on congruent evolution of sequence and structure and insist on analogous base pairs being formed by homologous nucleotides.

Incongruent evolution calls for a novel formal framework that enables capturing incongruent evolutionary changes mathematically, which serves as a basis for developing algorithmic approaches to systematically study this phenomenon.

2 Theory

2.1 Bi-alignments

We embrace the idea that the evolution of sequence and structure of two RNAs **a** and **b** is properly modeled by a pair of alignments $\mathbb{U}$ and $\mathbb{V}$, where $\mathbb{U}$ is intended to represent sequence homology, while $\mathbb{V}$ represents structural similarities. If the sequence and structure evolve congruently, then $\mathbb{U}$ and $\mathbb{V}$ coincide, i.e. both have the same sequence of (mis)match, insertion, and deletion columns, respectively. In the incongruent case, the gap pattern sequences in $\mathbb{U}$ and $\mathbb{V}$ differ. This is captured by an alignment $\mathbb{W}$ of the two alignments $\mathbb{U}$ and $\mathbb{V}$ of **a** and **b**. We call the triple $(\mathbb{U}, \mathbb{V}, \mathbb{W})$ a *bi-alignment* of **a** and **b**. It is a well-known fact that an alignment of a pair of pairwise alignments is a four-way alignment (e.g., see [1] for a formal discussion of the (de)composition of alignments). Thus we can think of a bi-alignment also as a corresponding four-way alignment $\mathbb{A} \simeq (\mathbb{U}, \mathbb{V}, \mathbb{W})$. Each column of $\mathbb{A}$ consists of a column of $\mathbb{U}$ and a column of $\mathbb{V}$ (match or mismatch column of $\mathbb{W}$), or a column of $\mathbb{U}$ or $\mathbb{V}$ padded by a pair of gaps (insertion or deletion in $\mathbb{W}$), see Fig. 3.

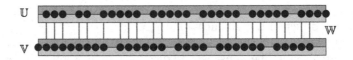

Fig. 3. A bi-alignment consists of two pairwise alignments $\mathbb{U}$ and $\mathbb{V}$ of the same sequences (shown as the two colored bars). Their columns (shown as black balls) in turn arranged in a pairwise alignment $\mathbb{W}$ indicated by the red lines. (Color figure online)

A shift between $\mathbb{U}$ and $\mathbb{V}$ occurs whenever an alignment column of $\mathbb{W}$ consumes a different number of sequence positions in **a** or **b** (or both)—thus, changing the lengths of aligned prefixes differently in the respective copies of **a** or **b**.

In other words, a shift in a given alignment column of $\mathbb{A}$ corresponds to a difference in the gap patterns of the two copies of $\mathbf{a}$ (or $\mathbf{b}$). We write the gap patterns of pairwise alignments mnemonically as $\binom{\bullet}{\bullet}$ (mismatch), $\binom{\bullet}{-}$ (deletion), $\binom{-}{\bullet}$ (insertion), and $\binom{-}{-}$ (gap column). Algebraically, we interpret them as differences between the prefix length of the current and previous alignment columns, i.e., as $\binom{1}{1}$, $\binom{1}{0}$, $\binom{0}{1}$, and $\binom{0}{0}$, respectively. For an alignment column $\binom{c_U}{c_V}$ we can therefore write the number of shifts as $\|c_U - c_V\|$, i.e., 0 if the gap patterns coincide, 1 if there is a difference for only one $\mathbf{a}$ or $\mathbf{b}$, and 2 if there is difference for each of the sequences. The possible combinations are

$$
\begin{array}{c|cccc}
 & \binom{\bullet}{\bullet} & \binom{\bullet}{-} & \binom{-}{\bullet} & \binom{-}{-} \\
\hline
\binom{\bullet}{\bullet} & 0 & 1 & 1 & 2 \\
\binom{\bullet}{-} & 1 & 0 & 2 & 1 \\
\binom{-}{\bullet} & 1 & 2 & 0 & 1 \\
\binom{-}{-} & 2 & 1 & 1 & - \\
\end{array}
\tag{1}
$$

The combination $\binom{-}{-}/\binom{-}{-}$ corresponds to an all-gap column in $\mathbb{A}$ and therefore does not appear. Each column of $\mathbb{A}$ that involves $\binom{-}{-}$ in either $\mathbb{U}$ or $\mathbb{V}$ corresponds to an insertion or deletion in $\mathbb{W}$, otherwise the column is a (mis)match in $\mathbb{W}$. Thus each column of $\mathbb{W}$ is scored by the shift penalty $-\Delta\|c_U - c_V\|$. The sum of the shift penalties over all columns of $\mathbb{W}$ defines the natural score $s_*(\mathbb{W})$ of $\mathbb{W}$. Finally, we define the *bi-alignment problem* as finding a triple $(\mathbb{U}, \mathbb{V}, \mathbb{W})$ that optimizes the sum $s_1(\mathbb{U}) + s_2(\mathbb{V}) + s_*(\mathbb{W})$.

In the four-way alignment $\mathbb{A} \simeq (\mathbb{U}, \mathbb{V}, \mathbb{W})$ we evaluate the two constituent alignments $\mathbb{U}$ and $\mathbb{V}$ (i.e., the first and second pair of rows, respectively), based on (mis)match scores μ, which in practice differ for $\mathbb{U}$ and $\mathbb{V}$ and can be position-specific. Furthermore, gaps in either alignment are linearly penalized based on the gap cost γ. The shift penalty of the column $\binom{c_U}{c_V}$ of $\mathbb{A}$ is $-\Delta\|c_U - c_V\| = -\Delta(|c_1 - c_3| + |c_2 - c_4|)$. Therefore, shift cost in $\mathbb{A}$ can be interpreted as a kind of linear gap cost Δ for indels in the pairwise alignments of the two copies of $\mathbf{a}$ and $\mathbf{b}$ (as contained in $\mathbb{A}$). The remaining two pairs of rows 1 & 4 and 2 & 3, resp., are not scored at all. In consequence, assuming additive cost models for both $\mathbb{U}$ and $\mathbb{V}$, the components $s_1(\mathbb{U})$, $s_2(\mathbb{V})$, and $s_*(\mathbb{W})$ of the scoring function are additively composed from the column-wise scores, depending on the 15 possible gap patterns

$$
\begin{array}{c|ccc|cccccccccccc}
\mathbf{a} & \bullet & \bullet & - & \bullet & \bullet & \bullet & - & \bullet & - & \bullet & - & \bullet & - & - & - \\
\mathbf{b} & \bullet & - & \bullet & \bullet & \bullet & - & \bullet & \bullet & \bullet & - & - & - & \bullet & - & - \\
\mathbf{a} & \bullet & \bullet & - & \bullet & - & \bullet & \bullet & - & \bullet & - & \bullet & - & - & \bullet & - \\
\mathbf{b} & \bullet & - & \bullet & - & \bullet & \bullet & \bullet & - & - & \bullet & \bullet & - & - & - & \bullet \\
\hline
s_1(\mathbb{U}) & \mu & \gamma & \gamma & \mu & \mu & \gamma & \gamma & \mu & \gamma & \gamma & 0 & \gamma & \gamma & 0 & 0 \\
s_2(\mathbb{V}) & \mu & \gamma & \gamma & \gamma & \gamma & \mu & \mu & 0 & \gamma & \gamma & \mu & 0 & 0 & \gamma & \gamma \\
s_*(\mathbb{W}) & 0 & 0 & 0 & \Delta & \Delta & \Delta & \Delta & 2\Delta & 2\Delta & 2\Delta & 2\Delta & \Delta & \Delta & \Delta & \Delta \\
\end{array}
\tag{2}
$$

While in the first three of these 15 cases both alignments move "in sync" and thus incur no shift penalty, all remaining cases induce shifts. In four of them, the alignments move out of sync simultaneously for both input sequences, thus incurring twice the shift penalty.

Let $M(x)$, with $x = (x_1, x_2, x_3, x_4)$ denote the score of the optimal four-way alignment of the prefixes $\mathbf{a}[1..x_1]$, $\mathbf{b}[1..x_2]$, $\mathbf{a}[1..x_3]$, and $\mathbf{b}[1..x_4]$. Depending on the gap pattern $c = (c_1, c_2, c_3, c_4)$ of the last alignment column, the prefix lengths up to the previous column are $x - c$. Following [9,11], M therefore satisfies the recursion

$$M(x) = \max_c M(x - c) + s(x, c) \quad \text{and} \quad M(0) = 0, \tag{3}$$

where $s(x, c)$ refers to the scoring function defined in Eq. (2), and c runs over the 15 possible gap patterns (not including the all-gap column). This simple dynamic programming recursion can be evaluated in quartic time and memory.

2.2 Limited Shifting and Complexity

Large numbers of shifts, i.e., large numbers of indels in the "self-alignments" of $\mathbf{a}$ and $\mathbf{b}$, i.e., row pairs 1 & 3 and 2 & 4, resp., are unlikely to be of interest in practice. Shifts thus can rather be strongly restricted, e.g. to 3 positions. This restriction can be easily realized in the algorithm by limiting index differences, such that one evaluates only a 'band' of the 4-dimension DP matrix. Consequently, one evaluates the recursions in quadratic time (w.r.t. the input length). An alternative would be to employ Carillo-Lipman-style bounds [4] to restrict the computational effort.

It might be tempting to simplify the comparison of $\mathbb{U}$ and $\mathbb{V}$ by using either $\mathbf{a}$ and $\mathbf{b}$ as a "reference" and to consider the implied alignment of the two copies of $\mathbf{b}$ (or $\mathbf{a}$, respectively) to assess the shifts. Optimizing over such three-way alignments, however, cannot replace the optimization over the (four-way) bi-alignments. To provide a brief argument, note that the three-way alignments with 'reference' $\mathbf{a}$ ($\mathbf{b}$, analogously) can be represented as—and scored like—bi-alignments without shifts between the copies of $\mathbf{a}$. Consequentially maximizing the score over the three-way alignments, yields a lower bound on the bi-alignment score. The existence of optimal bi-alignments that require shifts between the respective copies of a and b, which can be constructed easily for appropriate scoring schemes, shows that this bound is generally not tight.

A further lower bound could be obtained by constructing the bi-alignment progressively—first constructing alignments $\mathbb{U}^*$ and $\mathbb{V}^*$ (optimizing our respective alignment scores u and v) that, second, are aligned by a pairwise alignment optimizing the shift score. Denote with L the minimum of these lower bounds. In contrast, $u(\mathbb{U}^*) + v(\mathbb{V}^*)$ yields an upper bound. Immediately, this allows bounding the number of shifts $\#s$ in optimal alignments by

$$\#s \leq \frac{L - u(\mathbb{U}^*) - v(\mathbb{V}^*)}{\Delta}. \tag{4}$$

It remains open how strongly these bounds improve performance in an Carillo-Lipman approach to bi-alignment.

The (secondary) structure similarity of RNAs typically depends on similarities between corresponding (aligned) base pairs, which introduces a dependency between pairs of alignment columns. Moreover, in many cases the secondary structure of the RNAs is unknown, such that it must be inferred during the alignment. Both issues are addressed by computationally more complex algorithms often following the idea of Sankoff. As algorithmic short cut, one breaks the column dependencies and approximates structure similarity (resembling [3] and `stral` [5]) by a match similarity

$$\mu^S(i,j) = \sqrt{p_1^u(i)p_2^u(j)} + \sqrt{p_1^<(i)p_2^<(j)} + \sqrt{p_1^>(i)p_2^>(j)}, \tag{5}$$

where the $p_k^{\bullet}(i)$ denote the respective probabilities that position i is unpaired (u), paired upstream ($<$), or downstream ($>$) in the ensemble of RNA $k \in \{1, 2\}$).

2.3   Implementation

We implemented our bi-alignment algorithm in Python 3 as a free, open source software tool `BiAlign`; this work describes version 0.2. Our implementation evaluates the structure similarity following Eq. (5). It provides a convenient command line interface and, alternatively, can be integrated as Python module. Both interfaces support full parametrization of the alignment scores and the maximum shift between the sequence copies in the two alignments. Note that setting the maximum shift to zero provides a shift-free base line. Moreover, to facilitate by-eye inspection, the tool highlights conserved sequence and structure and indicates shift events in its output (see Fig. 4).

3   A Survey for Incongruent RNA Evolution

The `Rfam` database provides a large set of curated alignments for structured RNAs. For many of these families the alignments have been created with an emphasis on the consensus structure. In order to identify `Rfam` families in which incongruent evolution may have played a rule, we compare `Rfam` 14.1 seed alignments to their `MAFFT` [8] re-alignments. Employing a simple scoring function that considers the sequence positions observed in each column of the alignments, we determined to what extent `Rfam` and `MAFFT` alignments disagree. Incongruent evolution can be ruled out whenever `Rfam` and `MAFFT` alignments are nearly identical. Families for which the alignments disagree strongly, on the other hand, are likely to have experienced incongruent evolution. To limit the computational efforts, we focused on `Rfam` families with small and medium-width seed alignments (≤ 10 sequences, ≤ 120 columns), leaving us with 1181 of 3016 families in `Rfam` 14.1. Out of these we identify 709 cases where the `Rfam` alignment differs from the `MAFFT` realignments.

A more reliable indication for evolutionary shift events is an increase of the combined sequence and structure similarity in the bi-alignment compared to the shift-free baseline. To this end we used our bi-alignment algorithm twice: once to compute bi-alignments with at most three shifts, and once with forbidden shifts, thus enforcing congruent evolution. We chose *ad hoc*, but plausible parameters for assessing similarity: the sequence similarity in our bi-alignments is simply composed from scoring identical matching nucleotides with 100 (and mismatches with 0). For assessing structure similarity, we distinguish two cases: for *a priori* unknown structure, we use (mis)match scores defined in Eq. 5; for known structures, we simply count the matched symbols in the dot-bracket structure strings. In order have comparable weights for sequence and structure, structure scores and counts were multiplied by 100. All Indels are scored with -200 and each shift is penalized by -250.

In a second step, we computed the score difference of bi-alignment and shift-free baseline for all 10137 pairs of RNA sequences from the 709 alignments. The optimal bi-alignment exhibits at least one evolutionary shift event in 143 cases from 72 different Rfam families. Figure 4 shows one example of a very plausible evolutionary shift event. Naturally, the number and significance of predicted shifts strongly depend on the scoring parameters (in particular, shift costs).

4   Multiple Bi-alignments and Poly-alignments

The notion of bi-alignments can be generalized to a pair $(\mathbb{U}, \mathbb{V})$ of multiple alignments of the sequences $\mathbf{a^1}, \mathbf{a^2}, \ldots, \mathbf{a^k}$ that are aligned by a pairwise alignment $\mathbb{W}$ (see l.h.s. panel in Fig. 5). Denote the gap pattern in a column of the $\mathbb{W}$-alignment of $\mathbb{U}$ and $\mathbb{V}$ by c and d, resp. The total discrepancy between the gap patterns in the two alignments, i.e., the shift incurred in this column, is $s := \sum_{j=1}^{k} |c_j - d_j|$, where c_j or d_j is the 0-vector (all gaps) for insertions and deletions in $\mathbb{W}$. Note that this coincides with the sum of the number of indels $|c_j - d_j|$ observed in the projected pairwise alignments of the two copies of $\mathbf{a^j}$ with each other.

Let us assume that the k-way alignments $\mathbb{U}$ and $\mathbb{V}$ are scored with a sum-of-pair scores. The corresponding bi-alignment $(\mathbb{U}, \mathbb{V}, \mathbb{W})$ can then be represented as $2k$-way alignment with a scoring function of the form

$$\sigma = s_1(\mathbb{U}) + s_2(\mathbb{V}) + \frac{k-1}{2} s_*(\mathbb{W}) \tag{6}$$

where s_* is again defined by Eq. (2) and the pre-factor $(k-1)/2 = \frac{1}{k}\binom{k}{2}$ accounts for the fact that both $s_1(\mathbb{U})$ and $s_2(\mathbb{V})$ are sums of over $\binom{k}{2}$ pairwise alignments scores, while the shift score is a sum over k contributions. If s_1 and s_2 are additive scoring models, then the $2k$-way alignment also has an overall additive scoring model.

Although the most natural application for bi-alignments is the comparison of sequence and structure, one may want to consider $\ell > 2$ alignments $\mathbb{U}^{(i)}$, $1 \le i \le \ell$ of $\mathbf{a}$ and $\mathbf{b}$ with different scoring schemes. It is then natural to consider a *multiple* alignment $\mathbb{W}$ whose "rows" are the $\mathbb{U}^{(i)}$. We call this a

Rfam:

```
alpaca        ----UGGCUUCCUGCACAUCUGCC--UGUGGCU-AGGGUCACUCCGCGCGGGCACUGUGGCUUUUGUCUGCAUCUGCACAACUAGAUG
alpaca ss     ----.(((......(((((..(((((--((((...-((.....)).. .)))))))).))))......))).. (((((((......))))))
elephant      GGUGCGGUUUCCUGUAAGUCGGCCCGUGUGGCUGAGGGUCACUCCGCGCGGGCACUGCAGCUUUUGUCAGCAUCUGGACA--CCUGCCU
elephant ss   ((((.(((.(((((((.((..(((((((((((.(((...)))..)))))))))))))))))((((......))))....))).))-)))).
consensus     .....GG.UUCCUG.A. .UC.GCCCGUGUGGCU.AGGGUCACUCCGCGCGGGCACUG. .GCUUUUGUC.GCAUCUG.ACA.C. .G...
consensus ss  ....................[[[.................]]]....................
```

BiAlign:

```
alpaca ss     .(-((---......((--((..((((((((...-((.....))...)))))))).)))).......))).((((((......))))))
alpaca        UG-GC---UUCCUGCA--CAUCUGCCUGUGGCU-AGGGUCACUCCGCGCGGGCACUGUGGCUUUUGUCUGCAUCUGCACAACUAGAUG
consensus     .G.GC...UUCCUG.A........C.UGUGGCU.AGGGUCACUCCGCGCGGGCACUG..GCUUUUGUC.GCAUCUG.ACA.CU.G...
elephant      GGUGCGGUUUCCUGUAAGUCGGCCCGUGUGGCUGAGGGUCACUCCGCGCGGGCACUGCAGCUUUUGUCAGCAUCUGGACACCU-GCCU
elephant ss   ((((.(((.(((((((.((..((((((((((.(((...)))..))))))))))))))))((((......))))....))).)))-))).
consensus ss  ...........................................................................

alpaca        UGGCU---UCCUGCAC--AUCUGCCUGUGUGGCUA-GGGUCACUCCGCGCGGGCACUGUGGCUUUUGUCUGCAUCUGCACAACUA-GAUG
alpaca ss     .(((.---.....((((((((...-(-(.....))...)))))).)))......))).((((((......))-))))
consensus ss  ...............[[[.....[[[[[[[[.-(-(.....))...]]]]]]]].]]]
elephant ss   ((((.(((.((((((.((..((((((((((.(((...)))..)))))))))))))))((((......))))....))).)))-))).
elephant      GGUGCGGUUUCCUGUAAGUCGGCCCGUGUGGCUGAGGGUCACUCCGCGCGGGCACUGCAGCUUUUGUCAGCAUCUGGACACCU-GCCU
consensus     .G.....U.C...........CC....G.....GG......C.....GG.........UUU...............AC..-G...
```

Fig. 4. **Rfam** alignment and bi-alignment of conserved region 1 of the long non-coding RNA Six3os1 (RF02246) from alpaca (ABRR01379223.1/276-356) and elephant (AAGU03061906.1/13962-14048). The alignments are based on the minimum free energy structures of the sequences, which are annotated next to their corresponding sequences (marked with 'ss'). The consensus sequence string ('**consensus**') indicates conserved nucleotides by capital letters. Matched base pairs appear in the consensus structure strings ('**consensus ss**') as balanced '[]' pairs. The minimum free energy structures of both sequences are essentially characterized by a large hairpin. The **Rfam**-based alignment (top) reveals high sequence conservation while the corresponding consensus structure only recovers three base pairs of the putative analogous hairpin structure. In contrast, the bi-alignment by **BiAlign** (below) reconciles the incongruent sequence homology and structure analogy by introducing shifts. Its sequence-based alignment (first part) recovers almost all of the conserved nucleotides found in the **Rfam**-based alignment (the two respective consensus sequences are highlighted in red). At the same time, in its structure-based alignment part, it matches a larger hairpin, which shows up in the consensus structure. Unlike the weak consensus structure of the Rfam alignment, this suggests an incongruently evolved, analogous helix (consensus structures highlighted in red). (Color figure online)

poly-alignment, l.h.s. panel in Fig. 5. Writing $\mathbb{W}_{ij}$ for the restriction of $\mathbb{W}$ to the two "rows" i and j, a natural score is

$$\sigma = \sum_{i=1}^{\ell} s_i(\mathbb{U}^{(i)}) + \frac{2}{\ell-1} \sum_{i<j} s_*(\mathbb{W}_{ij}). \tag{7}$$

The poly-alignment problem then consists in simultaneously optimizing $\mathbb{U}^{(1)}$, $\mathbb{U}^{(2)}, \ldots, \mathbb{U}^{(\ell)}$, and $\mathbb{W}$. The normalization factor $2/(\ell-1)$ compensates for the $\binom{\ell}{2}$ shift contributions in relation to the ℓ pairwise alignments. For additive cost functions, the poly-alignment problems can be seen as a 2ℓ-way alignment problem with an additive cost function.

Of course it is also possible to consider poly-alignments consisting of ℓ multiple alignments of k input sequences. In the additive case this becomes a $k\ell$-way

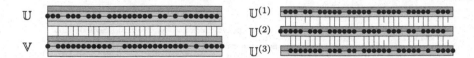

Fig. 5. Generalizations of bi-alignments. L.h.s.: bi-alignment of two multiple (four-way) alignments of the same quadruple of sequences. Bullets denote columns that are present in the constituent alignments, empty spaces indicate insertions/deletions, (i.e. all-gap columns) introduced by the alignment of alignment(s). R.h.s.: poly-alignment comprising three pairwise alignments of the same sequence pair. Red lines indicate the columns of $\mathbb{W}$, the constitutent alignments $\mathbb{U}^{(i)}$ are shown as blocks with a color for each sequence.

alignment problem with a scoring scheme taking into account each multiple alignment as well as the indels between the copies of the same sequence. For additive s_i, the poly-alignment again reduces to a $k\ell$-way alignment problem with additive scores. All these generalizations are thus amenable to dynamic programming algorithms and Carillo-Lipman-type sparsification [4] is applicable.

5   Conclusion and Outlook

Incongruent evolution of sequence and structure cannot be captured by the existing RNA alignment methods, which require that consensus structures are formed by homolgous nucleotides. Instead of performing a single common alignment, the sequence and the structure alignment therefore need to be represented separately to account for incongruencies. Bi-alignments appear to be a well-suited mathematical construction for this purpose. Here, we have shown that bi-alignments can be treated as four-way alignments with a scoring function that separately evaluates the two constituent alignments and the shifts between them, i.e., the alignment of the two alignments. The notion of bi-alignments can be generalized naturally to bi-multi-alignments in which both constituent alignment are k-way multiple alignments, and to poly-alignments in which more than two pairwise alignments of the same objects are aligned. In either case one obtains again a multiple alignment with an additive cost model if all constituent alignments use additive costs.

Limiting the total amount of shifts between sequence and structure alignment, the computational cost exceeds the individual alignment problems only by a constant factor. Consequently, bi-alignments are not only of conceptual interest but are also computationally feasible. We provide a Python implementation that can readily be applied to larger data sets.

A survey on a large part of Rfam already provides strong indications that incongruent evolution of RNA sequence and secondary structure is not a very rare phenomenon. Shifts of structure relative to sequence seem to have affected at least a few percent of the Rfam families. Given that Rfam is dedicated to RNA families with well-defined consensus structures, it is plausible to expect that the phenomenon is even more common in general. It would also be interesting to investigate whether a similar phenomenon can be detected in proteins at the level of secondary structures.

For the conceptual nature of this work, we made two simplifying design choices. Firstly, we chose *ad hoc* parameters for base similarities, indel and shift costs. Although Δ was chosen to penalize shifts more heavily than indels, it would be of great interest to train a more realistic scoring model. Ideally, this could be done in a probabilistic setting as a log-odds ratio, provided a sufficiently large set of shifts can be recovered from empirical data.

Secondly, we follow previous work on sequence-like alignments of structures using scoring models of the form of Eq. (5) [3,5]. In this way, structure-based alignments are modeled as sequence alignments incorporating secondary structures only indirectly via the probabilities of pairing towards the 5'- or 3'-side. This approximation makes the scoring function column-wise additive and thus make it possible to solve the bi-alignment problem by simple recursion (3). Since this model can only approximate the structure-induced dependencies between sequence positions, it does not yield the accuracy of true sequence-structure alignment approaches (e.g. based on Sankoff's algorithm).

We briefly describe the possible extension of our approach to more powerful RNA structure alignment: The (bi-)alignment model presented here corresponds to a regular grammar $A \rightarrow Ac|\epsilon$, where c is one of the 15 different possible gap patterns in a four-way alignment [12]. For Sankoff-style structural alignments we additionally need context-free grammar rules of the form $A \rightarrow Ac|A(A)|\epsilon$. The alternative production refers to a base pair in the consensus structure. More precisely, this production is of the form $\binom{A}{B} \rightarrow \binom{A}{B}\binom{u}{(}\binom{A}{B}\binom{v}{)}$ where the first coordinate refers to the sequence-based alignment and the second coordinate denotes the structural alignment. Here we only allow the insertion of a consensus base pair if both x and y support a base pair at the matching position. In the sequence part we may have $\binom{u}{v} = \binom{\bullet}{\bullet}$, $\binom{\bullet}{-}$, $\binom{-}{\bullet}$, or $\binom{-}{-}$. The four-way version of Sankoff's algorithm has a space complexity of $O(n^8)$ and a time complexity of $O(n^{12})$, which can be reduce to quadratic time and space requirements using the heuristics introduced in LocARNA [14] together with a restriction of the shifts. We shall describe Sankoff-style bi-alignments in detail in a forthcoming contribution.

Availability. The software package BiAlign is freely available at https://github.com/s-will/BiAlign.

Acknowledgments. Partial financial support by the German Federal Ministry of Education and Research (BMBF, project no. 031A538A, de.NBI-RBC and 031L0164C, RNAproNET) and the Austrian science fund (FWF project I 2874 "Prediction of RNA-RNA interactions" and doctoral college W 1207 "RNA Biology") is gratefully acknowledged.

References

1. Berkemer, S., Höner zu Siederdissen, C., Stadler, P.F.: Alignments as compositional structures (2018, submitted). arXiv:1810.07800

2. Bernhart, S.H., Hofacker, I.L., Will, S., Gruber, A.R., Stadler, P.F.: RNAalifold: improved consensus structure prediction for RNA alignments. BMC Bioinformatics **9**, 474 (2008)
3. Bonhoeffer, S., McCaskill, J.S., Stadler, P.F., Schuster, P.: RNA multi-structure landscapes. A study based on temperature dependent partition functions. Eur. Biophys. J. **22**, 13–24 (1993)
4. Carrillo, H., Lipman, D.: The multiple sequence alignment problem in biology. SIAM J. Appl. Math. **48**, 1073–1082 (1988)
5. Dalli, D., Wilm, A., Mainz, I., Steger, G.: STRAL: progressive alignment of non-coding RNA using base pairing probability vectors in quadratic time. Bioinformatics **22**, 1593–1599 (2006)
6. Hoechsmann, M., Voß, B., Giegerich, R.: Pure multiple RNA secondary structure alignments: a progressive profile approach. IEEE/ACM Trans. Comput. Biol. Bioinform. **1**, 53–62 (2004)
7. Kalvari, I., et al.: Rfam 13.0: shifting to a genome-centric resource for non-coding RNA families. Nucleic Acids Res. **46**, D335–D342 (2017)
8. Katoh, K., Standley, D.M.: MAFFT multiple sequence alignment software version 7: improvements in performance and usability. Mol. Biol. Evol. **30**, 772–780 (2013)
9. Lipman, D.J., Altschul, S.F., Kececioglu, J.D.: A tool for multiple sequence alignment. Proc. Natl. Acad. Sci. USA **86**, 4412–4415 (1989)
10. Sankoff, D.: Simultaneous solution of the RNA folding, alignment and protosequence problems. SIAM J. Appl. Math. **45**, 810–825 (1985)
11. Setubal, J.C., Meidanis, J.: Introduction to Computational Molecular Biology. PWS Publications, Boston (1997)
12. Höner zu Siederdissen, C., Hofacker, I.L., Stadler, P.F.: Product grammars for alignment and folding. IEEE/ACM Trans. Comput. Biol. Bioinform. **12**, 507–519 (2015)
13. Waldl, M., Will, S., Wolfinger, M.T., Hofacker, I.L., Stadler, P.F.: Bi-alignments as models of incongruent evolution and RNA sequence and structure. In: CIBB (2019). bioRxiv: https://doi.org/10.1101/631606
14. Will, S., Missal, K., Hofacker, I.L., Stadler, P.F., Backofen, R.: Inferring non-coding RNA families and classes by means of genome-scale structure-based clustering. PLoS Comput. Biol. **3**, e65 (2007)
15. Yao, Z., Weinberg, Z., Ruzzo, W.L.: CMfinder-a covariance model based RNA motif finding algorithm. Bioinformatics **22**, 445–452 (2006)

Label Core for Understanding RNA Structure

Michela Quadrini[1,2]($\boxtimes$) (iD), Emanuela Merelli[1] (iD), and Riccardo Piergallini[1] (iD)

[1] School of Science and Technology, Department, University of Camerino,
Via Madonna delle Carceri, 9, 62032 Camerino, Italy
{michela.quadrini,riccardo.piergallini}@unicam.it
[2] Department of Information Engineering, University of Padova,
Via Gradenigo 6/A, 35131 Padova, Italy
michela.quadrini@unipd.it

Abstract. The RNA structure, the main predictor of biological function, is the result of the folding process. While the nucleotides in the RNA sequence rapidly coupled forming weak bonds, the spatial arrangement is a slow process. Although many computational approaches have been proposed to study the folding process of RNA, most of them do not consider the hierarchical aspect existing among the bonds. In this work, we propose to collapse nucleotides and bonds underpinning the primary and secondary structure of RNA in a unique *label core* congruent with the spatial configuration. A label core is represented as a term of generalized context-free grammar properly defined to support RNA structural reduction and analysis.

Keywords: Generalized context-free grammar · Label core grammar · RNA structure reduction

1 Introduction

Ribonucleic acid (RNA) is a single stranded polymer, with a preferred 5′-3′ direction, made of four different types of nucleotides, known as Adenine (A), Guanine (G), Cytosine (C) and Uracil (U). Each nucleotide is linked to the previous one by a phosphodiester bond, referred to as a **strong bond**. Moreover, it can interact with at most another non-contiguous one, establishing a hydrogen bond, called a **weak bond**. Such a process, known as the **folding process**, induces complex *three-dimensional structure* (or shape). Such a shape is tied to its biological function. Discovering the relationships among nucleotides sequence, shape, and biological function has been considered one of the challenges in biology. RNAs play a variety of roles in cellular processes and are directly involved in the diseases for their ability to turn genes on and off. Disregarding the spatial configuration of the molecules and reducing nucleotides to dots, the molecule is abstracted in terms of *secondary structure*. Such an abstraction represents an intermediate level between the sequence and the shape of the molecule. It is both tractable

© Springer Nature Switzerland AG 2020
P. Cazzaniga et al. (Eds.): CIBB 2019, LNBI 12313, pp. 171–179, 2020.
https://doi.org/10.1007/978-3-030-63061-4_16

from a computational point of view and relevant from a biological perspective. As an example, under the action of antibiotics, many 16S ribosomal RNAs preserve the nucleotides sequence but change the shape. Such changes are detected by secondary structure. This structure can be formalized as an *arc diagram*, where the nucleotides are represented by vertices on a straight line and the base pairs are drawn as arcs in the upper half-plane, see Fig. 1 for an example. An RNA secondary structure is said to be *pseudoknot-free* if the arc diagram does not present crossing among arcs, as illustrated in Fig. 1-**a**, otherwise it is called *pseudoknotted*, as depicted in Fig. 1-**b**.

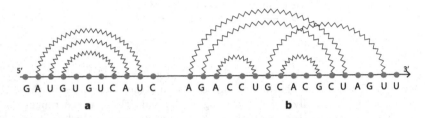

Fig. 1. An example of secondary structure represented as arc diagram. In **a**, the zigzagged arcs do not cross while, in **b**, pseudoknots are clearly visible as crossings of arcs

Here we propose an approach based on formal grammar to study the relationships between RNA structures and functions.

In the literature different computational approaches have been exploited to face such a problem. Maestri and Merelli studied the relationships between RNA structure and functions by resorting to process calculi [9], while an algebraic language has been defined for representing and comparing secondary structures with arbitrary pseudoknots in polynomial time [10,11]. Andersen *et al.* have exploited a combinatorial approach [1], and Yousef *et al.* have proposed an approach, able to differentiate among species when the evolutionary distance increases [14]. Recently, Quadrini *et al.* have defined a context-free grammar to identify common substructures considering both the primary and secondary structures [12].

In this work, we introduce the concept of the **label core** of each molecule. For each secondary structure represented as an arc diagram, the label core is obtained by collapsing groups of consecutive unpaired nucleotides into a single one, and consecutively parallel arcs are collapsed into a single one. Two arcs, identified by the pairs (i_1, j_1) and (i_2, j_2) are consecutively parallel if $i_1 = i_2 - 1 < j_2 = j_1 + 1$. In the literature, two similar concepts, the core and the shape, have been introduced by Reidys [13]; the shape is determined by removing nucleotides and arcs that do not cross, and collapsing the parallel arcs of the structure into single arcs, while the definition of the core preserves arcs that do not cross.

The label core permits both the classification of the RNAs in terms of equivalent class of secondary structures such as the core or the shape, and the consideration of the nucleotides sequence. Moreover, it is a particular arc diagram, where each vertex represents a finite ordered set of nucleotides rather than a single one, and it is unique for each arc diagram. As an example, the label core of the structures in Fig. 1 is shown in Fig. 2.

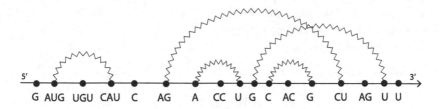

Fig. 2. The core of the structure shown in Fig. 1

To gain the ability to syntactically representing each RNA secondary structure, we define a generalized context-free grammar, namely **Label Core Grammar**, that uniquely represents the label core of each molecule as a string. Such a string consists of sequences of characters, which represent the nucleotides, enclosed in special symbols, $\langle$, $\rangle$, and equipped with natural numbers. Each sequence enclosed by the special characters represents a block of consecutive nucleotides. Moreover, if the special symbol is followed by a number h, it means that the consider block of nucleotides performs weak bonds with another block enclosing other h blocks.

This representation of the structure provides several advantages, among which it allows us to analyze substructures taking into account the nucleotides synthesis and the local folding.

This paper is organized as follows. In Sect. 2, we give the necessary background on formal grammars. In Sect. 3, we define a grammar, called label core grammar, that describes a language corresponding to the label core of RNA secondary structure. In Sect. 4, we use the words generated by the language to analyze the RNA structures to understand the relations between the sequence and its secondary structure. In particular, we use two different measures, CG-skew and AU-skew, to analyze the structures. The paper ends with some conclusions and future perspective, Sect. 5.

2 Generalized Context-Free Grammars

In this section, we give the background on formal grammar needed for this work. Context-free grammars are not expressive enough to model RNA pseudoknotted structures, as formally proved by Ogden's Lemma, a generalization of the pumping lemma for context-free languages [5]. A natural extension of context-free grammars is generalized context-free grammars that deals with tuples of strings.

Such formalism has the same generative capacity of Type 0 grammars. For a more detailed description, the interested reader can refer to [7].

Definition 1. *A generalized context-free grammar $\mathcal{G}$ is a 5-tuple (V, T, F, P, S) where:*

- *V is a finite set of nonterminal symbols or variables;*
- *T is a set of n-tuples, $n \geq 1$, over a finite set of symbols;*
- *F is a finite set of partial functions from $T \times T \times \cdots \times T$ to T.*
- *P is a finite set of rewriting rules; and*
- *$S \in V$ is the start symbol.*

Each rewriting rule in P has the form

$$A \rightarrow f[A_1, A_2, \ldots, A_q]$$

where A and $A_1, A_2, \ldots, A_q$ are nonterminal symbols and $f \in F$ is a function from T^q to T. A rewriting rule is called terminating rule, if $q = 0$ holds, i.e. there is no terminal symbol in the right-hand side of this rewriting rule.

Definition 2. *Let $\mathcal{G} = (V, T, F, P, S)$ be a generalized context-free grammar. For each nonterminal $A \in V$, the set $\mathcal{L}_{\mathcal{G}}(A)$ is defined as the smallest set that satisfies the following two conditions:*

- *if a rule $A \rightarrow \theta$ of P is a terminating rule, then θ belongs to the language accepted by A, denoted $\theta \in L_{\mathcal{G}}(A)$;*
- *if $\theta_i \in L_{\mathcal{G}}(A_i)$ for $i = 1, \ldots, q$, $A \rightarrow f[A_1, A_2, \ldots, A_q] \in P$ and $f[\theta_1, \theta_2, \ldots, \theta_q]$ is defined, then $f[\theta_1, \theta_2, \ldots, \theta_q] \in \mathcal{L}_{\mathcal{G}}(A)$.*

The language generated by $\mathcal{G}$, denoted by $\mathcal{L}_{\mathcal{G}}$, is $\mathcal{L}_{\mathcal{G}}(S)$.

Definition 3. *A derivation tree of a generalized context-free grammar $\mathcal{G} = (V, T, F, P, S)$ is defined as follows:*

- *for a terminating rule $A \rightarrow \theta$, the tree whose root is labelled with A and has only one child labelled with θ is a derivation tree of θ;*
- *if T_i, for $i = 1, \ldots, q$, is a derivation tree of $\theta_i \in \mathbb{T}$ whose root is labelled with A_i, $A \rightarrow f[A_1, A_2, \ldots, A_q]$ is in P and $f[\theta_1, \theta_2, \ldots, \theta_q]$ is defined, then a tree such that*
  - *the root is labelled with A;*
  - *the root has q children;*
  - *for all $i = 1, \ldots, q$, the subtree rooted at the ith child is isomorphic to T_i is a derivation tree of $f[\theta_1, \theta_2, \ldots, \theta_q]$;*
- *there is no other derivation tree.*

3   Label Core Grammar

RNA molecules can be formally described hierarchically. An RNA is a finite sequence, called *primary structure*, represented as a word over the alphabet of the nucleotide, A, G, C and U. By folding back onto itself, an RNA molecule forms a complex shape, also called *tertiary structures*. The tertiary structure of a molecule represents the spatial configuration of the molecule, based on the relative position of atoms, obtained by NMR spectroscopy or X-Ray diffraction [6,8]. Moreover, it can be abstracted in terms of *secondary structure*, which consists of a nucleotides sequence and a set of disjoint base pairs that correspond to the weak bonds. As mentioned in Sect. 1, it can be represented as an arc diagram by putting the dots representing the nucleotides and the strong bonds (the backbone of the molecule) on the x-axis and realizing the weak bonds as semi-circular zigzag arcs in the upper half-plane. Formally, the arc diagram is a labelled graph over the vertex set $[\ell] = \{1, \ldots, \ell\}$, in which each vertex has degree ≤ 3, and the edges are all the segments. We introduce the concept of **label core**. For each RNA secondary structure represented as an arc diagram, its *label core* is obtained by collapsing groups of consecutive unpaired nucleotides into a single one, while consecutively parallel arcs are collapsed into a single one. Since two arcs, identified by the pairs (i_1, j_1) and (i_2, j_2) are consecutively parallel if $i_1 = i_2 - 1 < j_2 = j_1 + 1$, a core is unique for each arc diagram. Moreover, it can be considered as a particular arc diagram, where each vertex represents a finite string over the alphabet of nucleotides rather than the single character, as shown in Fig. 2. To gain the ability to syntactically represent the label core of each RNA secondary structure, we define a generalized context-free grammar. Each term of the generated language u core. It is a word formed by sequences of characters enclosed by " $\langle$ " and " $\rangle$ ", which can be followed by a natural number h. Each subsequence represents a set of consecutive nucleotides and each natural number h indicates that the considered sequence is linked with a previous one overtaking h of them. In order words, the two linked elements are separated by other h elements. For brevity, as used in [3] by Giegerich *et al.* for tree grammars, we add a *lexical level* to the generalized context-free grammar concept, allowing strings in place of single symbols over $\mathcal{A}$. Let $\mathcal{A}_{RNA} = \{\text{A}, \text{U}, \text{G}, \text{C}\}$ be the alphabet of RNA. The *label core grammar* is a $\mathcal{C}_{RNA} = (V, T, F, P, S)$, where $V = \{S, S'\}$, $T = \mathcal{A}_{RNA} \cup \{\langle, \rangle\} \cup \{(,)\} \cup \{1, \ldots, N\}$, $F = \{f\}$ and the set of rewriting rules P is defined as follows:

$$
\begin{aligned}
S &::= \epsilon & emptystructure \\
&\mid S' & non-emptystructure \\
S' &::= S\langle b \rangle & primarystructure \\
&\mid f(S', (h, b)) & secondarystructure
\end{aligned}
$$

with $b \in \mathcal{A}_{RNA}^+$ and $h \in \{1, \ldots, N\}$. The partial function f depends on $M = \nu(S)$ and $\psi(S')$ defined as follows, respectively

$$
\nu(S) = \begin{cases} \emptyset & \text{if } S = \epsilon \\ \{\ell\} \cup \{\ell - h - 1\} \cup \nu(S') & \text{if } S = S'\langle b_h \rangle \\ \nu(S') & \text{if } S = S'\langle b \rangle \end{cases}
$$

and

$$
\psi(S') = \begin{cases} 1 & \text{if } S' = b, S' = b_h \\ 1 + \psi(S) & \text{if } S' = S\langle b \rangle, S' = S\langle b_h \rangle \end{cases}
$$

The partial function f is defined as follows

$$\bar{f}(S', (h, b)) = \begin{cases} S'\langle b_h \rangle & \text{if } h < \psi(S') = \ell \text{ and } \ell - \bar{h} - 1 \notin M \text{ and} \\ & |b_h| = |S\langle \ell - h - 1 \rangle| \\ \bot & \text{otherwise} \end{cases}$$

The partial function f, when it is defined, depends on an integer h that does not correspond to the nucleotides included into the weak bond, but it represents the number of components composed of collapsed nucleotides or weak bonds that are overtaken by the considered weak bond. Moreover, such weak bond links an ordered set of nucleotides with another one. For this reason, the partial function f depends on a further condition, i.e., the cardinality of the two linked stacks must be the same. We describe, step by step, the unique way to represent the core of the pseudoknotted component represented in Fig. 1 using the productions of the grammar. In other words, we illustrate a procedure analogous to a deterministic recursive descent parser for $\mathcal{C}_{RNA}$ grammar. The result is that we can determine uniquely a derivation tree. The first step consists in recognizing the vertex of the structure applying the production $S' := S\langle b \rangle$ of the grammar. In this case, $b = \text{U}$. The second step is to apply the production $S' := f(S', (h, \langle b \rangle))$ to represent the bond determined by the second to last stack of nucleotides, made by only one nucleotide, U. In this case, the value of h is 5, because there are 5 stacks between the two considered stacks of nucleotides. The third step consists of the formalization of the stack AG, using production $S' := S\langle b \rangle$. Proceeding this way, we obtain the string

$$\langle \text{G} \rangle \langle \text{AUG} \rangle \langle \text{UGU} \rangle \langle \text{CAU} \rangle_1 \langle \text{C} \rangle \langle \text{AG} \rangle \langle \text{A} \rangle \langle \text{CC} \rangle \langle \text{U} \rangle_1 \langle \text{G} \rangle \langle \text{C} \rangle \langle \text{AC} \rangle \langle \text{G} \rangle_1 \langle \text{CU} \rangle_7 \langle \text{AG} \rangle \langle \text{U} \rangle_5 \langle \text{U} \rangle$$

Such scheme works in general to give a unique algebraic expression of each motif of RNA secondary structure. This observations yields the following:

Theorem 1. *The Label Core Grammar, $\mathcal{C}_{RNA}$, generates uniquely all RNA label core.*

4   RNA Structural Analysis

The grammar described in the previous section generates a language whose words uniquely represents the label core of RNA secondary structures. Here we use two different measures, *CG-skew* (cytosine-guanine ratio) and *UG-skew* (adenine-uracil ratio) for analyzing the words of the language. The two measures were introduced to identify particular sequences in a genome at which replication starts and ends [4].

Definition 4 (CG-skew). *The CG-skew of a strand is a measure between -1 and 1 for dominance in the occurrence of cytosine compared to guanine:*

$$CG\text{-}skew = \frac{C - G}{C + G}$$

where G and C represent the frequency of occurrence in the considered sequence.

Definition 5 (AU-skew). *The AU-skew of a strand is a measure between* -1 *and* 1 *for dominance in the occurrence of cytosine compared to guanine:*

$$AU\text{-}skew = \frac{A - U}{A + U}$$

where A *and* U *represent the frequency of occurrence in the considered sequence.*

To understand the nucleotides sequence contribution in the weak bonds formation, we analyze the word generated by the grammar using the two measures. Since such strings contain numbers, which indicate the formation of weak bonds between the associated block and another previous one, we can analyze substructures taking into account the synthesis of the nucleotides. It is equivalent to analyze subwords of the form $\omega[1:i]$, i.e., subwords from the first character to the i-th one that corresponds a substructure formed by the first i nucleotides. The value of i depends on the presence of weak bonds. As an example, we consider the word generated by the grammar $\mathcal{C}_{RNA}$ that identifies the label core of the structure illustrated in Fig. 1

$$\langle G \rangle \langle AUG \rangle \langle UGU \rangle \langle CAU \rangle_1 \langle C \rangle \langle AG \rangle \langle A \rangle \langle CC \rangle \langle U \rangle_1 \langle G \rangle \langle C \rangle \langle AC \rangle \langle G \rangle_1 \langle CU \rangle_7 \langle AG \rangle \langle U \rangle_5 \langle U \rangle$$

as determined in Sect. 3. For this term, we analyze the six subwords, i.e., GAUGUGUCAU, GAUGUGUCAUCAGACCU, GAUGUGUCAUCAGACCUGCACG, GAUGUGUCAUC AGAC CUGCACGCU, GAUGUGUCAUCAGACCUGCACGCUAGU, and GAUGUGUCAUCAGACCUGC ACGCUU. As a consequence, we are able to analyze substructures taking into account the nucleotides synthesis and the local folding. Following this schema, we have selected two different sets of molecules from the RNA STRAND database [2]. A group (Group A) is composed of 13 molecules of 5S ribosomal RNA that we have downloaded from the RNA STRAND Databases selecting all molecules of this class validated by MNR or X-ray, the other one (Group B) is composed of 31 molecules of 16S. The relative lists are reported in Appendix A. Moreover, each of them has been validated by MNR or X-ray. From the analysis, it follows that every time the first block of weak bonds is created, there is an imbalance of the presence of nucleotides of cytosine with respect to guanines which is approximately double of the whole sequence, while no evidence for the imbalance of the presence of nucleotides of uracil with respect to adenines. Although this result has no statistical value, it represents a starting point for further analysis.

5   Conclusion

In this work, we have introduced the concept of the *label core* of RNA molecules. Considering an RNA secondary structure represented as an arc diagram, the label core is determined by collapsing consecutive nucleotides and arcs parallel arcs. We have defined a generalized context-free grammar, called *label core grammar*, to represent syntactically the label core of a molecule. The words generated by the grammar consists of a sequence of characters, which represent nucleotides, equipped with the information related to weak bonds. Such representation has favored the analysis of RNA substructures taking into account the nucleotides synthesis and local folding. In particular, we have applied two different measures, CG-skew and AU-skew, over two different sets of molecules, and we consider the results are a starting point for further analysis. Our long-term goal is to use the information on nucleotides sequence contribution, combined with environmental conditions, to predict the formation of intra and inter molecule weak bonds. In particular, we intend to define stochastic grammar to predict RNA secondary structures. Moreover, the sequence of characters divided into blocks allows us to identify the unpaired nucleotides that can interact with other RNAs. We also intend to use this theoretical result to represent the RNA primary and secondary structures as a vector space to apply machine learning techniques to identify targets of RNA-RNA interactions. Understand the targets and how the mechanism of RNA-RNA interactions is a fundamental task to link the structure and biological function, both healthy and sick cells.

Acknowledgments. We acknowledge the financial support of the Future and Emerging Technologies (FET) programme within the Seventh Framework Programme (FP7) for Research of the European Commission, under the FET-Proactive grant agreement TOPDRIM (www.topdrim.eu), number FP7-ICT- 318121.

Appendix A

In this appendix we list the molecules of the two group that we have downloaded from the RNA STRAND database.

List of molecules of Group A.

PDB_00004	PDB_00030	PDB_00048	PDB_00158	PDB_00249
PDB_00250	PDB_00252	PDB_00253	PDB_00288	PDB_00339
PDB_00570	PDB_00752	PDB_01100		

List of molecules of Group B.

PDB_00069	PDB_00227	PDB_00228	PDB_00409	PDB_00456
PDB_00457	PDB_00458	PDB_00459	PDB_00478	PDB_00589
PDB_00645	PDB_00703	PDB_00769	PDB_00771	PDB_00773
PDB_00775	PDB_00777	PDB_00791	PDB_00793	PDB_00933
PDB_00935	PDB_01107	PDB_01220	PDB_01222	PDB_01224
PDB_01226	PDB_01228	PDB_01230	PDB_01232	PDB_01234
PDB_01241	PDB_01243	PDB_01245	PDB_01247	PDB_01282
PDB_01284				

References

1. Andersen, J.E., Huang, F.W., Penner, R., Reidys, C.: Topology of RNA-RNA interaction structures. J. Comput. Biol. **7**(19), 928–943 (2012)
2. Andronescu, M., Bereg, V., Hoos, H.H., Condon, A.: RNA STRAND: the RNA secondary structure and statistical analysis database. BMC Bioinform. **9**(1), 340 (2008)

3. Giegerich, R., Steffen, P.: Implementing algebraic dynamic programming in the functional and the imperative programming paradigm. In: Boiten, E.A., Möller, B. (eds.) MPC 2002. LNCS, vol. 2386, pp. 1–20. Springer, Heidelberg (2002). https://doi.org/10.1007/3-540-45442-X_1

4. Grigoriev, A.: Analyzing genomes with cumulative skew diagrams. Nucl. Acids Res. **26**(10), 2286–2290 (1998)

5. Harrison, M.A.: Introduction to Formal Language Theory. Addison-Wesley Longman Publishing Co., Inc., Boston (1978)

6. Holbrook, S.R., Kim, S.H.: RNA crystallography. Biopolym.: Orig. Res. Biomol. **44**(1), 3–21 (1997)

7. Kasami, T., Seki, H., Fujii, M.: Generalized context-free grammars and multiple context-free grammars. Sys. Comput. Jpn. **20**(7), 43–52 (1989)

8. Kjems, J., Egebjerg, J.: Modern methods for probing RNA structure. Curr. Opin. Biotechnol. **9**(1), 59–65 (1998)

9. Maestri, S., Merelli, E.: Process calculi may reveal the equivalence lying at the heart of RNA and proteins. Sci. Rep. **9**(559), 1–9 (2019)

10. Quadrini, M., Tesei, L., Merelli, E.: An algebraic language for RNA pseudoknots comparison. BMC Bioinform. **20**(4), 161 (2019)

11. Quadrini, M., Merelli, E.: Loop-loop interaction metrics on RNA secondary structures with pseudoknots. In: Proceedings of the 13th International Joint Conference on Biomedical Engineering Systems and Technologies - BIOINFORMATICS, pp. 29–37 (2018)

12. Quadrini, M., Merelli, E., Piergallini, R.: Loop grammars to identify RNA structural patterns. In: Proceedings of the 12th International Joint Conference on Biomedical Engineering Systems and Technologies - BIOINFORMATICS, pp. 302–309. SciTePress (2019)

13. Reidys, C.: Combinatorial Computational Biology of RNA. Springer, New York (2011)

14. Yousef, M., Khalifa, W., Acar, I.E., Allmer, J.: MicroRNA categorization using sequence motifs and k-mers. BMC Bioinform. **18**(1), e170 (2017)

Modification of Valiant's Parsing Algorithm for the String-Searching Problem

Yuliya Susanina[1,2](✉)(iD), Anna Yaveyn[1](iD), and Semyon Grigorev[1,2](iD)

[1] Saint Petersburg State University, Universitetskaya nab. 7/9,
St. Petersburg 199034, Russia
jsusanina@gmail.com, anya.yaveyn@yandex.ru
[2] JetBrains Research, Primorskiy Prospekt 68-70, Building 1,
St. Petersburg 197374, Russia
semyon.grigorev@jetbrains.com

Abstract. Some string-matching problems can be reduced to parsing: verification whether some sequence can be derived in the given grammar. To apply parser-based solutions to such area as bioinformatics, one needs to improve parsing techniques so that the processing of large amounts of data was possible. The most asymptotically efficient parsing algorithm that can be applied to any context-free grammar is a matrix-based algorithm proposed by Valiant. This paper presents a modification of the Valiant's algorithm, which facilitates efficient utilization of modern hardware in highly-parallel implementation. Moreover, the modified version significantly decreases the number of excessive computations, accelerating the search of substrings.

Keywords: Context-free grammar · Parsing · Valiant's algorithm · String-matching · Secondary structure

1 Introduction

The secondary structure of RNA is tightly related to biological functions of organisms and plays an important role in classification and recognition problems. One of the approaches to analyze the secondary structure of RNA is based on formal language methods. Namely, one can process RNA sequence as a string over 4-letter alphabet $\{G, A, C, U\}$ and use formal language methods to describe properties of this string and then analyze strings w.r.t. described properties by parsing them.

Context-free languages (CFL-s) are the most prominent in this area, while probabilistic context-free grammars are widely used for secondary structure description and related tasks [3,8]. But more expressive language classes are required to describe some important features of secondary structure. For example, pseudoknots cannot be expressed in terms of a context-free grammar, but

---

The research was supported by the Russian Science Foundation, grant No. 18-11-00100.

P. Cazzaniga et al. (Eds.): CIBB 2019, LNBI 12313, pp. 180–192, 2020.
https://doi.org/10.1007/978-3-030-63061-4_17

can be expressed with a conjunctive grammar [14] which were proposed by
Okhotin [10] and are the natural extension of CFG.

For some problems, it is necessary to find all derivable substrings of a given
string [4]. This case is the string-matching problem also known as a string-
searching problem. The classical example of it is to find substrings matched
by the given regular expression (or regular template). But if one tries to find
a substring with a specific secondary structure, then it is necessary to use at
least context-free template (context-free grammar) and, as a result, utilize an
appropriate parsing algorithm.

Most CFG-based approaches suffer the same issue: the computational complexity is poor. Traditionally used CYK [7,13] runs with a cubic time complexity
and demonstrates poor performance on long strings or big grammars [9]. We
argue that more efficient algorithms are needed in fields such as bioinformatics
where large amount of data is common.

Asymptotically most efficient parsing algorithm is Valiant's algorithm [12]
which is based on matrix multiplication. Okhotin generalized this algorithm
to conjunctive and boolean grammars [11]. Moreover, in comparison to CYK,
Valiant's algorithm simplifies the utilization of parallel techniques to improve
performance by offloading critical computations onto matrices multiplication.
However, this algorithm is not suitable for the string-matching problem because
it computes a huge amount of redundant information.

In this paper we present the modification of Valiant's algorithm which
improves the utilization of GPGPU and parallel computations by processing
some submatrices products concurrently. Also, the proposed algorithm can be
easily utilized for the string-matching problem. We also prove the correctness
of our algorithm and analyze its time complexity. The proposed solution was
implemented using fast matrix multiplication algorithms and parallel techniques.
Evaluation of the implementation shows that GPGPU version is up to 20 times
faster then GPGPU-based implementation of the Valiant's algorithm.

2   Background

We start by introducing some basic definitions from the formal language theory.
Using these definitions, we describe Valiant's parsing algorithm on which we
base our modification.

2.1   Formal Languages

An *alphabet* Σ is a finite nonempty set of symbols. Σ^* denotes the set of all finite
strings over Σ. A *context-free grammar* G_S is a quadruple (Σ, N, R, S), where
Σ is a finite set of *terminals*, N is a finite set of *nonterminals*, R is a finite set of
productions of the form $A \to \beta$, where $\Sigma \cap N = \varnothing$, $A \in N, \beta \in V^*$, $V = \Sigma \cup N$
and $S \in N$ is a start nonterminal. Context-free grammar $G_S = (\Sigma, N, R, S)$
is said to be in *Chomsky normal form* if all productions in R are of the form:
$A \to BC$, $A \to a$, or $S \to \varepsilon$, where $A, B, C \in N, a \in \Sigma, \varepsilon$ is an empty string.

For each context-free grammar G of length N one can construct an equivalent grammar in Chomsky normal form with length N^2 [6].

$L_G(S) = \{\omega \mid S \xrightarrow[G_S]{*} \omega\}$ is a *language specified by the grammar* $G_S = (\Sigma, N, R, S)$, where $S \xrightarrow[G_S]{*} \omega$ means that ω can be derived in a finite number of rules applications from the start symbol S.

2.2 Valiant's Parsing Algorithm

Tabular parsing algorithms construct a matrix T, cells of which are filled with nonterminals from which the corresponding substring can be derived. These algorithms usually work with the grammar in Chomsky normal form. For $G_S = (\Sigma, N, R, S)$, $T_{i,j} = \{A \mid A \in N, a_{i+1} \ldots a_j \in L_G(A)\}$ $\forall i < j$.

The parsing matrix T are filled successively starting with diagonal elements $T_{i-1,i} = \{A \mid A \to a_i \in R\}$. Then, $T_{i,j} = f(P_{i,j})$, where $P_{i,j} = \bigcup_{i<k<j} T_{i,k} \times T_{k,j}$ (here, $\times$ is the Cartesian product of two sets) and $f(P) = \{A \mid \exists A \to BC \in R : (B, C) \in P\}$. Finally, the input string $a_1 a_2 \ldots a_n$ belongs to $L_G(S)$ iff $S \in T_{0,n}$.

If all cells are filled sequentially, the time complexity of this algorithm is $O(n^3)$. Valiant proposed to offload the most intensive computations to the Boolean matrix multiplication. The most time-consuming is computing $\bigcup_{i<k<j} T_{i,k} \times T_{k,j}$ and Valiant's idea is to compute $T_{i,j}$ by multiplication of submatrices of T.

Multiplication of two submatrices of parsing table T is defined as follows. Let $X \in (2^N)^{m \times l}$ and $Y \in (2^N)^{l \times n}$ be two submatrices of the parsing table T. Then, denote $X \times Y = Z$, where $Z \in (2^{N \times N})^{m \times n}$ and $Z_{i,j} = \bigcup_{1 \le k \le l} X_{i,k} \times Y_{k,j}$.

Note that the computation of $X \times Y$ can be replaced by the $|N|^2$ multiplication of $|N|$ Boolean matrices (for each nonterminal pair). Denote the matrix corresponding to the pair $(B, C) \in N \times N$ as $Z^{(B,C)}$, then $Z_{i,j}^{(B,C)} = 1$ iff $(B, C) \in Z_{i,j}$. It should also be noted that $Z^{(B,C)} = X^B \times Y^C$. Each Boolean matrix multiplication can be computed independently. Following these changes, time complexity of this algorithm is $O(|G|\text{BMM}(n)log(n))$ for an input string of length n, where $\text{BMM}(n)$ is the number of operations needed to multiply two Boolean matrices of size $n \times n$.

Valiant's algorithm written as described by Okhotin is presented in Listing 1. All elements of T and P are initialized by empty sets. Then, the elements of these two table are successively filled by two recursive procedures.

The procedure $compute(l, m)$ computes values of $T_{i,j}$ for all $l \le i < j < m$. The procedure $complete(l, m, l', m')$ constructs the submatrix $T_{i,j}$ for all $l \le i < m, l' \le j < m'$. This procedure assumes $T_{i,j}$ for all $l \le i < j < m, l' \le i < j < m'$ are already constructed and the current value of

$$P_{i,j} = \{(B, C) \mid \exists k, (m \le k < l'), a_{i+1} \ldots a_k \in L(B), a_{k+1} \ldots a_j \in L(C)\}$$

for all $l \le i < m, l' \le j < m'$. The submatrix partition during the procedure call is shown in Fig. 1.

Listing 1: Parsing by Matrix Multiplication: Valiant's Algorithm

---

Input: Grammar $G = (\Sigma, N, R, \mathcal{S}), w - a_1 \ldots a_n, n \geq 1, a_i \in \Sigma$, where $n + 1 = 2^p$

1 main():
2   compute(0, n + 1);
3   accept iff $S \in T_{0,n}$
4 compute(l, m):
5 if $m - l \geq 4$ then
6   compute(l, $\frac{l+m}{2}$);
7   compute($\frac{l+m}{2}$, m)
8 complete(l, $\frac{l+m}{2}$, $\frac{l+m}{2}$, m)
9 complete(l, m, l', m'):
10 if $m - l = 4$ and $m = l'$ then $T_{l,l+1} = \{A \mid A \to a_{l+1} \in R\}$;
11 else if $m - l = 1$ and $m < l'$ then $T_{l,l'} = f(P_{l,l'})$;
12 else if $m - l > 1$ then
13     $leftgrounded = (l, \frac{l+m}{2}, \frac{l+m}{2}, m), rightgrounded = (l', \frac{l'+m'}{2}, \frac{l'+m'}{2}, m')$,
14     $bottom = (\frac{l+m}{2}, m, l', \frac{l'+m'}{2}), left = (l, \frac{l+m}{2}, l', \frac{l'+m'}{2})$,
15     $right = (\frac{l+m}{2}, m, \frac{l'+m'}{2}, m'), top = (l, \frac{l+m}{2}, \frac{l'+m'}{2}, m')$;
16   complete(bottom);
17     $P_{left} = P_{left} \cup (T_{leftgrounded} \times T_{bottom})$;
18   complete(left);
19     $P_{right} = P_{right} \cup (T_{bottom} \times T_{rightgrounded})$;
20   complete(right);
21     $P_{top} = P_{top} \cup (T_{leftgrounded} \times T_{right})$;
22     $P_{top} = P_{top} \cup (T_{left} \times T_{rightgrounded})$;
23   complete(top)

A simple example of the several first steps of Valiant's algorithm execution is presented in Fig. 3. Only several steps are shown, but it is enough to compare our version with the original algorithm.

3   Modified Valiant's Algorithm

In this section we propose a way to rearrange submatrices handling in the algorithm. The different order improves the independence of submatrices handling and facilitates the implementation of parallel submatrix processing.

3.1   Layered Submatrices Processing

We propose to divide the parsing table into layers of disjoint submatrices of the same size (see Fig. 2). Such division is possible because the derivation of a substring of the fixed length does not depend on either left or right contexts. Each layer consists of square matrices which size is a power of 2. The layers are computed successively in the bottom-up order. Each matrix in the layer can be handled independently, which facilitates parallelization of layer processing.

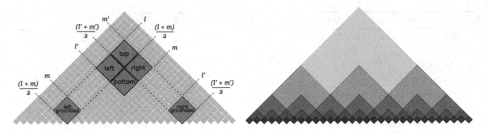

Fig. 1. Matrix partition used in procedure *complete(l, m, l', m')*

Fig. 2. Matrix partition on V-shaped layers used in modification

Fig. 3. An example of the first steps of Valiant's algorithm

Figure 4 demonstrates the modified algorithm. The lowest layer (submatrices of size 1) has already been computed. The second layer is filled in during steps 1–2. So the same part of parsing matrix (as in Fig. 3) can be computed in just two steps using parallel computation of submatrix products.

The modified version of Valiant's algorithm is presented in Listing 2. The procedure *main()* computes the lowest layer ($T_{l,l+1}$), and then divides the table into layers, and computes them with the *completeVLayer()* function. Thus, *main()* computes all elements of parsing table T.

We define *rightgrounded(subm)*, *leftgrounded(subm)*, *left(subm)*, *right(subm)*, *top(subm)* and *bottom(subm)* functions which return the submatrices for matrix $subm = (l, m, l', m')$ according to the original Valiant's algorithm (Fig. 2).

The procedure *completeVLayer(M)* takes an array of disjoint submatrices M which represents a layer. For each $subm = (l, m, l', m') \in M$ this procedure computes *left(subm)*, *right(subm)*, *top(subm)*. It is assumed in the procedure that the elements of *bottom(subm)* and $T_{i,j}$ for all i and j such that $l \leq i < j < m$ and $l' \leq i < j < m'$ are already constructed. Also it is assumed that the current value of $P_{i,j} = \{(B,C) \mid \exists k, (m \leq k < l'), a_{i+1} \ldots a_k \in L_G(B), a_{k+1} \ldots a_j \in L_G(C)\}$ for all i and j such that $l \leq i < m$ and $l' \leq j < m'$.

The procedure *completeLayer(M)* takes an array of disjoint submatrices M, but unlike the previous one, it computes $T_{i,j}$ for all $(i,j) \in subm$. This procedure requires the same assumptions on $T_{i,j}$ and $P_{i,j}$ as in the original algorithm.

In other words, *completeVLayer(M)* computes the entire layer M and *completeLayer(M_2)* is a helper function which is necessary for computation of smaller square submatrices $subm_2 \in M_2$, where M_2 is a sublayer of M.

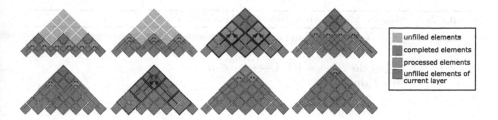

Fig. 4. Steps of the modification of Valiant's algorithm

Finally, the procedure *performMultiplication(tasks)*, where *tasks* is an array of triples of submatrices, performs the basic step of the algorithm: matrix multiplication. It is worth mentioning that in this procedure $|tasks| \geq 1$ and each task can be computed independently, while the original algorithm handles one *task* per step sequentially. Thus, the practical implementation of this procedure can easily utilize different techniques of parallel array processing.

3.2   Correctness and Complexity

We provide the proof of correctness and time complexity of the proposed modification in this section.

Lemma 1. *Let M be a layer. If for all $(l, m, l', m') \in M$:*

1. $T_{i,j} = \{A \mid a_{i+1} \ldots a_j \in L_G(A)\}$ *for all i and j such that $l \leq i < j < m$ and $l' \leq i < j < m'$;*
2. $P_{i,j} = \{(B,C) \mid \exists k, (m \leq k < l') : a_{i+1} \ldots a_k \in L_G(B), a_{k+1} \ldots a_j \in L_G(C)\}$ *for all $l \leq i < m$ and $l' \leq j < m'$.*

Then the procedure completeLayer(M), returns correctly computed sets of $T_{i,j}$ for all $l \leq i < m$ and $l' \leq j < m'$ for all $(l, m, l', m') \in M$.

Proof. We only sketch the proof of this lemma, as *completeLayer()* and *completeVLayer()* are two mutually recursive functions, so this lemma and the theorem 1 overlap and is proven similarly.

As basis we consider the matrix of size $m - l = 1$ (and then the elements of layer M are correctly computed in lines 10–11) and thereafter the lemma can be proved by induction on $m - l$.   □

Theorem 1. *Algorithm from Listing 2 correctly computes $T_{i,j}$ for all i and j, thus an input string $a = a_1 a_2 \ldots a_n \in L_G(S)$ if and only if $S \in T_{0,n}$.*

Proof. Primarily to prove the theorem, we show by induction that all layers of the parsing table T are computed correctly.

Basis: layer of size 1×1. Parsing table T consists of one layer of size 1 and its elements are correctly computed in lines 2–3 in Listing 2.

Listing 2: Parsing by Matrix Multiplication: Modified Version

---

Input: $G = (\Sigma, N, R, S), w = a_1 \ldots a_n, n \geq 1, n + 1 = 2^p, a_i \in \Sigma$

1 main():

2 **for** $l \in \{1, \ldots, n\}$ **do** $T_{l,l+1} = \{A | A \to a_{l+1} \in R\}$;

3 **for** $1 \leq i \leq p - 1$ **do**

4     $layer = constructLayer(i)$;

5     $completeVLayer(layer)$

6 accept iff $S \in T_{0,n}$

7 constructLayer(i):

8 $\{(k \cdot 2^i, (k+1) \cdot 2^i, (k+1) \cdot 2^i, (k+2) \cdot 2^i) \mid 0 \leq k < 2^{p-i} - 1\}$

9 completeLayer(M):

10 **if** $\forall (l, m, l', m') \in M \quad (m - l = 1)$ **then**

11   **for** $(l, m, l', m') \in M$ **do** $T_{l,l'} = f(P_{l,l'})$;

12 **else**

13     $completeLayer(\{bottom(subm) \mid subm \in M\})$;

14     $completeVLayer(M)$

15 completeVLayer(M):

16 $multiplicationTask_1 =$
    $\{left(subm), leftgrounded(subm), bottom(subm) \mid subm \in M\} \cup$
    $\{right(subm), bottom(subm), rightgrounded(subm) \mid subm \in M\}$;

17 $multiplicationTask_2 = \{top(subm), leftgrounded(subm), right(subm) \mid subm \in M\}$;

18 $multiplicationTask_3 = \{top(subm), left(subm), rightgrounded(subm) \mid subm \in M\}$;

19 $performMultiplications(multiplicationTask_1)$;

20 $completeLayer(\{left(subm) \mid subm \in M\} \cup \{right(subm) \mid subm \in M\})$;

21 $performMultiplications(multiplicationTask_2)$;

22 $performMultiplications(multiplicationTask_3)$;

23 $completeLayer(\{top(subm) \mid subm \in M\})$

24 performMultiplications(tasks):

25 **for** $(m, m1, m2) \in tasks$ **do** $P_m = P_m \cup (T_{m1} \times T_{m2})$;

Inductive step: assume any layer of size less than or equal to $2^{r-2} \times 2^{r-2}$ are computed correctly.

Define layer of size $2^{r-1} \times 2^{r-1}$ as M. Hereinafter $subm = (l, m, l', m')$ is a typical element of layer M.

Consider $completeVLayer(M)$ call.

First, $performMultiplications(multiplicationTask_1)$ adds to each $P_{i,j}$ all pairs (B, C) such that $\exists k, (\frac{l+m}{2} \leq k < l'), a_{i+1} \ldots a_k \in L_G(B), a_{k+1} \ldots a_j \in L_G(C)$ for all $(i, j) \in leftsublayer(M)$ and (B, C) such that $\exists k, (m \leq k < \frac{l'+m'}{2})$, $a_{i+1} \ldots a_k \in L_G(B), a_{k+1} \ldots a_j \in L_G(C)$ for all $(i, j) \in rightsublayer(M)$. Now $completeLayer(leftsublayer(M) \cup rightsublayer(M))$ can be called and it returns the correctly computed $leftsublayer(M) \cup rightsublayer(M)$.

Then *performMultiplications* called with arguments *multiplicationTask2* and *multiplicationTask3* adds pairs (B, C) such that $\exists k, \ (\frac{l+m}{2} \leq k < m)$, $a_{i+1} \dots a_k \in L_G(B)$, $a_{k+1} \dots a_j \in L_G(C)$ and pairs (B, C) such that $\exists k$, $(l' \leq k < \frac{l'+m'}{2})$, $a_{i+1} \dots a_k \in L_G(B)$, $a_{k+1} \dots a_j \in L_G(C)$ to each element $P_{i,j}$ for all $(i, j) \in topsublayer(M)$. So as $m = l'$ (from the construction of the layer), condition for elements of matrix P are fulfilled. Now *completeLayer(topsublayer(M))* can be called and it returns the correctly computed *topsublayer(M)*.

All $T[i, j] \ \forall (i, j) \in M$ are computed correctly.

Thus, *completeVLayer(M)* returns correct $T_{i,j}$ for all $(i, j) \in M$ for any layer M of parsing table T and lines 4–6 in Listing 2 return all $T_{i,j} = \{A \mid A \in N, \ a_{i+1} \dots a_j \in L_G(A)\}$. □

Lemma 2. *Let callsᵣ be a number of the calls of completeVLayer(M) where for all $(l, m, l', m') \in M$ with $m - l = 2^{p-r}$.*

- *for all $r \in \{1, \dots, p-1\}$ $\sum_{n=1}^{calls_r} |M|$ is exactly $2^{2r-1} - 2^{r-1}$;*
- *for all $r \in \{1, \dots, p-1\}$ products of submatrices of size $2^{p-r} \times 2^{p-r}$ are calculated exactly $2^{2r-1} - 2^r$ times.*

Proof. Prove the first statement by induction on r.

Basis: $r = 1$. $calls_1$ and $|M| = 1$. So, $2^{2r-1} - 2^{r-1} = 2^1 - 2^0 = 1$.

Inductive step: assume that $\sum_{n=1}^{calls_r} |M|$ is exactly $2^{2r-1} - 2^{r-1}$ for all $r \in \{1, .., q\}$.

Let us consider $r = q + 1$.

Firstly, note that function *costructLayer(r)* returns $2^{p-r} - 1$ matrices of size 2^r, so in the call of *completeVLayer(costructLayer(p - r))* *costructLayer(p - r)* returns $2^r - 1$ matrices of size 2^{p-r}. Secondly, *completeVLayer()* is called 3 times for the left, right and top submatrices of size $2^{p-(r-1)}$. Finally, *completeVLayer()* is called 4 times for the bottom, left, right and top submatrices of size $2^{p-(r-2)}$, except $2^{r-2} - 1$ matrices which were already computed.

Then,

$$\sum_{n=1}^{calls_r} |M| = 2^r - 1 + 3 \times (2^{2(r-1)-1} - 2^{(r-1)-1})$$
$$+4 \times (2^{2(r-2)-1} - 2^{(r-2)-1}) - (2^{r-2} - 1)$$
$$= 2^{2r-1} - 2^{r-1}.$$

Now we know $\sum_{n=1}^{calls_{r-1}} |M|$ is $2^{2(r-1)-1} - 2^{(r-1)-1}$ and we can calculate the number of products of submatrices of size $2^{p-r} \times 2^{p-r}$. During these calls *performMultiplications* runs 3 times, $|multiplicationTask1| = 2 \times 2^{2(r-1)-1} - 2^{(r-1)-1}$ and $|multiplicationTask2| = |multiplicationTask3| = 2^{2(r-1)-1} - 2^{(r-1)-1}$. So, the number of products of submatrices of size $2^{p-r} \times 2^{p-r}$ is

$$4 \times (2^{2(r-1)-1} - 2^{(r-1)-1}) = 2^{2r-1} - 2^r.$$

□

Theorem 2. *Let $|G|$ be the length of the description of the grammar G and let n be a length of an input string. Then algorithm in Listing 2 calculates matrix T in $\mathcal{O}(|G|BMM(n)\log n)$ where $BMM(n)$ is the number of operations needed to multiply two Boolean matrices of size $n \times n$.*

Proof. The proof is almost identical to the proof of the Theorem 1 given by Okhotin [11], because the modified algorithm computes the same number of products of submatrices just like the Valiant's algorithm (lemma 2), so the time complexity of our algorithm is the same as of the original one.   □

To summarize, we proved the correctness of the modification and shown that the time complexity remained the same as in Valiant's version.

3.3   Algorithm for Substrings

Next, we show how our modification can be applied to the string-matching problem. To find all substrings of size s, which can be derived from the start symbol for an input string of size $n = 2^p - 1$, we need to compute layers with submatrices of size not greater than 2^r, where $2^{r-2} < s \le 2^{r-1}$.

Let $r = p - (m - 2)$ and consequently $(m - 2) = p - r$. For any $m \le i \le p$ products of submatrices of size 2^{p-i} are calculated exactly $2^{2i-1} - 2^i$ times and each of them imply multiplying $C = \mathcal{O}(|G|)$ Boolean submatrices. Let $BMM = n^\omega f(n)$, where $\omega \ge 2$ and $f(n) = n^{o(1)}$. Now we estimate the number of operations needed to find all substrings:

$$
C \cdot \sum_{i=m}^{p} 2^{2i-1} \cdot 2^{\omega(p-i)} \cdot f(2^{p-i}) = C \cdot 2^{\omega r} \sum_{i=2}^{r} 2^{(2-\omega)i} \cdot 2^{2(p-r)-1} \cdot f(2^{r-i}) \le
$$
$$
C \cdot 2^{\omega r} f(2^r) \cdot 2^{2(p-r)-1} \sum_{i=2}^{r} 2^{(2-\omega)i} = BMM(2^r) \cdot 2^{2(p-r)-1} \sum_{i=2}^{r} 2^{(2-\omega)i}
$$

Thus, time complexity for searching all substrings of size not greater than 2^r is $O(2^{2(p-r)-1} |G| \, BMM(2^r)(r-1))$ where the appeared factor meet the number of matrices in the last completed layer, while time complexity for the whole input string is $O(|G|BMM(2^p)(p-1))$. The Valiant's algorithm completely calculates at least 2 triangle submatrices of size $\frac{n}{2}$, as shown in Fig. 5, thus the minimum asymptotic complexity is $O(|G| \, BMM(2^{p-1})(p-2))$. Thus we can conclude that the modification is asymptotically faster than the original algorithm for substrings of size $s \ll n$.

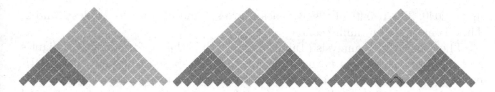

Fig. 5. Valiant's algorithm: it is necessary to calculate at least 2 triangle submatrices of size $\frac{n}{2}$ even for short substrings finding

4   Evaluation

In this section we present the results of experiments whose purpose is to demonstrate the practical applicability of the proposed algorithm. All tests were run on a PC with the following characteristics: OS: Linux Mint 19.1, CPU: Intel i5-8250U, 3400 MHz, RAM: 8 GB, GPU: NVIDIA GeForce GTX 1050 MAX-Q.

We implement two different versions of Valiant's algorithm and its modification in C++ programming language[1]:

- **CPU-based solutions** (valCPU and modCPU). One of the most efficient implementations of the "Method of the Four Russians" [2] from the library M4RI [1] is used for Boolean matrix multiplication.
- **GPU-based solutions** (valGPU and modGPU). A naive Boolean matrix multiplication in CUDA C with Boolean values treated as bits and packed into uint_32 is implemented.

We evaluate these implementations on context-free grammars D_2:

$$s \to s\,s \mid A\,s\,U \mid C\,s\,G \mid \varepsilon$$

and BIO:

$$
\begin{aligned}
s &\to \quad \mathrm{stem}\langle s0\rangle \\
\mathrm{any\_str} &\to \quad \mathrm{any\_smb} * [2..10] \\
s0 &\to \quad \mathrm{any\_str} \mid \mathrm{any\_str}\ \mathrm{stem}\langle s0\rangle\ s0 \\
\mathrm{any\_smb} &\to \quad A \mid U \mid C \mid G \\
\mathrm{stem1}\langle s1\rangle &\to \quad A\ s1\ U \mid G\ s1\ C \mid U\ s1\ A \mid C\ s1\ G \\
\mathrm{stem2}\langle s1\rangle &\to \quad \mathrm{stem1}\langle \mathrm{stem1}\langle s1\rangle\rangle \\
\mathrm{stem}\langle s1\rangle &\to \quad A\ \mathrm{stem}\langle s1\rangle\ U \mid U\ \mathrm{stem}\langle s1\rangle\ A \mid C\ \mathrm{stem}\langle s1\rangle\ G \\
&\quad\ \mid G\ \mathrm{stem}\langle s1\rangle\ C \mid \mathrm{stem1}\langle \mathrm{stem2}\langle s1\rangle\rangle
\end{aligned}
$$

We want to look at two grammars of different sizes. Grammar D_2 is chosen because grammars that describe well-balanced sequences of brackets are often used in string analysis in bioinformatics. Grammar BIO applies to the tRNA classification problem in paper [5]. We test both synthetic strings and real RNA subsequences with length n up to 8191 and search substrings with length *subs*

---

[1] The source code is available on GitHub: https://github.com/SusaninaJulia/PBMM.

up to 2040. The results of the evaluation are summarized in the Tables 1 and 2. Time is measured in milliseconds.

The comparative analysis (Table 1) shows that the performance of Valiant's and modified algorithms is the same for the CPU-based solutions. The GPU-based implementation of Valiant algorithm is slower for grammar D_2 than the CPU-based one. It is probably because of processing a large amount of small matrix multiplication which cannot be computed concurrently. But it also shows that the modified algorithm is more efficient to use parallel techniques. In other cases, GPU-based solution provides significant performance improvement, especially for our modification, where utilization of parallelism facilitates parallel multiplication of the matrices itself and matrices in a layer.

Table 1. Comparison of the Valiant's algorithm and the modification

n	Grammar D_2				Grammar BIO			
	valCPU	modCPU	valGPU	modGPU	valCPU	modCPU	valGPU	modGPU
127	78	76	195	105	1345	1339	193	106
255	289	292	523	130	5408	5488	525	140
511	1212	1177	1909	250	21969	22347	1994	256
1023	4858	4779	7878	540	88698	90318	7890	598
2047	19613	19379	33508	1500	363324	374204	34010	1701
4095	78361	78279	140473	4453	1467675	1480594	141104	5472
8191	315677	315088	-	13650	-	-	-	18039

To adapt our algorithm for the string-matching problem the, $main()$ function takes an additional argument sub—the maximum length of strings we want to find, so the modification has no need to compute all layers as shown in Sect. 3.3. The corresponding implementations are named as adpCPU and adpGPU.

Table 2. Modified algorithm evaluation on the string-searching for the BIO grammar

Sub	n	adpCPU	adpGPU
250	1023	2996	242
	2047	6647	255
	4095	13825	320
	8191	28904	456
510	2047	12178	583
	4095	26576	653
	8191	56703	884
1020	4095	48314	1590
	8191	108382	1953
2040	4095	197324	5100

The results of the second evaluation (Table 2) shows that the modified version of algorithm can find all derivable substrings much faster than the Valiant's algorithm, thus it can be efficiently applied to the string-searching problem.

5 Conclusion and Future Works

We presented a modification of the Valiant's algorithm which makes it possible to process each matrix in layer independently and use parallel computations more efficiently. This new algorithm can efficiently handle the problem of finding all substrings of a specified length. The proposed algorithm is accompanied by the proof of correctness and computational complexity analysis. We demonstrated practical applicability of our modification. Concurrent processing of matrices in layer significantly increases the performance of GPU-based solution. Also, the modification can find all substrings much faster than Valiant's algorithm through the possibility to stop parsing matrix filling.

The directions for future research is to extend the proposed algorithm to handle conjunctive and boolean grammars. It is useful for complex secondary structure features processing.

References

1. Albrecht, M., Bard, G., Hart, W.: Algorithm 898. ACM Trans. Math. Softw. **37**(1), 1–14 (2010). https://doi.org/10.1145/1644001.1644010
2. Arlazarov, V.L., Dinitz, Y.A., Kronrod, M., Faradzhev, I.: On economical construction of the transitive closure of an oriented graph. In: Doklady Akademii Nauk, vol. 194, pp. 487–488. Russian Academy of Sciences (1970)
3. Dowell, R.D., Eddy, S.R.: Evaluation of several lightweight stochastic context-free grammars for RNA secondary structure prediction. BMC Bioinform. **5**(1), 71 (2004)
4. Durbin, R., Eddy, S., Krogh, A., Mitchison, G.: Biological Sequence Analysis. Cambridge University Press, Cambridge (1996)
5. Grigorev., S., Lunina., P.: The composition of dense neural networks and formal grammars for secondary structure analysis. In: Proceedings of the 12th International Joint Conference on Biomedical Engineering Systems and Technologies - Volume 3 BIOINFORMATICS: BIOINFORMATICS, pp. 234–241. INSTICC, SciTePress (2019). https://doi.org/10.5220/0007472302340241
6. Hopcroft, J.E.: Introduction to Automata Theory, Languages, and Computation. Pearson Education India, Boston (2008)
7. Kasami, T.: An efficient recognition and syntax-analysis algorithm for context-free languages. Coordinated Science Laboratory Report no. R-257 (1966)
8. Knudsen, B., Hein, J.: RNA secondary structure prediction using stochastic context-free grammars and evolutionary history. Bioinformatics (Oxford, England) **15**(6), 446–454 (1999)
9. Liu, T., Schmidt, B.: Parallel RNA secondary structure prediction using stochastic context-free grammars. Concurr. Comput. Pract. Exp. **17**(14), 1669–1685 (2005)
10. Okhotin, A.: Conjunctive grammars. J. Autom. Lang. Comb. **6**(4), 519–535 (2001)

11. Okhotin, A.: Parsing by matrix multiplication generalized to Boolean grammars. Theor. Comput. Sci. **516**, 101–120 (2014). https://doi.org/10.1016/j.tcs.2013.09.011

12. Valiant, L.G.: General context-free recognition in less than cubic time. J. Comput. Syst. Sci. **10**(2), 308–315 (1975). https://doi.org/10.1016/S0022-0000(75)80046-8

13. Younger, D.H.: Context-free language processing in time n3. In: Proceedings of the 7th Annual Symposium on Switching and Automata Theory (Swat 1966). SWAT 1966, pp. 7–20. IEEE Computer Society, Washington, DC, USA (1966). https://doi.org/10.1109/SWAT.1966.7

14. Zier-Vogel, R., Domaratzki, M.: RNA pseudoknot prediction through stochastic conjunctive grammars. In: Computability in Europe 2013. Informal Proceedings, pp. 80–89 (2013)

On Secondary Structure Analysis by Using Formal Grammars and Artificial Neural Networks

Polina Lunina[1,2]([✉]) [iD] and Semyon Grigorev[1,2] [iD]

[1] Saint Petersburg State University, 7/9 Universitetskaya nab.,
St. Petersburg 199034, Russia
lunina_polina@mail.ru, s.v.grigoriev@spbu.ru
[2] JetBrains Research, Primorskiy prospekt 68-70, Building 1,
St. Petersburg 197374, Russia
semyon.grigorev@jetbrains.com

Abstract. A way to combine formal grammars and artificial neural networks for biological sequences processing was recently proposed. In this approach, an ordinary grammar encodes primitive features of the RNA secondary structure, parsing is utilized for features extraction and artificial neural network—for processing of the extracted features. Parsing is a bottleneck of the solution: input sequences should first be parsed before processing with a trained model which is a time-consuming operation when working with huge biological databases. In this work, we solve this problem by employing staged learning and limiting parsing to be used only during network training. We also compare networks which represent the parsing result in two different ways: by a vector and a bitmap image. Finally, we evaluate our solution on tRNA classification tasks.

Keywords: DNN · CNN · Machine learning · Secondary structure · Genomic sequences · Formal grammars · Parsing

1 Introduction

Development of effective computational methods for genomic sequences analysis is an open problem in bioinformatics. While the existing algorithms for sequences classification and subsequences detection adopt different concepts and approaches, most of them share one idea: the secondary structure of genomic sequences contains important information about the biological functions of organisms. There are different ways to handle secondary structure, for example, probabilistic grammars [6,13] and covariance models [7,8].

Real-world biological data commonly contains different mutations, noise and random variations. This issue requires some sort of probability estimation while modeling the secondary structure. Probabilistic grammars and covariance models

---

Supported by the Russian Science Foundation grant 18-11-00100.

P. Cazzaniga et al. (Eds.): CIBB 2019, LNBI 12313, pp. 193–203, 2020.
https://doi.org/10.1007/978-3-030-63061-4_18

provide such functionality, are expressive and handle long-distance connections. They are successfully used in practical tools, such as Infernal [15], but building and training accurate grammar or model for predicting the whole secondary structure involves theoretical and practical difficulties. On the other hand, artificial neural networks are a common way to process noisy data and find complex structural patterns. Moreover, the efficiency of neural networks for genetic data processing has already been shown in some works [10,14,19–21].

An approach for biological sequences processing which employs the combination of ordinary formal grammars and artificial neural networks was proposed in [9]. The key idea is to use an ordinary (not probabilistic) context-free grammar to describe only basic secondary structure features and leave the entire sequence analysis along with probabilistic estimation to the neural network which takes parsing-provided data as an input and solves some given task. This approach was proven to be applicable for real-world data processing, however, we faced some problems during the experimental research. Firstly, it appears that the choice of parsing-provided data representation affects the accuracy and performance of the solution, so, it should somehow preserve data locality and secondly, the time costs required for parsing can become crucial while working with huge amounts of data. In this paper, we improve this solution. We provide some ideas that are aimed to optimize its quality and performance and solve the described above problems.

Namely, we do the following contribution in this work.

- We represent parsing matrices as bitmap images to preserve data locality and to process such images using convolutional layers. A dense neural network is used in the original solution, so parsing matrix vectorization is required to fit data format. As a result, information about contacts arrangement is implicit which makes training harder.
- We propose a way to train a network which handles sequences instead of parsing results. A network which handles parsing result is proposed in the original work. But parsing is a very time-consuming step. We use parsing only for network training, and as a result, improve the performance of the final solution.
- We provide an evaluation of the both proposed improvements on tRNA classification tasks. We show that, first of all, convolutional layers utilization improves the quality of our solution in terms of networks accuracy, precision and recall, and decreases the total training time. Also, we show that parsing elimination significantly improves the performance of the final solution without the degradation of its quality.

2   Ordinary Context-Free Grammars and Artificial Neural Networks for Secondary Structure Analysis

In this section we provide a brief description of the approach for genomic sequences secondary structure analysis which is proposed in [9].

The secondary structure of RNA sequences can be viewed as a composition of stems [16]. So, one can create an ordinary context-free grammar which describes a set of such compositions and use it to extract the actual features of the given sequence. Grammar G_0 presented in Fig. 1 is an example of such grammar. This grammar is used in [9] as well as in the present work. This grammar considers only the conventional base pairs (line **5**) and describes the recursive composition of stems which are at least three base pairs in height (lines **7-12**). Stems may be connected by an arbitrary sequence of length from 2 up to 10, and loops also have length from 2 up to 10 (line **2**). These parameters were tuned manually as a result of several experiments and can be changed to provide a better grammar for some specific goal.

```
1    s1: stem<s0>
2    any_str : any_smb*[2..10]
3    s0: any_str | any_str stem<s0> s0
4    any_smb: A | U | C | G
5    stem1<s>: A s U | G s C | U s A | C s G
6    stem2<s>: stem1< stem1<s> >
7    stem<s>:
8          A stem<s> U
9        | U stem<s> A
10       | C stem<s> G
11       | G stem<s> C
12       | stem1< stem2<s> >
```

Fig. 1. Context-free grammar G_0 for RNA secondary structure features description

The result of a parsing algorithm for an input string w and a fixed grammar non-terminal N (start nonterminal) is an upper-triangular boolean matrix M_N, where $M_N[i,j] = 1$, iff the substring $w[i, j-1]$ is derivable from N. This means that, for the grammar G_0, a matrix contains 1 in a cell iff a correspondent substring folds to a stem of height at least 3. Such stem results in a diagonal chain of one-s in the matrix. Figure 2 presents the parsing result for a sequence

$$w_1 = CCCCATTGCCAAGGACCCCACCTTGGCAATCCC$$

w.r.t the grammar G_0. Colored boxes map a substring which folds to a stem to correspondent cells in the matrix. Besides, this matrix contains other non-zero cells, because parser detects all possible foldings for all possible substrings. It can be either noise or some important information about the secondary structure. One of the tasks that the neural network should perform is to process such matrices and filter all the insignificant contacts between the nucleotides.

The parsing result in a form of a matrix can be linearized, compressed into a byte or integer vector, and be further handled by a dense neural network, as described in [9]. Unfortunately, linearization breaks data locality: a diagonal chain of one-s, which signifies a high stem, is local in a matrix, but is broken

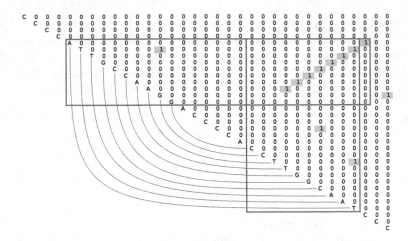

Fig. 2. Parsing result for sequence which should fold to stem

apart during its linearization. We see it to be an argument to investigate the applicability of convolutional networks for parsing result handling, as a boolean matrix can be converted to a black-white bitmap image. In this paper, we provide an empirical comparison of networks which handle vectors and images.

Another problem is a bad performance of the earlier solution. Since the trained network handles parsing result, each input sequence should first be parsed. Parsing is a very time-consuming step: context-free parsing has cubic complexity in terms of the input length. Even if we use matrix-based parsing algorithm [3] which utilizes GPGPU, performance is insufficient. We believe it would be better to avoid the parsing step at the final stage of the solution. In this work, we propose a way to solve this problem by building a network which handles raw sequences, not parsing results.

3   Convolutional Neural Network Utilization

First, we describe how to use a convolutional network for parsing result processing. Parsing result is a boolean matrix that contains the information about sequence secondary structure features and we utilize the artificial neural network to detect sufficient features and find patterns in their appearance. Therefore, we need to transform these boolean matrices to some data structure acceptable by the neural network. Currently, we came up with two possible ways: vectorization and conversion to a black and white image.

The first way is to drop out the nullary bottom left triangle, vectorize the top right triangle row by row and transform it into a byte vector. This approach reduces the input size, but it requires all the input sequences to have equal length. Thus we propose to either cut sequences to be of some predefined length or to pad them up with some blank symbols. Vectorization breaks data locality which makes learning harder: the network should restore back the relations broken during linearization. This also means that learning takes more time.

The second way is to represent the matrix as an image: the false bits of the matrix as white pixels and the true bits as black ones. This approach makes it possible to process sequences of different lengths since the images are easily transformed to a specified size. Data locality is also preserved: the information about relative positions of extracted basic features does not get lost which should improve learning.

The architecture of the neural network that takes vectorized data as an input is described in [9] and it consists of the long sequence of interchangeable dense and dropout layers with aggressive batch normalization. To handle images, we propose to use a network which consists of a small number of convolutional layers, linearization, and dense network which has a similar architecture as for vectorized data. An example of the proposed architecture is provided in Fig. 3 (network N1).

4   Parsing Step Elimination

Another improvement that we came up with concerns parsing elimination in the context of our solution. The idea is to create a model which can handle original sequences instead of the parsing matrices. For that, we propose to use two-staged learning: first, a network which solves a subtask is trained and then it is used as pretrained layers in the training of the resulting network. In our solution, we first train a neural network to handle parsing results which performs classification according to a problem at hands. We create two networks in order to compare different architectures: one of them handles vectorized parsing result, the other handles parsing result represented as a bitmap image. After that, we extend these neural networks by several input layers that take the initial nucleotide sequence as an input and convert it to the parsing result which is handled appropriately by the pretrained layers.

Figure 3 represents the detailed description of these three neural networks architectures. Here N1 is a network which handles images, N2 is a network which handles vectorized parsing results, and N0 is an additional block which converts the input sequence into a set of features which can be handled by using N1 or N2. So, firstly we train N1 and N2 on parsed data. After that, for vector-based network we combine the extension N0 and the whole original sequence of layers and for image-based network we use the similar architecture, except we remove the convolutional layer from the extended model, thus, the first layer at the junction of the blocks corresponds to the linearized image.

To sum up, we developed a technique to process parsing matrices as images by convolutional neural networks. Also, we built a model that handles sequences and requires parsing only for training the network it is based on. This removes the parsing step from the usage of the trained model.

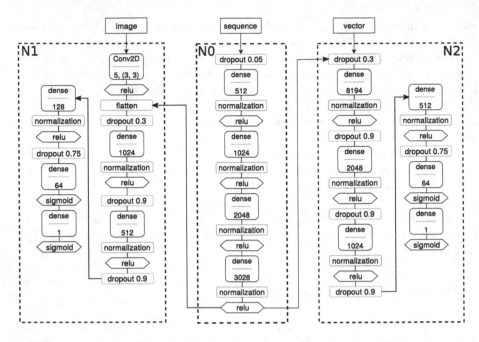

Fig. 3. Neural networks architectures

5 Evaluation

We evaluated the proposed approach with the described above modifications on two tRNA sequences analysis tasks. The first one was a classification of tRNA into two classes: eukaryotes and prokaryotes, while the second was a classification into four classes: archaea, bacteria, plants and fungi.

We used the neural networks shown in the Fig. 3 each of them having the following hyperparameters.

- Batch size – 64.
- Cross entropy loss function.
- Adagrad optimizer (adaptive gradient) with learning rate 0.05.

We took sequences from tRNA databases GtRNAdb[1] [4] and tRNADB-CE[2] [2] for these experiments. We used the parsing algorithm implemented by means of the YaccConstructor[3] platform and Keras library [5] with Tensorflow

---

[1] GtRNAdb tRNA database Web page: http://gtrnadb.ucsc.edu/. Access date: 05.06.2019.
[2] The tRNADB-CE tRNA database Web page: http://trna.ie.niigata-u.ac.jp/cgi-bin/ trnadb/index.cgi. Access date: 05.06.2019.
[3] YaccConstructor is an SDK for syntax analysis tools development. Project repository on GitHub: https://github.com/YaccConstructor/YaccConstructor. Access date: 07.03.2020.

framework [1] for neural networks training and testing. All models, as well as parsing tool, were run on GPU NVIDIA GeForce GTX 1070. We selected the equal number of samples (single tRNA molecule sequences) for each class for both classification tasks. Each sample was parsed w.r.t. the grammar G_0 and then both vectorized and transformed into an image. After that, we trained two neural networks: the first handles the representation of the parsing result as vectors, and the second—as images. Finally, we trained the extended neural network. It consists of a block which takes an initial tRNA sequence as an input and transforms it into the parsing result, and the block of pretrained layers: either the vector- or the image-based model from the previous step.

All extended neural networks were trained, validated (by hold-out validation) and tested on the same datasets as the corresponding base ones. The trained models for two classes (EP) and for four classes (ABFP) classification tasks were estimated by using classical machine learning metrics: accuracy, precision and recall.

Accuracy metrics for each problem for the test datasets are presented in the Table 1, where base model is a model which handles parsing result (image or vector respectively) and extended model handles tRNA sequences and extends the corresponding base model. Also, this table shows the total time spent on two stages of training (base network + extended network) for both problems and types of data.

Table 1. Base and extended models test results by accuracy metrics

Classifier	EP		ABFP	
Approach	Vector-based	Image-based	Vector-based	Image-based
Base model accuracy	94.1%	96.2%	86.7%	93.3%
Extended model accuracy	97.5%	97.8%	96.2%	95.7%
Total training time	30000s	4600s	31800s	3600s
Samples for train:valid:test	20000:5000:10000 (57%:14%:29%)		8000:1000:3000 (67%:8%:25%)	

The estimations by precision and recall metrics for extended models for both classifiers on the same samples as in the Table 1 are presented in the Table 2.

The results show that our approach is applicable to tRNA classification tasks and both vector- and image-based models can be used along with dense and convolutional layers in neural networks architectures. While the differences in results for extended models are insignificant, for base models image-based network demonstrates slightly better results (see Table 1). We believe that the reason of this effect lays in a better locality of features in the image-based representation of parsing result: chain of one-s which means a high stem is local in terms

Table 2. Extended models test results by precision and recall metrics for each class

Classifier	Class	Vector-based approach		Image-based approach	
		Precision	Recall	Precision	Recall
EP	Prokaryotic	95.8%	99.4%	96.2%	99.4%
	Eukaryotic	99.4%	95.6%	99.4%	99.5%
ABFP	Archaeal	91.1%	99.2%	91.6%	98.5%
	Bacterial	96.6%	95.1%	95.2%	95.5%
	Fungi	98.5%	94.9%	97.5%	94.3%
	Plant	99.4%	95.7%	99.2%	94.7%

of picture but is broken during linearization. Also, we analyzed the time spent on all the models training (Table 1) and, although some of these numbers could probably be decreased by more detailed networks tuning, we can state that the image-based networks learn much faster than the vector-based ones. The current model for images classification uses a single convolutional layer. Whether it is possible to utilize deep convolutional networks for secondary structure analysis in the discussed approach is a question for future research.

The idea of the extended model that handles sequences instead of parsing results is proved to be applicable in practice and it demonstrates even higher quality than the original parsing-based model, as illustrated by Table 1. We can conclude that it is possible to use parsing only for network training without decreasing the network quality.

To demonstrate the advantage of this technique in practical use in comparison with the classical way (when sequences should first be parsed) we took 100 tRNA sequences from two classes: eukaryotes and prokaryotes and used all four of the trained models to predict their classes. While using base models each sequence was parsed, transformed to correspondent format (image or vector) and fed to the neural network. Extended networks run on original sequences, so the parsing step was skipped. We measured the total time required to output predicted class for all the considered sequences in each case. In the Table 3 the results of this experiment are provided and it is clear that the time spent for parsing is crucial relative to the total working time. So, parsing elimination significantly improves the performance of our solution.

Table 3. Time measurements for 100 sequences processing

Step	Vector based approach		Image based approach	
	Base	Extended	Base	Extended
Parsing	307.6 s	–	310.5 s	–
Weights loading	0.2 s	0.2 s	0.1 s	0.3 s
Class predicting	0.2 s	0.2 s	0.2 s	0.3 s
Total	308.0 s	0.4 s	310.8 s	0.6 s

6   Conclusion

We describe the modifications of the proposed approach [9] for biological sequences analysis using the combination of formal grammars and neural networks. We show that it is possible to improve the quality of the solution by representing parsing result as an image and handling it by using convolutional layers while processing it with a neural network. Also, we provide a technique that removes the parsing step from the trained model use and allows to run models on the original RNA sequences. As a result, the performance of the solution is significantly improved. We demonstrate the applicability of the proposed modifications for real-world problems[4].

We can provide several directions for future research. First of all, it is necessary to investigate the applicability of the proposed approach for other sequences processing tasks such as 16s rRNA processing and chimeric sequences filtration.

Another possible application is a secondary structure prediction. We plan to investigate the possibility of creating a network which generates the most possible contact map for the given sequence. It is necessary to compare this approach with both classical approaches and tools for secondary structure prediction [11,12,17,18] and artificial neural network-based ones [14,20].

The image-based model demonstrates a higher quality. We believe that it is caused by a better locality of features. If so, it should be possible to create a deep convolutional network for secondary structure analysis: further investigation is needed.

Finally, it is important to find a theoretical base for grammar tuning. It is important to adopt the theoretical results on secondary structure description by using formal grammar, such as [16] to find the optimal grammar for our approach.

References

1. Abadi, M., et al.: TensorFlow: Large-scale machine learning on heterogeneous systems (2015). http://tensorflow.org/. Software available from tensorflow.org
2. Abe, T., Inokuchi, H., Yamada, Y., Muto, A., Iwasaki, Y., Ikemura, T.: TRNADB-CE: TRNA gene database well-timed in the era of big sequence data. Front. Genet. **5**, 114 (2014)
3. Azimov, R., Grigorev, S.: Context-free path querying by matrix multiplication. In: Proceedings of the 1st ACM SIGMOD Joint International Workshop on Graph Data Management Experiences & Systems (GRADES) and Network Data Analytics (NDA). GRADES-NDA 2018, Association for Computing Machinery, New York, NY, USA (2018). https://doi.org/10.1145/3210259.3210264
4. Chan, P.P., Lowe, T.M.: GTRNADB 2.0 an expanded database of transfer RNA genes identified in complete and draft genomes. Nucleic Acids Res. **44**(D1), D184–D189 (2016)

---

[4] Project description is available at the project page: https://research.jetbrains.org/groups/plt_lab/projects?project_id=43. Source code and documentation are published at GitHub: https://github.com/LuninaPolina/SecondaryStructureAnalyzer. Access date: 07.03.2020.

5. Chollet, F., et al.: Keras (2015). https://keras.io
6. Dowell, R.D., Eddy, S.R.: Evaluation of several lightweight stochastic context-free grammars for RNA secondary structure prediction. BMC Bioinform. **5**, 71 (2004). https://doi.org/10.1186/1471-2105-5-71. https://pubmed.ncbi.nlm.nih.gov/15180907
7. Durbin, R., Eddy, S.R., Krogh, A., Mitchison, G.: Biological Sequence Analysis: Probabilistic Models of Proteins and Nucleic Acids. Cambridge University Press, Cambridge (1998). https://doi.org/10.1017/CBO9780511790492
8. Eddy, S.R., Durbin, R.: RNA sequence analysis using covariance models. Nucleic Acids Res. **22**(11), 2079–2088 (1994). https://doi.org/10.1093/nar/22.11.2079
9. Grigorev, S., Lunina, P.: The composition of dense neural networks and formal grammars for secondary structure analysis. In: Proceedings of the 12th International Joint Conference on Biomedical Engineering Systems and Technologies - Volume 3 BIOINFORMATICS: BIOINFORMATICS, pp. 234–241. INSTICC, SciTePress (2019). https://doi.org/10.5220/0007472302340241
10. Higashi, S., Hungria, M., De O. C. Brunetto, M.A.: Bacteria classification based on 16s ribosomal gene using artificial neural networks. In: Proceedings of the 8th WSEAS International Conference on Computational Intelligence, Man-Machine Systems and Cybernetics, pp. 86–91. CIMMACS 2009, World Scientific and Engineering Academy and Society (WSEAS), Stevens Point, Wisconsin, USA (2009)
11. Jabbari, H., Condon, A.: A fast and robust iterative algorithm for prediction of RNA pseudoknotted secondary structures. BMC Bioinform. **15**(1), 147 (2014)
12. Jabbari, H., Condon, A., Pop, A., Pop, C., Zhao, Y.: HFold: RNA Pseudoknotted Secondary Structure Prediction Using Hierarchical Folding. In: Giancarlo, R., Hannenhalli, S. (eds.) WABI 2007. LNCS, vol. 4645, pp. 323–334. Springer, Heidelberg (2007). https://doi.org/10.1007/978-3-540-74126-8_30
13. Knudsen, B., Hein, J.: RNA secondary structure prediction using stochastic context-free grammars and evolutionary history. Bioinform. **15**(6), 446–454 (1999). https://doi.org/10.1093/bioinformatics/15.6.446
14. Lu, W., et al.: Predicting RNA secondary structure via adaptive deep recurrent neural networks with energy-based filter. BMC Bioinform. **20**(25), (2019). https://doi.org/10.1186/s12859-019-3258-7
15. Nawrocki, E.P., Eddy, S.R.: Infernal: 1.1 100-fold faster RNA homology searches. Bioinform. **29**(22), 2933–2935 (2013)
16. Quadrini, M., Merelli, E., Piergallini, R.: Loop grammars to identify RNA structural patterns. In: Proceedings of the 12th International Joint Conference on Biomedical Engineering Systems and Technologies, vol. 3, pp. 302–309. BIOINFORMATICS INSTICC, SciTePress (2019). https://doi.org/10.5220/0007576603020309
17. Sato, K., Hamada, M., Asai, K., Mituyama, T.: CentroidFold: a web server for RNA secondary structure prediction. Nucleic Acids Res. **37**(suppl.2), W277–W280 (2009)
18. Sato, K., Kato, Y., Hamada, M., Akutsu, T., Asai, K.: IPknot: fast and accurate prediction of RNA secondary structures with pseudoknots using integer programming. Bioinformat. **27**(13), 85–93 (2011). https://doi.org/10.1093/bioinformatics/btr215
19. Sherman, D.J.: Humidor: microbial community classification of the 16s gene by training cigar strings with convolutional neural networks (2017)

20. Singh, J., Hanson, J., Paliwal, K., Zhou, Y.: RNA secondary structure prediction using an ensemble of two-dimensional deep neural networks and transfer learning. Nat. Commun. 10(1) (2019). https://doi.org/10.1038/s41467-019-13395-9
21. Steeg, E.W.: Neural Networks, Adaptive Optimization, and RNA Secondary Structure Prediction, pp. 12–160. American Association for Artificial Intelligence, USA (1993)

Intelligence Methods for Molecular Characterization and Dynamics in Translational Medicine

Integration of Single-Cell RNA-Sequencing Data into Flux Balance Cellular Automata

Davide Maspero[1,2,6], Marzia Di Filippo[4], Fabrizio Angaroni[1],
Dario Pescini[4,5], Giancarlo Mauri[1,5], Marco Vanoni[3,5],
Alex Graudenzi[1,6(✉)], and Chiara Damiani[3,5(✉)]

[1] Department of Informatics, Systems and Communication,
University of Milan-Bicocca, Milan, Italy
alex.graudenzi@unimib.it
[2] Fondazione IRCCS Istituto Nazionale dei Tumori, Milan, Italy
[3] Department of Biotechnology and Biosciences,
University of Milan-Bicocca, Milan, Italy
chiara.damiani@unimib.it
[4] Department of Statistics and Quantitative Methods,
University of Milan-Bicocca, Milan, Italy
[5] SYSBIO Centre of Systems Biology, University of Milan-Bicocca, Milan, Italy
[6] Institute of Molecular Bioimaging and Physiology, CNR, Segrate, Milan, Italy

Abstract. FBCA (Flux Balance Cellular Automata) has been recently proposed as a new multi-scale modeling framework to represent the spatial dynamics of multi-cellular systems, while simultaneously taking into account the metabolic activity of individual cells. Preliminary results have revealed the potentialities of the framework in enabling to identify and analyze complex emergent properties of cellular populations, such as spatial patterns phenomena and synchronization effects. Here we move a step forward, by exploring the possibility of integrating real-world data into the framework. To this end, we seek to customize the metabolism of individual cells according to single-cell gene expression profiles. We investigate the effect on cell metabolism of the interplay between: (a) the environmental conditions determined by nutrient diffusion dynamics; (b) the activation or deactivation of metabolic pathways determined by gene expression.

Keywords: Flux Balance Analysis · Cellular Potts Model · Single-cell RNA-seq

1 Scientific Background

The alteration of cellular metabolism plays a significant role in tumor origin and development. While the reprogramming of cancer metabolism potentially opens

Supported by SYSBIO and ITFOC.

P. Cazzaniga et al. (Eds.): CIBB 2019, LNBI 12313, pp. 207–215, 2020.
https://doi.org/10.1007/978-3-030-63061-4_19

new therapeutic opportunities, heterogeneity of cancer metabolism hinders the identification of effective treatments. Tumors with different tissue and cell type origin present extensive genetic and phenotypic variability, resulting in a differentiated aggressiveness and sensitivity to cytotoxic therapies. Such an inter-tumor heterogeneity is accompanied by intra-tumor heterogeneity, since multiple subclones having different genetic, epigenetic, and phenotypic features can characterize distinct regions of the same primary tumor. A complex metabolic interplay occurs among cancer cells, the host stroma, and cells of the immune system. Malignant cells may extract high-energy metabolites (e.g., lactate and fatty acids) from adjacent cells, contributing to treatment resistance. Therefore, effective therapeutic strategies should incorporate knowledge of cooperation and competition phenomena within cancer cell populations.

Many efforts have been devoted to investigate inter-tumor metabolic heterogeneity [1], which rely on tissue-specific steady-state modeling to unravel distinctive metabolic alterations of multiple cancer types that may represent potential targets for the development of personalized therapies. Fewer efforts have been instead put forward to investigate intra-tumor metabolic heterogeneity [2], which however require knowledge *a priori* of the composition of the cell population. This limit can be overcome by exploiting the information of gene expression at the single cell level today fully enabled by RNA-seq. In this regard, we have recently proposed the single-cell Flux Balance Analysis framework (scFBA) [3], which allows to portray a snapshot of the single-cell metabolic phenotypes within a cell population at a given moment, by relying on unsupervised integration of scRNA-seq data. scFBA does not explicitly model spatial organization and dynamics. Yet, cancer (sub)population evolve and compete in a (micro)environment with usually limited resource (e.g., oxygen and nutrients) and with specific spatial properties, which significantly differs in distinct tissues and organs. Therefore, modeling the spatio-temporal dynamics of heterogeneous cell populations may assist the development of strategies able to investigate processes and phenomena involving populations of interacting cells at different time/space scales. The simulation of the spatial/morphological dynamics of multicellular systems, such as tissues and organs, has recently progressed [4]. As a first attempt to combine the spatial/morphological dynamics of a multicellular model simulated via Cellular Potts Model (CPM) [5], and the metabolic activity of its constituting single cells computed via Flux Balance Analysis (FBA) [6], we developed FBCA (Flux Balance with Cellular Automata) [7,8,10]. In [9], we extended the previous framework by modeling the metabolic communication among cells, via diffusion of metabolites over the tissue according to a local spatial gradient.

In this work, we introduce the integration of single-cell RNA-seq (scRNA-seq) data into FBCA, aiming to evaluate the impact of different single-cell fluxomes on the overall cell population spatial dynamics. Metabolic behavior(s) characterizing every single cell in the population will emerge from the combination of the corresponding intracellular constraints dictated by single-cell transcriptomic data, together with its nutritional constraints. As a proof of principle, we applied the methodology to the lung adenocarcinoma patient derive xenograft scRNA-seq data, already used in [3].

2 Materials and Methods

2.1 The **FBCA** framework

FBCA is a computational methodology which combines the spatial dynamics representation of the system through Cellular Potts Model (CPM) [5], with a model of cell metabolism by means of Flux Balance Analysis (FBA) [6]. For an extensive and formal description of the FBCA framework, the reader is referred to [7,8]. Here we just recall the general idea: the morphology of a generic tissue is modeled via CPM, in which biological cells are represented by sets of contiguous lattice sites, which evolve via flip attempts driven by a Hamiltonian function that accounts for the Differential Adhesion Hypothesis [5] and for the growth tendency of each cell due to the accumulation of biomass.

Accumulation of biomass is determined according to the biomass production rate of the metabolic network associated with each cell, according to the corresponding nutrient availability. A metabolic network is defined as the set of metabolites and chemical reactions taking place in a cell. The biomass production rate of a given metabolic network is computed at each Monte Carlo simulation Step (MCS) by means of FBA, according to the corresponding nutrient availability. The uptake rate of extracellular nutrients are constrained to be lower than the sum of the corresponding concentration values in the lattice sites belonging to that cell neighborhood. The concentration of nutrients in the lattice sites in turns depend on the diffusion of nutrients which is updated at each simulation step.

Given the constraints on uptake boundaries, FBA solves a Linear Programming problem to identify the flux distribution $v = (v_1, \ldots, v_R)$ that maximizes or minimizes the biomass reaction flux v_b of the cell, given a steady-state assumption for the abundance of each metabolite, i.e., $S\,v = 0$, and boundaries for allowed reaction fluxes.

At each time step s, the biomass synthesis rate is added to the current value of the biomass so far accumulated by a given cell. As described in [8], to avoid an uncontrolled raising of the cell density a limitation of biomass accumulation was introduced.

Each cell is assigned an initial equal area. When it reaches twice its initial area, it splits into two daughters, along a randomly chosen horizontal or vertical direction. Daughter cell inherit their parents' properties.

The nutrient diffusion process, as discussed in [9] is implemented by averaging the nutrient concentrations in a neighborhood of each lattice site. Different diffusion coefficients can be set for different nutrients.

At each simulation time step, the upper bounds of the nutrients uptake rates of each cell are set proportionally to its area.

2.2   Integration of ScRNA-Seq Data

In order to customize the cells according to scRNA-seq data, we first classify genes as "on" (1) if the corresponding Transcript Per Kilobase Milion (TPM) is greater than 0, as "off" (0) otherwise. We then solve the logic Gene-Protein-Reaction (GPR) rules included in the model, which determine the set of proteins that must be present for the reaction to carry flux, in a Boolean fashion, by considering the binary gene expression values. If the GPR is not satisfied (i.e., false), we limit the flux capacity of the reaction by setting the upper/lower bound as 0.01% of the original value.

We note that the Boolean simplification can be effective, to a first approximation, in investigating how the simple activation/inactivation of genes may influence the complex metabolic interplay involving cell subpopulations. Nevertheless, more refined data integration strategies, such as the employment of discretized or continues normalized expression values may be effective to this end, and will be included in further extensions of the method.

It is worth mentioning that, to reduce the computation time, we did not embrace the philosophy used in scFBA of setting the flux capacity as a continuous function of the expression of the associated genes, as this would require to perform a computationally demanding Flux Variability Analysis for each cell at each simulation step [3]. By doing so, we maintain the simulation time reported in [9].

It is also worth mentioning that we did not exploit the strategy used in scFBA of solving a unique mass balance problem for the entire population, given constraints on the uptake and consumption rate of the bulk, but each cell was modeled separately. Because space is explicitly modeled in FBCA, each cell must indeed have its own constraints, which depend in turn on the diffusion of nutrients in its neighborhood.

In our framework, nutrients diffusion and cells evolution take place with interleaved dynamics of different speeds (a MCS every ten diffusion steps). In order to avoid the case that cells starve at first steps of the simulation we let nutrients diffuse for 1500 steps before seeding the lattice with cells and letting them to enter in the dynamics process. Moreover, to guarantee the representativeness of each experimental cell phenotype, we seed the lattice with ten copies of each cell.

Datasets. In this work, we used the Lung Adenocarcinoma (LUAD) scRNA-seq dataset LCPT45Re obtained from the NCBI Gene Expression Omnibus (GEO) data repository under accession number GSE69405. The dataset is composed of 43 cells acquired from a xenograft, obtained by sub-renal implantation in mice of a surgical resection of a 37-mm irregular primary lung lesion in the right middle lobe of a 60-year-old untreated male patient.

2.3   Experimental Setting

We considered a rectangular lattice space of 150 × 100 sites to represent a tissue-like environment. The lattice is closed on all sides to avoid the washing out of cells, which can disappear form the systems only as a result of cell death. The metabolism of each cell is simulated by using a core model of human central carbon metabolism [1]. As in [9], we considered a single cellular type and defined an Hamiltonian function to mimic the tendency of cells to fill the empty space if available. All simulation parameters are set as in [9], unless otherwise specified. In the initialization phase, the lattice is populated with ~ 500 cells with different and randomly assigned initial areas, in the range [25, 50], as originally proposed in [8].

In order to introduce biophysically plausible nutrient sources in the modelling framework, in [9] we described two scenarios involving projections of three-dimensional blood vessels into the two-dimensional space of the lattice. Here we opted for the *Cross section* scenario, which represents a transversal section of a generic tissue that includes 5 squared nutrient sources positioned on the lattice (see Fig. 1).

The nutrient diffusion process is then simulated as follows. As in [9], at each diffusion step, the concentration of each nutrient in each lattice site of the nutrient source area is set equal to a specific value, i.e., oxygen is equal to $100[fmol]$, glucose is equal to $50[fmol]$, and glutamine is equal to $50[fmol]$. The lactate is not supplied in the extracellular environment, but it may be just produced and then exchanged with other cells. If nutrients are not consumed by cells, they are removed from the simulated lattice through a constant flux value when the corresponding edges of the lattice itself are reached.

In [9] we explored two different descriptions of nutrient diffusion across cellular membranes: permeable versus impermeable cells, and we concluded that, given that the population dynamics is only slightly affected by this choice, in order to speed up the computation time the second option is more convenient. Therefore, here we assume cell permeability, meaning that the diffusion process takes place "independently" from the cells positions. Cell and nutrients move on two distinct overlapping layers: the cell matrix, in which cells evolve, overlays the nutrient matrix, where metabolites freely diffuse disregarding the presence of the cells. In this configuration, cells have access to the nutrients and metabolites of each lattice site over which they "float" (See Fig. 1). More details can be found in [9].

In this work, we decided to perform 10 nutrient update steps at each step of the spatial dynamics (MCS), and to account for different diffusion rates of the distinct nutrients according to experimental estimations in literature. We also allow the nutrients to diffuse over the lattice for 1500 update steps before positioning the cells and start running the spatial simulation.

3   Results

As previously mentioned, we allow nutrients to diffuse for 1500 updates before positioning the cells. We also verified that after 1500 time steps all nutrients have diffused all over the lattice.

We consider this situation as our starting point (MCS=0). At MCS=0 we place the cells and we let them evolve. We ran 9 distinct simulations with the same experimental setting. An example of simulation run is depicted in Fig. 1 for MCS 1, 500, 1000 and 2000. Cells are colored according to emerging metabolic properties: green cells consume lactate, where orange cells secrete it. It can be noticed that at time MCS 1, no cells consume lactate as it is not supplied with blood but it can just be secreted by cells in the lattice. It can be observed that at time 500 cells have lost their initial (unrealistic) squared shape and have filled up the lattice. Remarkably, at this point cells that consume lactate appear. As it is apparent also in later evolution steps, these cells tend to be confined at the extremities of the lattice, where nutrients are less abundant.

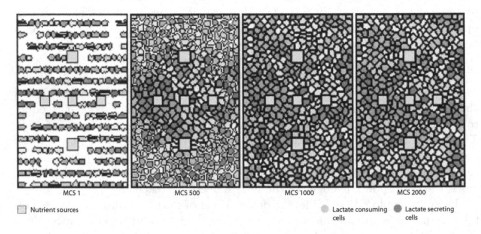

Nutrient sources   Lactate consuming cells   Lactate secreting cells

Fig. 1. Snapshots of four selected simulation steps (at time: $MCS = 1$, $MCS = 500$, $MCS = 1000$ and $MCS = 2000$) of a single simulation run. The tissue is represented from a cross section and the cells have permeable membrane. The yellow lattice sites represent the nutrient sources in the lattice. Biological cells are differently colored according the corresponding lactate metabolism, i.e., green if lactate is produced and orange is lactate is consumed. (Color figure online)

It is interesting to investigate whether this patterning is determined merely by nutrient gradients or also by the underling metabolic network, which should reflect the gene expression patterns. At this aim, we need to classify cells differently according to their metabolic network. As a first approximation, we considered as an indicator of differences in the metabolic network the differences in the theoretical capacity of the corresponding cell to make biomass given (the same) availability of all nutrients. We refer to this value as Optimal Biomass Production (OBP). Following the distribution of OPB values, we could easily divide cells into three non-overalaping groups: low OPB; medium OBP and high OBP (data not shown here).

In Fig. 2A cells are labeled with different colors according to the OBP group. The dots in Fig. 2 correspond to the barycenter of each cell in each of the 9 simulation runs at a given MCS. When observing MCS 1 (first panel from the left) it is apparent that most cells fall in the high OBP group. Cells belonging to the tree groups are uniformly distributed across space. As the simulation proceed, it can be observed that cells tend to cluster according to their OBP, and hence to their gene expression. High OBP cells tend to colonize the surroundings of the blood vessels, whereas low and medium OBP cells are confined to the corners of the lattice. This interesting result proves that the metabolic behaviour of a cell is the result of an interplay involving gene expression patterns and the properties of the environment, in this case in terms of proximity to the nutrient source and availability of space. To further investigate such behaviour, we counted the number of cells included in each OBP group and placed at a given distance from the nutrient sources – computed in terms of lattice sites. In particular, we computed the proportion of cells of each group having distance d: $0 \leq d < 3$, $3 \leq d < 6, 6 \leq d < 9, \ldots$ As shown in Fig. 2B, at $MCS = 1$, the prevalence of the distinct groups is homogeneous with respect to the distance from the nutrient sources. However, as the simulation continues, the regions close to the nutrient sources get progressively colonized by the high OBP cells.

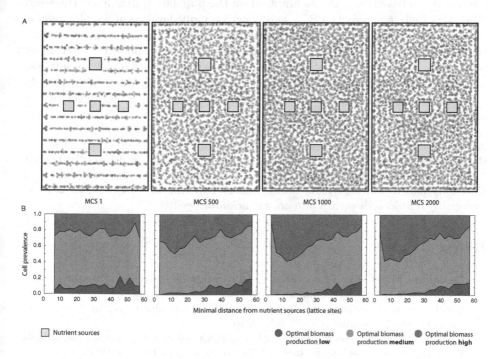

Fig. 2. A. The barycenters of the cells present in the lattice. B. Variation of the prevalence of the OBP cell populations at each time point. The distance is computed by considering for each cell only the nearest nutrient source. The plots correspond to time $MCS = 1$, $MCS = 500$, $MCS = 1000$ and $MCS = 2000$ in all 9 simulation runs are displayed. The colours are related the Optimal Biomass Production – OBP – groups, as defined in the main text: blue, green and red, corresponding to low, medium and high OBP, respectively. (Color figure online)

4 Conclusion

The results presented in Figs. 1 and 2 prove that FBCA has the potential to unravel the complex interplay between gene expression and nutrient gradients. Results are still preliminary, but suggest how a similar approach could be used to group cells in homogeneous clusters due to the projection of single-cell RNAseq data into cell-specific metabolic models. In [3] we showed how the scFBA methodology based exclusively on steady state modeling can be exploited to cluster cells with the same data, according to growth rate and metabolic phenotype. FBCA allows to refine such analysis by taking into account their spatial properties and the interaction with other cells and the environment. Of course, the former approach requires much less assumption and parameters, as scFBA needs only constraints on the rate of consumption/secretion of nutrients of the overall population, which can be promptly measured with current methodologies. On the contrary, FBCA requires information on the nutritional constraints of each single cell and is based on many assumption on the population dynamics. However, due to its high expressivity FBCA can describe multi-level complex phenomena that emerge specifically due to interaction and competition of cells in an environment with limited space and resources. This provides a powerful instrument to investigate intra-tumor metabolic heterogeneity in distinct in-silico scenarios, and surely deserves further investigations. We finally specify that a user-friendly tool for the simulation of the FBCA framework is currently under development and will be released in the near future.

Acknowledgements. The institutional financial support to SYSBIO.ISBE.IT within the Italian Roadmap for ESFRI Research Infrastructures and the FLAG-ERA grant ITFoC are gratefully acknowledged. Financial support from the Italian Ministry of University and Research (MIUR) through grant Dipartimenti di Eccellenza 2017 to University of Milano Bicocca is also greatly acknowledged.

References

1. Di Filippo, M., et al.: Zooming-in on cancer metabolic rewiring with tissue specific constraint-based models. Comput. Biol. Chem. **62**, 60–69 (2016)
2. Conde, M., do Rosario, P., Sauter, T., Pfau, T.: Constraint based modeling going multicellular. Front. Mol. Biosci. **3**:3 (2016)
3. Damiani, C., et al.: Integration of single-cell RNA-seq data into population models to characterize cancer metabolism. PLoS Comput. Biol. **15**(2), e1006733 (2019)
4. Graudenzi, A., Caravagna, G., De Matteis, G., Antoniotti, M.: Investigating the relation between stochastic differentiation, homeostasis and clonal expansion in intestinal crypts via multiscale modeling. PLoS ONE **9**(5), e97272 (2014)
5. Scianna, M., Preziosi, L.: Cellular Potts Models: Multiscale Extensions and Biological Applications. CRC Press, Boca Raton (2013)
6. Orth, J.D., Thiele, I., Palsson, B.: What is flux balance analysis? Nat. Biotechnol. **28**(3), 245 (2010)

7. Graudenzi, A., Maspero, D., Damiani, C.: Modeling Spatio-Temporal Dynamics of Metabolic Networks with Cellular Automata and Constraint-Based Methods. In: Mauri, G., El Yacoubi, S., Dennunzio, A., Nishinari, K., Manzoni, L. (eds.) ACRI 2018. LNCS, vol. 11115, pp. 16–29. Springer, Cham (2018). https://doi.org/10.1007/978-3-319-99813-8_2

8. Maspero, D., et al.: Synchronization Effects in a Metabolism-Driven Model of Multi-cellular System. In: Cagnoni, S., Mordonini, M., Pecori, R., Roli, A., Villani, M. (eds.) WIVACE 2018. CCIS, vol. 900, pp. 115–126. Springer, Cham (2019). https://doi.org/10.1007/978-3-030-21733-4_9

9. Maspero, D., et al.: The influence of nutrients diffusion on a metabolism-driven model of a multi-cellular system. Fundam. Inform. **171**(1–4), 279–295 (2020)

10. Graudenzi, A., Maspero, D., Damiani, C.: FBCA, a multiscale modeling framework combining cellular automata and flux balance analysis. J. Cell. Automata **15** (1/2), 75–95 (2020)

Machine Learning in Healthcare
Informatics and Medical Biology

Characterizing Bipolar Disorder-Associated Single Nucleotide Polymorphisms in a Large British Cohort Using Association Rules

Alberto Pinheira[1]([ID]), Rodrigo Dias[2][ID], Camila Nascimento[2][ID], and Inês Dutra[1][ID]

[1] Department of Computer Science, Faculty of Sciences, University of Porto, Porto, Portugal
alberto_pinheira@hotmail.com

[2] Bipolar Disorder Research Program, Department of Psychiatry, University of São Paulo Medical School, University of São Paulo, São Paulo, Brazil

Abstract. Bipolar Disorder (BD) is chronic and severe psychiatric illness presenting with mood alterations, including manic, hypomanic and depressive episodes. Due to the high clinical heterogeneity and lack of biological validation, both BD treatment and diagnostic are still problematic. Patients and clinicians would benefit from better clinical and biological characterization, ultimately opening a new possibility to distinct forms of treatment. In this context, we studied genome wide association (GWA) data from the Wellcome Trust Case Control Consortium (WTCCC). After an exploratory analysis, we found a higher prevalence of homozygous compared with heterozygous in different single nucleotide polymorphisms (SNPs) in genes previously associated with BD risk. Results from our association rules analysis indicate that there is a group of patients presenting with different groups of genotypes, including pairs or triples, while others present only one. We performed the same analysis with a control group from the same cohort (WTCCC) and found that although healthy subjects may present the same SNPs combinations, the risky alleles occur in a lower frequency. Moreover, no subject in the control group presented the same pairs or triples of genotypes found in the BD group, and if a pair or triple is found, the support and confidence are lower than in the BD group (<50%).

Keywords: Machine learning · Association rules · Apriori algorithm · Bipolar disorder · WTCCC · GWAS · SNPs

1 Scientific Background

Bipolar disorder (BD) is characterized by the presence of mood changes from mania to depression, associated with progressive functional impairment. Following the Global Burden of Diseases 2013 (GBD 2013) [12], BD is a mental illness

© Springer Nature Switzerland AG 2020
P. Cazzaniga et al. (Eds.): CIBB 2019, LNBI 12313, pp. 219–231, 2020.
https://doi.org/10.1007/978-3-030-63061-4_20

that affects around 48.8 million people (43.5–54.4 million) with peak of incidence between 20 and 34 years old. Life expectancy of BD patients is decreased 12.0–8.7 years and 10.6–8.3 years, for males and females aged 25 to 45 years, respectively [14].

Like other psychiatric conditions, BD can be the result of multiple factors: genetic, biological or environmental [24]. Due to the high clinical heterogeneity, bipolar can be misdiagnosed, which confers a poor treatment and prognostic outcomes [4]. Biological information, such as genomic variants, can be helpful in the case of complex brain disorders since their diagnostic rely only on clinical identification, which is highly heterogeneous. In this work, we are interested in categorizing different groups of BD patients. Our hypothesis is that distinction among genetic patterns in patients may offer improved treatment and clinician planning. We explored different Single Nucleotide Polymorphisms (SNPs) in seven genes previously associated with BD (ARPP21, CACNA1c, SYNE1, NCAN, GABRB1, SYN3 [18]) in data from the Wellcome Trust Case Control Consortium (WTCCC), a large British cohort. Genotype data was produced through blood samples using the Affymetrix 500K microarrays comprising probes for more than 500K SNPs. While there is a reasonable number of studies focusing on how each one of these SNPs affect BD risk [24], and how the same SNP can be involved in multiple mental disorders [16], to the best of our knowledge, investigation of the combination of these SNPs and their role on describing patients with BD has not yet been explored.

In our study, we applied association rules to the WTCCC data. In order to reduce computation time, we trained subjects and controls separately, using a non-supervised setting. We obtained rules and then analysed the resulting patterns for both groups. Results of our study indicate that: (1) most genotype patterns for bipolar disorder and controls are homozygous with a prevalence of homozygous in the controls for GABRB1 and SYN3; (2) CACNA1c (SNP rs4765914) very often occurs together with SYN3 (SNP rs9621532) for a group of patients; (3) some SNPs associated with BD can also appear in the genetic sequence of healthy patients, but the prevalence of risk alleles is higher for BD, as expected; (4) some combinations of genotypes found in BD patients either did not that appear in the controls or presented low frequency.

We organized this paper as follows. Next section provides a brief summary of works that have been studied in bipolar disease from machine learning and clinical standpoints, highlighting their main results. Section 3 describes the dataset used and our experimental methodology. Section 4 presents and discusses our findings. Finally, we conclude summarizing the main contributions and lessons learned from this work.

2   Related Work

We searched Google Scholar and Scopus and queried for "Bipolar Disorder'and "Machine Learning" and "Genetics". We selected only public papers. Other machine learning (ML) techniques have been used in the literature of BD. Moreira [17] used the k-means algorithm applied to the same genetic data with a selection of 7 chromosomes. Saylan *et al.* [20] tested several machine learning algorithms to classify BD and schizophrenia. They used k-nearest neighbors, Decision Trees and a Naive Bayes Network on a very small dataset of genetic data. Results are not very conclusive since the sample was too small (seven individuals, 4 controls and 3 bipolar). Hajek *et al.* [11] used ML to identify individuals at genetic high risk for BD based on brain structure. With a large-scale genome-wide association (GWA) data, Chuang and Kuo [3] built a genetic risk model using random forests on the STEP-BD and GAIN-BD datasets consisting of subjects and controls and found genetic overlaps between the two datasets.

From a clinical standpoint, some works investigated the relation of BD genes and loci with other diseases. For example, Green *et al.* discovered that the bipolar disorder risk allele at CACNA1c also confers risk of recurrent major depression and of schizophrenia [10]. Stahl *et al.* [23] discovered that bipolar I disorder is strongly genetically correlated with schizophrenia, driven by psychosis, whereas bipolar II disorder is more strongly correlated with major depressive disorder.

3   Materials and Methods

Our BD data consists of a cohort of 1,998 subjects from the WTCCC repository [5]. For the controls, we used the database from the WTCCC1 project samples, the 1958 British Birth Cohort, which contains 1,502 individuals. The files, among other information, contain Genotypes: SNPs, the genotype present in the SNP and the respective Genetic Risk Score (GRS); and Meta-data: SNP identifier, position, the subjects and the frequency.

A file for each chromosome (25 in total) is associated with a patient and contains the SNP and patient identifiers, the genotype of that patient and the GRS, which is calculated by the Chiamo software [22]. Other file contains demographic data for each patient. Such information contains the gender (1 – Male, 2 – Female), cohort (in our case, BD), the supplier, the plate, the region of the patient and age of recruitment (1 – age between 10 and 19, 2 – age between 20 and 29, and so on).

Our methodology follows the steps below:

1. Search the literature for genomic variants (SNPs) related to BD and select those that are in our dataset;
2. Collect statistics and frequencies of pairs of genotypes;
3. Visualize homozygous and heterozygous with respective GRS, and the number of genotypes per patient;
4. Create the dataset for the apriori algorithm containing each genotype per SNP and its genetic risk score, with one row per patient;
5. Apply the Apriori algorithm for both controls and BD;
6. Cross the patterns found with the demographic data: region, gender and age.

Contrary to other datasets used in the literature and mentioned in Sect. 2, the WTCCC dataset does not provide any clinical information or type of bipolar disorder. It also does not have information about other mental illnesses the patients might have. Therefore, we concentrated our study in finding subgroups that potentially share combinations of SNPs. We started by performing an exploratory analysis and then used 10-fold cross-validation to validate our findings. In order to reduce computational time, we trained subjects and controls separately in a non-supervised setting using association rules. We used the apriori algorithm [1], popular for mining frequent patterns (rules). Rules are ranked according to their support, confidence and lift. The support of a rule measures how frequent that rule occurs in the dataset [21]. Lift measures how many times the items in the antecedent and in the consequent of the rule occur together in the transactions than would be expected if the itemsets at both antecedent and consequent were specifically independent. We tested two scenarios:

- smaller values of support trying to capture not very frequent itemsets (rare subgroups of items). We chose minsupport = 3% in order to produce as many diverse itemsets as possible. We chose minconfidence = 60% to compensate the choice of using very low support. We chose minlift = 3 in order to capture itemsets where items have high probability of appearing together and having a positive dependency;
- higher values of support to capture frequent itemsets, to find common subgroups of items. We chose minsupport = 50%, minconfidence = 50% and minlift = 1 to find independent items.

Table 1. Genes, SNPs and the respective chromosome regions found in the WTCCC BD cohort. Number is the encoding we used for each SNP.

Gene	SNP	Chromosome	Number
ARPP21	rs1523041	3	1
CACNA1c	rs1006737/rs4765914	12	2/3
SYNE1	rs9371601	6	4
NCAN	rs1064395	19	5
GABRB1	rs7680321	4	6
SYN3	rs9621532	22	7

Rules were generated for both subjects and controls. For the first scenario, we selected the top rules according to the algorithm generation order. All subsequent generated rules had all the same or less values of support, confidence and lift. For the second scenario, we sorted the rules from the highest to the lowest support.

4   Results

4.1   Descriptive Analysis

A summary of demographic data for the BD cohort is as follows. 751 patients are men (37.6%) and 1247 are women (62.4%). The majority of patients is from the

Midlands (475 – 24%), followed by Wales (414 – 20.7%), Scotland (199 – 10%), Northern (176 – 8.8%), London (133 – 6.6%), North Midlands (118 – 5.9%), Southern (116 – 5.8%), Southwestern (115 – 5.7%), Southeastern (95 – 4.7%), Northwestern (68 – 3.4%), Eastern (63 – 3.1%) and East + West Ridings (26 – 1.3%). Most of the patients are adults. 575 (28.7%) are in the range 40–49, 480 (24%) are in 50–59, 400 (20%) are in the range 30–39, 277 (13.8%) are in 60–69, 195 (9.7%) in the 20s, and 56 (2.5%) in the 70s. Teenagers and seniors are minority, 14 (1%) and 7 (0.3%), respectively. The summary for the controls is: 752 patients are male (50%) and 750 (50%), female. 161 are from Eastern (11%), 160 from Northwestern (10.6%), 153 from Scotland (10.2%), 141 from Midlands (9.3%), 128 from East + West Ridings (8.5%), 124 from Southwestern (8.3%), 119 from London (7.9%), 117 from Northern (7.7%), 115 from Southern (7.6%), 98 from North Midlands (6.5%) and 76 from Wales (5%).

Following step (1) of our methodology, we found the SNPs shown in Table 1 and their respective chromosomes. The column "Number" refers to the encoding we use for each SNP.

Table 2. BD: counts and scores of Homozygous and Heterozygous genotypes per SNP

Gene	ARPP21	CACNA1c	CACNA1c	SYNE1	NCAN	GABRB1	SYN3
SNP	rs1523041	rs1006737	rs4765914	rs9371601	rs1064395	rs7680321	rs9621532
Tot. Patients	1,998	1,998	1,998	1,998	1,998	1,998	1,998
Tot. Homo	1,091	1,075	1,276	1,069	1,444	1,641	1,796
Tot. Hetero	907	923	722	929	554	357	202
Gen. A	CC	AA	CC	TT	AA	CC	AA
# A	791	298	1,180	259	65	26	1,788
Gen. B	GG	GG	TT	GG	GG	TT	CC
#B	300	827	96	810	1,377	1,615	8
Gen. C	CG	AG	CT	GT	AG	CT	AC
#C	907	923	722	929	554	357	202
[0–0.2]	6	0	2	1	3	1	8
]0.2–0.4]	0	0	0	0	1	0	0
]0.4–0.6]	0	1	1	0	2	0	2
]0.6–0.8]	1	4	1	0	1	0	1
]0.8–0.9]	2	2	3	1	0	1	3
]0.9–1.0]	1,989	1,991	1,991	1,996	1991	1,996	1,984
Tot <0.9	9	7	7	2	7	2	14

Observing the distribution of these SNPs, we noticed that there is an allelic frequency in homozygotic and heterozygotic groups of BD patients per SNP. Table 2 shows those differences. The last six lines show counters for intervals of genetic risk score (GRS). Most genotype patterns for bipolar disorder are homozygous. We also see that genes SYN3 and GABRB1 have a greater imbalance between homozygous and heterozygous genotypes, (90% homozygous × 10% heterozygous and 82% homozygous × 18% heterozygous, respectively).

Table 3. Controls: counts and scores of Homozygous and Heterozygous genotypes per SNP

Gene	ARPP21	CACNA1c	CACNA1c	SYNE1	NCAN	GABRB1	SYN3
SNP	rs1523041	rs1006737	rs4765914	rs9371601	rs1064395	rs7680321	rs9621532
Tot. Subjects	1,502	1,502	1,502	1,502	1,502	1,502	1,502
Tot. Homo	775	875	1,015	826	1,086	1,278	1,370
Tot. Hetero	727	627	487	676	416	224	132
Gen A	CC	AA	CC	TT	AA	CC	AA
# Gen A	587	191	953	174	37	12	1367
Gen B	GG	GG	TT	GG	GG	TT	CC
# Gen B	188	684	62	652	1,049	1266	3
Gen C	CG	AG	CT	GT	AG	CT	AC
#Gen C	727	627	487	676	416	224	132
[0–0.2]	2	0	2	1	2	3	3
[0.2–0.4]	0	0	0	0	0	0	0
[0.4–0.6]	1	1	0	0	0	0	0
[0.6–0.8]	1	0	2	0	0	3	0
[0.8–0.9]	0	0	1	1	0	0	1
[0.9–1.0]	1,498	1,501	1,497	1,500	1,500	1,496	1,498
Tot <0.9	4	1	5	2	2	6	4

At the position rs4765914 of the gene CACNA1c, there is a high rate of genotype CC (92%) compared to TT (8%) among the homozygous patients, and SYN3 has a very high rate of CC (99%) compared to AA (1%), once again among the homozygous. Subjects with the gene SYN3 at the position rs9621532 are distinct from other patients as most of them (14) have Genetic Risk Score (GRS) below 0.9. A single patient with the gene SYNE1 at the position rs9371601 has a very low GRS compared to the other 1997 patients with the genotype GT (heterozygous). Another single patient with the gene GABRB1 at the position rs7680321 has a very low GRS when compared with the other 1,997 patients with genotype TT (homozygous).

Table 3 shows the same information for the controls. Again, most genotype patterns are homozygous. The same imbalance observed for genes SYN3 and GABRB1 in the BD patients is also observed in the controls. However, for these genes, the proportion of homozygous over heterozygous is higher for the controls than for the BD, 91% homozygous versus 9% for SYN3 and 85% homozygous and 15% heterozygous for GABRB1 (note that the controls group has a smaller number of individuals).

Figure 1 shows the ratio of homozygous and heterozygous for subjects and controls for all SNPs (the blue line corresponds to the subjects and the red corresponds to the controls).

Besides plotting the homozygous versus heterozygous ratio, we also plotted the ratio of different genotypes between BD patients and controls. Figure 2 shows the three ratio curves for the first homozygous pair (RatioA), the second homozygous pair (RatioB) and the heterozygous pair (RatioC). In the literature for BD, some risk alleles are reported. For example, for gene CACNA1c, SNP rs1006737, the risk allele is A and the curves show that the BD patients have a prevalence of A alleles over the controls (AA - RatioA and AC - RatioC). For gene SYNE1, SNP rs9371601, T is the risky allele, and TT and GT are prevalent in the BD patients. The same repeats for genes NCAN (prevalence of allele A for BD) and GABRB1 (prevalence of allele C for BD).

Next, we looked for associations between pairs of genes according to genotypes. First, we manually explored frequencies for all pairs. In Table 4, the

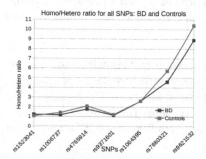 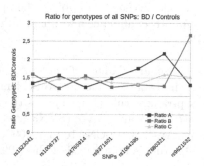

Fig. 1. Ratio between homozygous and heterozygous, for BD and Controls, for all SNPs, showing a prevalence of homozygous genotypes for the controls, particularly for SNPs GABRB1 and SYN3.

Fig. 2. Ratio between BD and Controls for each genotype, for all SNPs, showing in more detail the differences between BD and controls in SNPs GABRB1 and SYN3.

Table 4. Pairs of SNPs for all cases. This illustrates that some combinations of SNPs are more frequent than others, being the frequency for some pairs greater than for the controls.

	GG-1	CC-1	CG-1	AA-2	GG-2	AG-2	CC-3	TT-3	CT-3	GG-4	TT-4	GT-4	AA-5	GG-5	AG-5	TT-6	CC-6	CT-6
GG-1																		
CC-1																		
CG-1																		
AA-2	32	111	105															
GG-2	127	329	371															
AG-2	141	351	431															
CC-3	172	467	541	48	770	362												
TT-3	13	48	35	74	0	22												
CT-3	115	276	331	126	57	539												
GG-4	125	327	358	96	352	362	483	40	287									
TT-4	36	104	119	24	97	138	146	6	107									
GT-4	139	360	430	128	378	423	551	50	328									
AA-5	14	28	25	7	29	31	43	4	20	25	14	28						
GG-5	204	520	653	170	567	640	802	67	508	571	175	631						
AG-5	82	243	229	71	231	252	335	25	194	214	70	270						
TT-6	249	642	724	209	655	751	945	81	589	653	210	752	55	1114	446			
CC-6	6	12	8	1	14	11	20	0	6	13	4	9	1	22	3			
CT-6	45	137	175	38	158	161	215	15	127	144	45	168	11	241	105			
AA-7	270	722	796	216	745	827	1055	86	647	726	232	830	61	1235	492	1436	24	328
CC-7	3	2	3	0	5	3	6	1	1	3	0	5	1	5	2	5	1	2
AC-7	21	67	108	32	77	93	119	9	74	81	27	04	5	137	60	174	1	27

column and row names refer to the SNPs encoding (column "Number" in Table 1) and their genotypes. For example, GG-1 refers to SNP rs1523041, genotype GG. As we can observe, some of the pairs are very frequent. For example, TT-6 (gene GABRB1, SNP rs7680321) and GG-5 (gene NCAN, SNP rs1064395) or AA-7 (SYN3) and GG-5 (NCAN). However, these combinations are not so interesting as we are searching for subgroups.

4.2   Association Rules

We moved further and applied the apriori algorithm [3] with predefined values for support, confidence and lift in order to search for associations among several SNPs.

Case 1: Smaller Values of Support. Curiously, and contrasting with the numbers shown in Table 5, some not so evident associations were revealed. For example, the top-ranked rule (Rule 1) says that if the patient has both TT-3 (CACNA1c at rs4765914) and AA-7 (SYN3 at rs9621532) present, the probability of also having AA-2 (CACCNA1c at rs1006737) is high (confidence of 77% and lift of 6). This rule is true for 66 of the 1,998 patients.

Table 5. Pairs of SNPs for all controls. This illustrates that some combinations of SNPs are more frequent than others.

	GG-1	CC-1	CG-1	AA-2	GG-2	AG-2	CC-3	TT-3	CT-3	GG-4	TT-4	GT-4	AA-5	GG-5	AG-5	TT-6	CC-6	CT-6
GG-1																		
CC-1																		
CG-1																		
AA-2	24	81	86															
GG-2	80	270	334															
AG-2	84	236	307															
CC-3	119	372	462	39	648	266												
TT-3	6	25	31	54	0	8												
CT-3	63	190	234	98	36	353												
GG-4	87	251	314	78	300	274	422	26	204									
TT-4	19	72	83	21	83	70	107	9	58									
GT-4	82	264	330	92	301	283	424	27	225									
AA-5	6	10	21	7	17	13	24	4	9	10	6	21						
GG-5	117	425	507	134	472	443	668	41	340	454	121	474						
AG-5	65	152	199	50	195	171	261	17	138	188	47	181						
TT-6	156	486	624	157	582	527	805	53	408	555	146	565	31	896	339			
CC-6	2	6	4	1	4	7	5	0	7	4	2	6	0	8	4			
CT-6	30	95	99	33	98	93	143	9	72	93	26	105	6	145	73			
AA-7	171	540	656	173	628	566	867	52	442	601	150	616	32	959	376	1153	11	203
CC-7	0	0	3	0	2	1	2	0	1	2	0	1	0	3	0	1	0	2
AC-7	17	47	68	18	54	60	84	4	44	49	24	59	5	87	40	112	1	19

The second ranked rule (Rule 2) says that if the patient has both TT-3 (CACNA1c at rs4765914) and TT-6 (GABRB1 at rs9621532) present, the probability of having AA-2 (CACNA1c at rs1006737) is high with confidence of 80% and lift of 6. This rule is true for 65 of the 1,998 patients. Interestingly, this rule does not appear in any of the individuals of the control group with that confidence and lift, which may make it a strong case for BD. The combination TT-3,

AA-2, AA-7 (topmost ranked rule for BD) occurs in 56% of the controls while it occurs only 3% in the BD subjects, which may indicate that this combination of SNPs is not a good indicator for BD. Nevertheless, as this group of patients is distinct from others, it may be clinically interesting.

When intersecting both rules for the BD subjects, 58 patients have all four homozygous genotypes (TT-3, TT-6, AA-2, AA-7). 19 of these patients are male and the other 39 are female. All of the patients are more than twenty years old. The most frequent age intervals are 40–49 and 50–59 (with 15 patients in each subgroup). Also, most of the patients of this group are from the Midlands.

One subset of patients (8) from the subset of 66 covered by Rule 1 does not have TT-6, while another subset (7) from the 65 covered by Rule 2 does not have TT-3. From the first subset, 5 of the patients are female and the other 3 are male. All of the patients are more than twenty years old. In the male group, 2 of them are from Scotland and one is from Wales. And from the female group, 1 is from London, 2 are from Northern Midlands, 1 Northern and 1 Southeastern. From the second subset (7 patients out of the intersection), 4 of the patients are female and the other 3 are male. All of the patients are adults between twenty and sixty years old. From the male group, one of them is from Wales, one is from Midlands and one is from the Southern region. From the female group, two of them are from the Midlands, one is from Wales and one is from SouthWestern.

The amount of frequent itemsets for these rules is low, but significant, given the high confidence and lift. This finding may help the identification of rare groups of BD patients that possibly would need a different treatment from others or a different clinical and psychological/psychiatric evaluation.

Case 2: Higher Values of Support. The first rule says that if the patient has a GRS between [0.9–1.0]-3 (at SNP rs4765914, which gene is CACNA1c), the frequency of also having a GRS between [0.9–1.0]-5 (at rs1064395, which gene is NCAN) is high due to its high support (99.3%). The confidence is also high (99.7%) due to the high number of transactions.

The second rule says that if the patient has a GRS between [0.9–1.0]-1 (at SNP rs1523041, which gene is ARPP21), the frequency of also having a GRS between [0.9–1.0]-3 (at SNP rs4765914, which gene is CACNA1c) is high due to its support (99.3%). The confidence is high (99.8%), due to the high number of transactions.

None of these two rules appear among the top 10 rules obtained for the controls.

When intersecting both rules, three groups arise: 1,980 patients have all three GRS ([0.9–1.0]-1, [0.9–1.0]-3, [0.9–1.0]-5); 6 patients have [0.9–1.0]-3 and [0.9–1.0]-5 but don't have GRS between [0.9–1.0]-1; and 5 patients have [0.9–1.0]-1 and [0.9–1.0]-3 but don't have GRS between [0.9–1.0]-5;

By adding the demographic information, for the first group, from the 1,980 patients, 744 of them are male and 1236 are female. The most frequent age interval is 40–49 in both genders, and the majority of the patients is from the Midlands. For the second group, from the 6 patients, 3 patients are female and the other 3 are male. They have a 50/50 distribution in terms of age, however there are more men in their 30's and more women in their 50's. For the third group, from the 5 patients, 2 of them are female and the other 3 are male. There is a 50/50 distribution in gender when the patients are in their 40's, however there are cases where there are males in an age interval and no females and vice-versa. The majority of the patients come from the Midlands.

Although not shown here, since we focus on the exploratory study, results of cross-validation show the same findings for most of the 10 folds, using the same values of support, confidence and lift. For example, among the 10 folds, all of them had rules containing both TT-3 and AA-2, for the cases. A more detailed discussion about cross-validation results can be found in [19], for the BD cases.

5   Conclusion

It is well-known that BD presents a highly heterogeneous clinical course [4], with not a single gene or a bio-marker as diagnostic tool for this disorder, even with the high heritable component observed within families [8]. Computational methods may help in generating novel information regarding biological aspects, which could improve diagnostic and treatments choices. Our work explored genotypic and allelic characteristics of SNPs from genes previously associated with BD from a large British cohort including BD and controls. As far as we know this is the first work that applies association rules to BD in an unsupervised learning task, exploring how patients are categorized according to combinations of SNPs.

The main differences between our work and previously published data are: (1) the use of association rules in a non-supervised setting to find genetic patterns for BD; (2) categorization of BD patients in groups according to the combination of these genetic patterns; (3) comparison between combinations of SNPs for controls and BD subjects.

Our main findings can be summarized as follows. As we studied the behaviour of the GRS of each SNP per patient, we observed differences between homozygous and heterozygous in BD patients. There are some cases where even though the majority of the patients have a very high GRS in most of the SNPs, some cases exhibit a very low GRS for a given SNP. We did not find SNPs associated to any of the other genes related with BD reported in the literature (ANK3, ODZ4, TRANK1/LBA1, DISC1, ANKRD46, DUSP6 and GRIND2B) because they are not present in the WTCCC BD cohort.

From the apriori algorithm results, in both cases, we found subgroups presenting specific patterns of genetic characteristics in BD, in particular among the TT genotype at SNP rs7680321, the AA genotype at SNP rs9621532, the TT genotype at SNP rs4565914 and the AA genotype at SNP rs1006737. For the parameters we used in the apriori algorithm, one of the top-ranked rules found in the BD group does not show in the controls (combinations of the TT genotype at SNP rs7680321 with others). Other associations involve the GRS in the interval [0.9–1.0] in SNP rs476591, rs106439 and rs1523041, respectively. Because different age groups were clustered by specific SNPs, this would suggest that different age at onset for BD that is observed in the clinical settings would be associated to specific genetic variants. This information allied to clinical evaluation and other patient characteristics may help strength the effort in the precision medicine field devising better treatment.

While intersecting our results with the demographic data we observed that the female gender is more frequent than the male gender [7]. It could be because women usually seek more for medical care than men. Epidemiology data does not show higher prevalence of bipolar on women when compared to men. All of the patients are adults, and most of them older adults, it might be because most of the bipolar subjects have the diagnostic of the disease around 30 to 40 years old [6,9].

Regarding genotypes, studies with families, of adoption and of twins, indicate that BD has a high heritability (proportion of risk of disease in the population attributable to genetic variation), estimated between 79 and 83% [2]. Recent studies, with longitudinal follow-up and control for age and sex, demonstrated a heritability of 60% [13]. Family studies conducted to date suggest that first-degree relatives of BD patients have approximately 9% risk of developing the disorder, almost 10 times that of the general population [2]. Twin studies have shown that the greatest risk observed in family members of BD patients is largely due to genetic influence, since the agreement between monozygotic twins (38 to 43%) is significantly higher than between dizygotic (4 to 6%) [15]. Despite the high rate of heritability, genetic epidemiology studies suggest that the form of BD transmission is still unknown, and the contribution of genetic factors is complex, including interaction between multiple susceptibility and environmental influence genes [24].

Our work has some limitations, such as lack of clinical data and information regarding ancestry, which could be potential confounds. However, we studied a large cohort of British subjects, using novel techniques of machine learning, which had never been applied in this context. For this reason, we believe that our exploratory analysis may give new insights on biological differences within BD subjects, supporting the hypothesis that psychiatric disorders present multiple genes influencing in the disease's phenotype.

Future studies using clinical information may clarify whether the biological aggregation we observed are associated with clinical characteristics. Furthermore, validation of our results in other BD cohorts could confirm our findings. It would also be interesting to investigate the same genotypic information in other psychiatric disorders to assess the disease's specificity of the biological aggregation we showed here.

References

1. Agrawal, R., Srikant, R.: Fast algorithms for mining association rules in large databases. In: Proceedings of the 20th International Conference on Very Large Data Bases (VLDB 1994), pp. 487–499. Morgan Kaufmann Publishers Inc., San Francisco, CA, USA (1994). http://dl.acm.org/citation.cfm?id=645920.672836
2. Barnett, J.H., Smoller, J.W.: The genetics of bipolar disorder. Neuroscience 164(1), 331–343 (2009). https://doi.org/10.1016/j.neuroscience.2009.03.080

3. Chuang, L.C., Kuo, P.H.: Building a genetic risk model for bipolar disorder from genome-wide association data with random forest algorithm. PMC Sci. Rep. **7**, 39943 (2017). https://doi.org/10.1038/srep39943
4. Cohen, B.M., Öngür, D.: The urgent need for more research on bipolar depression. Lancet Psychiatry **5**(12), e29–e30 (2018). https://doi.org/10.1038/nrdp.2018.8
5. Craddock, N., et al.: Genome-wide association study of CNVs in 16,000 cases of eight common diseases and 3,000 shared controls. Nature **464**(7289), 713–720 (2010). https://doi.org/10.1038/nature08979
6. Dagani, J., et al.: Meta-analysis of the interval between the onset and management of bipolar disorder. Can J Psychiatry. **62**(4), 247–258 (2017). https://doi.org/10.1177/0706743716656607
7. Diflorio, A., Jones, I.: Is sex important? gender differences in bipolar disorder. Int. Rev. Psychiatry **22**(5), 437–452 (2010). https://doi.org/10.3109/09540261.2010.514601
8. Grande, I., Berk, M., Birmaher, B., Vieta, D.E.: Bipolar disorder. Lancet **387**(10027), 1561–1572 (2016). https://doi.org/10.1016/S0140-6736(15)00241-X
9. Grande, I., Berk, M., Birmaher, B., Vieta, D.E.: Is age of onset associated with severity, prognosis, and clinical features in bipolar disorder? a meta-analytic review. Bipolar Disord. Int. J. Psychiatry Neurosci. **18**(5), 389–403 (2016). https://doi.org/10.1111/bdi.12419
10. Green, E., et al.: The bipolar disorder risk allele at cacna1c also confers risk of recurrent major depression and of schizophrenia. Mol Psychiatry **15**(10), 1016–1022 (2010). https://doi.org/10.1038/mp.2009.49
11. Hajek, T., Cooke, C., Kopecek, M., Novak, T., Hoschl, C., Alda, M.: Using structural MRI to identify individuals at genetic risk for bipolar disorders: a 2-cohort, machine learning study. J Psychiatry Neurosci **40**(5), 316–324 (2015)
12. Global Burden of Disease. IHME, University of Washington, Seattle, WA (2015). http://www.healthdata.org/gbd. Accessed May 2020
13. Johansson, V., Kuja-Halkola, R., Cannon, T., Hultman, C., Hedman, A.: A population-based heritability estimate of bipolar disorder - in a swedish twin sample. Psychiatry Res. **278**, 180–187 (2019). https://doi.org/10.1016/j.psychres.2019.06.010. epub 2019 Jun 12
14. Kessing, L.V., Vradi, E., Andersen, P.K.: Life expectancy in bipolar disorder. Bipolar Disord. Int. J. Psychiatry Neurosci. **17**(5), 543–548 (2015). https://doi.org/10.1111/bdi.12296
15. McGuffin, P., Rijsdijk, F., Andrew, M., Sham, P., Katz, R., Cardno, A.: The heritability of bipolar affective disorder and the genetic relationship to unipolar depression. Arch. Gen. Psychiatry **60**(5), 497–502 (2003). https://doi.org/10.1001/archpsyc.60.5.497
16. Moon, A.L., Haan, N., Wilkinson, L.S., Thomas, K.L., Hall, J.: Cacna1c: association with psychiatric disorders, behavior, and neurogenesis. Schizophr. Bull. **44**(5), 958–965 (2018). https://doi.org/10.1093/schbul/sby096
17. Moreira, L.M.M.: Extraction of a Bipolar Disorder associated genetic pattern. Faculdade de Ciências da Universidade do Porto (2018)
18. Orrù, G., Carta, M.G.: Genetic variants involved in bipolar disorder, a rough road ahead. Clin. Pract. Epidemiol. Ment. Health: CP & EMH **14**, 37–45 (2018). https://doi.org/10.2174/1745017901814010037
19. Pinheira, A.: Genetic Analysis of Patients with Bipolar Disorder. Master's thesis, University of Porto, Porto, Portugal (2019)
20. Saylan, C.C., Yilancioglu, K.: Classification of schizophrenia and bipolar disorder by using machine learning algorithms. J. Neurobehavioral Sci. **3**(3), 1 (2016)

21. Shweta, M., Garg, D.K.: Mining efficient association rules through apriori algorithm using attributes and comparative analysis of various association rule algorithms. Int. J. Adv. Res. Comput. Sci. Software Eng. **3**, 306–312 (2013)

22. Spencer, C., Marchini, J., Donnelly, P., Teo, Y.: Chiamo. https://mathgen.stats.ox.ac.uk/genetics_software/chiamo/chiamo.html. Accessed May 2020

23. Stahl, E.A., et al.: Genome-wide association study identifies 30 loci associated with bipolar disorder. Nat. Genet. **51**, 793–803 (2019). https://doi.org/10.1038/s41588-019-0397-8

24. Vieta, E., et al.: Bipolar disorders. Nat. Rev. Dis. Primers **4**, 18008 (2018). https://doi.org/10.1038/nrdp.2018.8

Evaluating Deep Semi-supervised Learning for Whole-Transcriptome Breast Cancer Subtyping

Silvia Cascianelli[✉], Francisco Cristovao, Arif Canakoglu, Mark Carman, Luca Nanni, Pietro Pinoli, and Marco Masseroli

Dipartimento di Elettronica, Informazione e Bioingegneria, Politecnico di Milano, Milan, Italy
{silvia.cascianelli,francisco.cristovao,arif.canakoglu,mark.carman, luca.nanni,pietro.pinoli,marco.masseroli}@polimi.it

Abstract. We investigate the important clinical problem of predicting prognosis-related breast cancer molecular subtypes using whole-transcriptome information present in The Cancer Genome Atlas Project (TCGA) dataset. From a Machine Learning perspective, the data is both high-dimensional with over nineteen thousand features, and extremely small with only about one thousand labeled instances in total. To deal with the dearth of information we compare classical, deep and semi-supervised learning approaches on the subtyping task. Specifically, we compare a L_1-regularized Logistic Regression, a 2-hidden layer Feed Forward Neural Network and a Variational Autoencoder based semi-supervised learner that makes use of pan-cancer TCGA data as well as normal breast tissue data from a second source. We find that the classical supervised technique performs at least as well as the deep and semi-supervised learning approaches, although learning curve analysis suggests that insufficient unlabeled data may be being provided for the chosen semi-supervised learning technique to be effective.

Keywords: Deep learning · Breast cancer · Gene expression · Semi-supervised learning · Variational Autoencoder

1 Scientific Background

Over the last two decades, an accurate classification into prognostically relevant molecular subtypes has been recognized as crucial for a deeper understanding of breast cancer (BRCA) heterogeneity, improving patient outcome prediction, developing tailored treatments and supporting therapeutic decision making [3,12]. A significant body of evidence has confirmed the prognostic meaning and predictive ability of the intrinsic molecular subtypes: *Luminal A, Luminal B, Her2-enriched, Basal and Normal-like*, which were discovered in the early 2000s

S. Cascianelli and F. Cristovao—Co-first authors.

through unsupervised hierarchical clustering on BRCA microarray gene expression profiles [12] and confirmed by several other studies [5,11]. To date, the subtypes are commonly identified using the PAM50 method [10], which implements the Prediction Analysis for Microarrays (PAM) classification algorithm and examines specifically the differential expression of a signature of 50 genes. Yet, many other genes could play relevant roles in defining discriminant patterns of gene expression across intrinsic subtypes. Consequently, genome-wide analysis of RNA-Sequencing (RNA-Seq) data could yield a substantial contribution by taking advantage of larger gene expression spaces. Recently, the number of publicly available BRCA samples profiled with RNA-Seq has dramatically increased; although the PAM50 technique has been extensively adopted to categorize microarray and PCR-based gene expression data, only lately it has been applied to some RNA-Seq BRCA datasets [2]. Thus, only a small portion of available BRCA RNA-Seq gene expression data is labeled with intrinsic subtypes and hence usable for supervised learning. Furthermore, two main issues can affect classifier performance in learning intrinsic subtypes from these available RNA-Seq data: 1) the number of instances usable for training is always much smaller than the huge amount of genes in the feature space; and 2) the limits and uncertainties of the PAM50 method are inherited to some extent by any supervised method trained with PAM50 labeled data.

In such a complex scenario, we implemented two baseline supervised methods to perform RNA-Seq BRCA sample classification into intrinsic subtypes: an L_1-regularized Logistic Regression and Fully Connected Feed Forward Neural Networks, for which we examined several architectures. Furthermore, we considered semi-supervised learning techniques, both to leverage on available unlabeled RNA-Seq samples and to evaluate the possible gain from enlisting deep learning methods to tackle the BRCA intrinsic subtyping task (indeed, deep techniques have recently been adopted to improve accuracy in domain-related problems [4,8,13]). Particularly, we focused on Variational Autoencoders, since they can learn better continuous well-structured latent spaces by mixing deep learning with Bayesian inference. Hence, in this study we investigated the performance of this innovative deep approach compared with baseline methods, exploring different architectures, hyper-parameters and regularization techniques. Furthermore, for each approach we evaluated to what extent its accuracy is influenced by the dimension of the available labeled training samples, as to better assess the role and contribution of the semi-supervised learning methods.

2   Materials and Methods

2.1   Datasets

We used RNA-Seq data from the TCGA [9] and ARCHS4 [7] public datasets. The TCGA data comprises 11,284 RNA-Seq Version2 profiles of cancer samples from 33 cancer types, all present as raw counts. The ARCHS4 data was downloaded from the ARCHS4 project website and comprises 84,863 human samples as gene raw counts coming from HiSeq 2000, HiSeq 2500 and NextSeq 500 platforms;

however, only the ARCHS4 breast tissue samples were here considered, for a total of 4,123 RNA-seq samples. Notably, we used the expression values of the 19,036 genes that were common to both datasets.

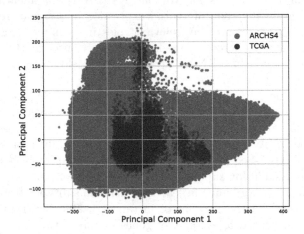

Fig. 1. First two components of Principal Component Analysis of the two datasets used in this study. The two datasets present a strong overlap, ruling out the presence of batch effects.

For all of the TCGA BRCA samples, subtype labels were traced on cBioPortal[1] and come from PAM50 classifications performed by the TCGA consortium at different stages, leading to a total of 1,053 labeled samples (546 *Luminal A*, 208 *Luminal B*, 179 *Basal*, 81 *Her2-enriched* and 39 *Normal-like*). The TCGA non-BRCA samples and the ARCHS4 breast tissue samples were used as unlabeled data for the pre-training of the semi-supervised models. In order to use the data from both sources, we computed the *reads per million* (RPM) of each gene g_i in each sample s_j as: $RPM = \frac{\#reads\ mapped\ to\ g_i}{total\ reads\ for\ sample\ s_j} * 10^6$, and then applied log2 and min-max normalization to make the training faster and reduce the chances of the algorithm getting stuck in local optima.

After normalization, we performed Principal Component Analysis (PCA) on the combined data (TCGA + ARCHS4), as to check their compatibility. Considering the overlap between the first two components, shown in Fig. 1, we assumed that no highly significant batch effects affect these two datasets. Small differences would probably be outweighed by the advantage of leveraging on much more data; therefore, for some of the assessed semi-supervised contexts, we used samples coming from the two different sources within the same experiment.

2.2 Supervised Learning

We first explored traditional supervised learning approaches, which by definition resort to labeled data in order to learn the model parameters. Particularly, a

---

[1] https://www.cbioportal.org/.

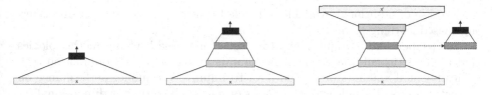

Fig. 2. L_1-regularized Logistic Regression (left); Fully-connected Feed Forward Neural Network (centre); Variational Autoencoder with additional Softmax layer (right).

subset of 817 TCGA BRCA samples labeled by Ciriello et al. in [2] was used and is henceforth indicated as the training set of all the supervised settings; conversely, the remaining 236 labeled TCGA BRCA samples were adopted as a test set. The following paragraphs describe the two supervised methods used.

L_1-regularized Logistic Regression: To deal with the very large number of features (19,036 genes) with respect to a very small number of training samples (817), we made use of multi-class Logistic Regression (LR) with a sparsity inducing L_1 penalty to prevent overfitting[2] (Fig. 2 on the left). The cost function to be optimised contains a regularization term with a shrinkage coefficient λ:

$$\mathcal{L}(\theta) = - \left[\sum_{i=1}^{n} \sum_{k=1}^{K} 1\left\{y^{(i)} = k\right\} \log \frac{\exp(\theta^{(k)\top} x^{(i)})}{\sum_{j=1}^{K} \exp(\theta^{(j)\top} x^{(i)})} \right] + \lambda \sum_{j=1}^{K} ||\theta^{(j)}||_1$$

where n and K denote the number of instances and classes (subtypes), $\theta^{(k)}$ is the vector of model parameters for class k, and $y^{(i)}$ & $x^{(i)}$ denote the class and feature vectors of the i-th sample, respectively.

Feed-Forward Neural Network: Techniques such as dropout, batch normalization and extensive use of validation data can help to control overfitting in over-parameterized Feed-Forward Neural Network (FFNN) models. We adopt them with various fully-connected FFNN structures. Each network has two-hidden layers (of varying width) and uses ReLu activation functions for hidden layers and Softmax for output neurons (Fig. 2, centre). The FFNN were implemented in Keras[3], and trained to minimize Categorical Cross Entropy loss:

$$\mathcal{L}(\theta) = - \sum_{i=1}^{n} \sum_{k=1}^{K} 1\left\{y^{(i)} = k\right\} \log \hat{p}(k|x^{(i)}; \theta)$$

where $\hat{p}(k|x^{(i)}; \theta)$ is the predicted probability for class k on instance i.

2.3 Semi-supervised Learning

Another way to mitigate the issue of having very few labeled gene expression instances to work with, compared to the huge feature space of profiled genes, is

---

[2] Implemented using Python's scikit-learn package, https://scikit-learn.org.
[3] https://www.tensorflow.org/guide/keras/.

making use of additional available unlabeled gene expression data, using *semi-supervised* learning techniques to perform BRCA subtyping.

Semi-supervised learning refers to a set of machine learning methodologies that leverage (usually) large quantities of unlabeled samples in conjunction with typically small amounts of labeled data to improve the model performance on predictive tasks. These learning methods often use clustering or dimensionality-reduction techniques to model the feature space, such that the class label information from the small number of labeled training instances can be more easily generalized to unlabeled parts of the sample space. The usefulness of semi-supervised learning is based on the assumption of *continuity* [1] among the data under investigation, a hypothesis of all supervised learning algorithms, which requires that data points lying nearby in the feature space tend to have the same label. This requirement is hence inherited also in semi-supervised context, involving both labeled and unlabeled data, with the extra underlying assumption that the unlabeled data is drawn from a similar distribution to the labeled data.

Variational Autoencoder: A popular deep learning method for modeling unlabeled data is the Variational Autoencoder (VAE) [6]. Autoencoders are neural networks trained to perform dimensionality reduction. The network is structured to map high-dimensional input data down into a low-dimensional representation and then back out to the original dimension. Weights are learned such that the reconstructed data is as close to the original input data as possible. *Variational Autoencoders* include an additional stochastic sampling step over the low-dimensional representation, before generating the output. This sampling process provides superior regularization and interpretability of the latent representation.

The motivation for using this type of architecture arose as a direct consequence of the high-dimensionality of the feature space; we aimed indeed at extracting a meaningful latent representation of our RNA-seq expression data, that could be helpful for the breast cancer subtyping task. Thus, we made use of *Variational Autoencoder* for semi-supervised learning as follows. We first trained the autoencoder in an unsupervised manner to minimize reconstruction error in terms of binary cross-entropy (experiments minimizing Mean Square Error were performed as well, but binary cross-entropy reached better prediction performances). We then added a Softmax layer to the low-dimensional representation, obtained as the concatenation of the mean and variance vectors (we also investigated sampling the hidden representation, but no performance improvement was observed) and trained the weights of the Softmax to maximise prediction performance on the labeled BRCA data (see the architecture on the right in Fig. 2). While learning the Softmax weights, we also fine-tuned the encoder component of the autoencoder on the supervised task since that was observed to improve performance markedly over keeping the encoder weights fixed[4].

---

[4] We leave investigating combined classification + autoencoder loss to future work.

3   Experiments

In the following subsections we discuss the experiments performed during this study and the collected results.

Table 1. Hyperparameter values of the best deep models

Architecture	Unlabeled data	Batch size	Learning rate	Dropout (Input)	Dropout (Hidden)
FFNN (19k → 300 → 100 → 5)	–	50	0.001	0.8	0.4
FFNN (19k → 100 → 20 → 5)	–	200	0.01	0.6	0.5
VAE + Softmax (19k → 300 → 100 → 5)	TCGA	50	0.001	0.4	0.4
VAE + Softmax (19k → 100 → 20 → 5)	TCGA	200	0.01	0.6	0.8
VAE + Softmax (19k → 300 → 100 → 5)	ARCHS4 Breast	50	0.001	0.6	0.4
VAE + Softmax (19k → 100 → 20 → 5)	ARCHS4 Breast	200	0.01	0.8	0.6

3.1   Experimental Settings

Models: In our tests we considered a total of five different models, namely: a simply L_1-regularized logistic regressor; a Feed Forward Neural Network with two hidden layers in two configurations, the first with 300 and 100 neurons and the second with 100 and 20 neurons; two Variational Autoencoders with two hidden layers in two different configuration, the first with 300 and 100 neurons and the second with 100 and 20 neurons. In all the mentioned classical and deep architectures, a grid search was done for each hyperparameter of interest, (i.e. regularization parameter, number of epochs, learning rate and dropout rate). Specifically for the regularization parameter of the Logistic Regression, different values ($\lambda = 10^i, i \in \{-3, -2, \ldots, 4\}$) were tested, whereas, in the case of deep models, different values for learning rate (0.0005, 0.001, 0.01), number of epochs (25, 50, 75, 100) and dropout rates (0, 0.2, 0.4, 0.6, 0.8) of input and hidden layers, were taken into consideration.

Evaluation Metrics: To evaluate the models on the validation and test data, we used the true positive (TP), true negative (TN), false positive (FP) and false negative (FN) counts for each class to compute the $Accuracy = \frac{TP+TN}{TP+FP+TN+FN}$, $Precision = \frac{TP}{TP+FP}$, $Recall = \frac{TP}{TP+FN}$ and $Specificity = \frac{TN}{TN+FP}$ for each subtype and then aggregated them in a weighted fashion across the five subtypes.

Cross-validation and Hyperparameter Tuning: All considered models were evaluated using stratified 5-fold cross-validation[5] to preserve the same percentages of subtypes of the whole training set, reporting as aggregated performance score the mean accuracy across folds. From grid-search and cross-validation the best

---

[5] We used the StratifiedKFold method from scikit-learn Python library.

performing L_1-regularized LR has $\lambda = 10$, while the best hyperparameter settings emerged for the deep models are summarized in Table 1.

Validation on the Held-Out Dataset: In addition to the cross-validation, we used the best trained models to predict the labels of the held-out dataset composed of the 236 independently labeled samples of TCGA. For this test we also reported the confusion matrices of the predictions (Table 4).

3.2   Experimental Results

Model Selection: Table 2 contains the cross-validation results for each considered model, using the corresponding best found hyperparameter setting. We note that the simpler L_1-regularized Logistic Regression matched or outperformed all of the other models by a small margin, for any metric present in the table. However, the performance for all other models lie within one standard deviation of this. Such a result can be explained by the small number of training samples available, making the more complex models less capable of extracting non-linear features that would allow them to outperform simpler models. Still, the FFNN with the hidden configuration (300, 100) performed almost in line with the L_1-regularized LR, matching the same Recall, albeit with a larger standard deviation. Also the VAE with the hidden configuration (300, 100) and trained with the ARCHS4 breast tissue subset behaved well, not falling much behind the L_1-regularized LR on all the three investigated metrics.

Table 2. Comparison of prediction performance on validation data for different models, architectures and datasets, using 5-fold cross-validation. Numbers in parenthesis denote the dimension of the layers in the neural network architectures.

Architecture	Unlabeled data	Labeled data	Accuracy	Precision	Recall
Logistic regression + L_1 (19k → 5)	–	TCGA BRCA	**0.885 ± 0.017**	**0.886 ± 0.042**	**0.879 ± 0.015**
FFNN (19k → 300→ 100 →5)	–	TCGA BRCA	0.879 ± 0.025	0.877 ± 0.028	0.879 ± 0.025
FFNN (19k → 100 → 20 → 5)	–	TCGA BRCA	0.859 ± 0.033	0.866 ± 0.036	0.878 ± 0.036
VAE + Softmax (19k → 300 → 100 → 5)	TCGA	TCGA BRCA	0.869 ± 0.023	0.863 ± 0.028	0.869 ± 0.023
VAE + Softmax (19k → 100 → 20 → 5)	TCGA	TCGA BRCA	0.858 ± 0.037	0.861 ± 0.040	0.858 ± 0.037
VAE + Softmax (19k → 300 → 100 → 5)	ARCHS4 Breast	TCGA BRCA	0.873 ± 0.019	0.876 ± 0.036	0.873 ± 0.020
VAE + Softmax (19k → 100 → 20 → 5)	ARCHS4 Breast	TCGA BRCA	0.873 ± 0.021	0.868 ± 0.030	0.873 ± 0.021

Sensitivity to Quantity of Training Data: In Fig. 3, we provide learning curves on the RNA-Seq validation data for the best four classifiers under comparison: an L_1-regularized Logistic Regression, a Feed-Forward Neural Network, and two Variational Autoencoder trained alternatively on the ARCHS4 breast tissue subset and on the pan-cancer TCGA dataset. The curves show the effect on accuracy of reducing the amount of labeled training data available to the algorithm, while keeping the same proportion of BRCA subtypes. We note that performance in all cases increases with the amount of labeled training data. More interesting, however, is that when in the presence of low quantities of training data the *relative* performance of the semi-supervised methods (particularly using the pan-cancer TCGA data) improves with respect to logistic regression.

A possible explanation for this effect is that for small amounts of labeled data, the information extracted and learned by the network during the unsupervised pre-training from the *relatively* larger amount of unlabeled data is able to provide a stronger regularization effect across the whole-transcriptome, which can be exploited by the (weaker) subsequent classifier. If the case, this would suggest that in the presence of a higher ratio between labeled and unlabeled training samples the semi-supervised techniques would become a more effective method to tackle the problem, hinting that larger quantities of unlabeled data may be needed to provide performance benefits from these techniques. However, this current lack of data should not be seen as an obstacle difficult to overcome and enough to disprove the deep semi-supervised methods; at the rate at which the available amount of this type of data is currently increasing, conditions to possibly validate this statement will be met in the near future [14].

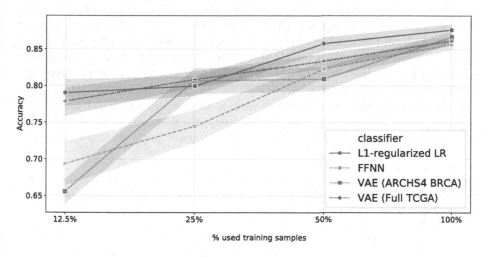

Fig. 3. Learning curves for the evaluated classifiers (the best performing ones) on the validation data.

Logistic Regression Feature Weights Analysis: Being the simplest model under study, the L_1-regularized Logistic Regression has the advantage of providing an easy interpretation of the learned model and its features (see Availability). This type of analysis is interesting since it allows us to bridge the gap between the two major fields of this work (Biology and Computational Intelligence), and so to give biological meaning to the mathematics behind the models. Table 3 provides information regarding the number of features (genes) with non-zero weights of the Logistic Regression model trained using RNA-seq data (the best performing one), together with the number of samples belonging to each class. One of the main takeaways is that, even for the class having the biggest value of non-zero weights (Luminal B), these features only account for around 5% of the initial set of 19,036 genes. Besides confirming the feature selection capabilities of the L_1-regularized Logistic Regression model, these results show the enormous amount of redundancy present in the information carried by most genes, and the irrelevance that most of them has for the task under study. These conclusions are in accordance with the work done to develop the PAM50 assay [10], currently a reference test for BRCA subtyping. Indeed, Parker et al. [10] analysed this problem starting from a space of about thousand genes and found a small gene panel, including the genes that carried more information for BRCA subtyping.

Hence, it becomes logical to compare the identified gene set with the genes belonging to the original PAM50 panel. The existence of this overlap confirms the presence of a biological meaning behind what the model has learned, and eases its interpretation. The last column of Table 3 includes all the PAM50 genes with non-zero weights for each class. Notably, the genes in bold are those that were identified as important for that particular subtype in the original PAM50 study [10]. Some of the PAM50 genes were not present in any of the classes[6] and may have been replaced by other genes providing similar information. We note that next-generation sequencing (NGS) data allows for investigating a wider set of genes. Indeed, the number of genes with non-zero coefficients for each class is around 10 times larger than the PAM50 panel. The use of whole-transcriptome subtyping approaches could provide new perspectives for biological investigations, aimed at assessing the relevance of many more genes than those hitherto considered significant. We expect that in the near future larger amounts of NGS data with long-term survival annotations will become available and useful for validating prognostically the role of all the genes involved and emerged as

Table 3. Number of training samples and of non-zero regression coefficients (LR model) for each subtype. PAM50 panel genes present with non-zero coefficients are also shown, with in **bold** those considered relevant (in PAM50) for that subtype.

Class	Samples	Coefficients	PAM50 panel genes present
Basal	136	501	GRB7, ERBB2, BAG1, BLVRA, **FOXC1**, MYC, CCNE1, NUF2, KIF2C, **MELK**, UBE2C, RRM2, BCL2, MAPT, MLPH
Her2-enriched	65	390	**SFRP1**, **GRB7**, FOXC1, EGFR, PHGDH, **CCNE1**, NUF2, **CDC20**, MYBL2, ORC6, SLC39A6, BCL2, ESR1, MAPT
Luminal A	415	716	SFRP1, KRT17, TYMS, NDC80, UBE2T, MELK, **CENPF**, ANLN, **UBE2C**, RRM2, CEP55, **NAT1**, BCL2, ESR1, GPR160
Luminal B	176	973	KRT14, KRT5, GRB7, ACTR3B, BLVRA, **CXXC5**, FOXC1, **EGFR**, **CDH3**, **PHGDH**, CCNE1, CDC20, TYMS, NDC80, UBE2T, CCNB1, EXO1, UBE2C, RRM2, NAT1, **BCL2**, GPR160, MLPH, FOXA1
Normal-like	371	25	**KRT5**, GRB7, **TYMS**, CENPF, ANLN, FOXA1

---

[6] Specifically: BIRC5, CDC6, FGFR4, MDM2, MIA, MKI67, MMP11, PGR, PTTG1, TMEM45B.

subtype-related; survival analysis and predictive models could thus be used to identify high or low risk patients and differentiate their treatments accordingly[7].

Performance on Held-Out Test Data: Having selected the most effective methods and architectures on the validation data, we evaluated them on the test set (Table 4), which was a held-out subset of the TCGA BRCA data, consisting of 236 samples. The higher accuracy results on the test data can be justified by minor differences on the distribution of the classes between the two subsets, marginally favoring the presence of classes that are easier to classify on the test set. However, since they were labeled at different stages using slightly different pipelines, we decided not to re-balance them. Other than avoiding the introduction of noise in the training data, this characteristic made both sets suitable to be used as training and test sets as it is.

Table 4. Comparison of prediction performance on test data for the most effective methods and architectures chosen in the validation phase.

Architecture	Unlabeled data	Labeled data	Accuracy	Average precision	Average accuracy	Average precision
Logistic regression + L_1 ($19k \rightarrow 5$)	–	TCGA BRCA	0.936	0.955	0.936	0.964
FFNN ($19k \rightarrow 300 \rightarrow 100 \rightarrow 5$)	–	TCGA BRCA	0.907	0.917	0.907	0.953
VAE + Softmax ($19k \rightarrow 300 \rightarrow 100 \rightarrow 5$)	TCGA	TCGA BRCA	0.911	0.920	0.911	0.950
VAE + Softmax ($19k \rightarrow 300 \rightarrow 100 \rightarrow 5$)	ARCHS4 Breast	TCGA BRCA	0.903	0.916	0.903	0.930

Concerning the gathered results, they are generally concordant with the ones obtained using 5-fold cross-validation, with the L_1-regularized Logistic Regression being the best performing across all of the explored models. For what concerns the semi-supervised techniques used, note that the most suitable model for classification is the one where the embedding is learned using non-BRCA TCGA. This may have happened because test data came from TCGA as well, avoiding possible issues caused by the data collection procedure that could hurt the results. Also, even if from breast tissue, the ARCHS4 samples are not cancer samples, which motivates the existence of relevant differences between the two.

Confusion Matrices: In Table 5, the confusion matrices for the models evaluated over the test set are presented. It can be verified that the *Basal* samples are the easiest to classify, having themselves the more characteristic molecular traits usually causing disease aggressiveness and poor prognosis. It is also shown that the wrongly classified *Luminal A* and *Luminal B* samples occur mostly among each other, with *Luminal A* samples being more frequently predicted as *Luminal B*. A likely reason for this to happen is the similarity that exists

---

[7] Here, clinical assessment was not performed due to insufficient annotated data.

Table 5. Confusion Matrices on the test data: LR with L_1-regularization (top left), FFNN (top right), VAE + Softmax trained with ARCHS4 breast (bottom left) and VAE + Softmax trained with TCGA (bottom right). Ba, H2, LA, LB and NL correspond to *Basal, Her2-enriched, Luminal A, Luminal B* and *Normal-like*, respectively.

LR + L_1 predicted labels

		Ba	H2	LA	LB	NL
	Ba	43	0	0	0	0
Actual	H2	0	16	0	0	0
labels	LA	0	1	126	4	0
	LB	0	0	2	30	0
	NL	0	3	4	1	6

FFNN predicted labels

		Ba	H2	LA	LB	NL
	Ba	43	0	0	0	0
Actual	H2	0	14	1	1	0
labels	LA	0	0	119	11	1
	LB	0	0	1	31	0
	NL	0	1	5	1	7

VAE (ARCHS4) + Softmax predicted labels

		Ba	H2	LA	LB	NL
	Ba	41	0	0	1	1
Actual	H2	0	13	2	1	0
labels	LA	0	0	123	8	0
	LB	0	0	3	29	0
	NL	0	0	7	0	7

VAE (TCGA) + Softmax predicted labels

		Ba	H2	LA	LB	NL
	Ba	42	0	0	1	0
Actual	H2	0	14	1	1	0
labels	LA	0	3	121	7	0
	LB	0	0	1	31	0
	NL	0	1	6	0	7

between these two classes from being both Luminal tumors, although *Luminal A* cases usually have a better prognosis. Finally, we can see that the *Normal-like* class is the one with the largest percentage of misclassified samples. It probably happens due to the very small number of training instances available, together with the fact that there are known similarities with the *Luminal A* class, having the biggest number of training samples. This unbalance greatly contributes to the difficulty in identifying the samples belonging to the *Normal-like* class, that should include mainly samples from "normal-like" tissues surrounding the primary breast tumor. Hence, besides being satisfactory, the results obtained in this analysis are overall in good accordance with the literature on the subject [3].

4   Conclusion

Multi-gene prognostic molecular tests play a crucial role in the study of breast cancer, enabling an accurate prognosis estimation that is of great use to help the physicians in providing tailored treatment for each individual. However, these tests are not able to leverage on the advantages that high-throughput sequencing technologies have brought to the area. By doing so, they would become much more reproducible and affordable, and consequently widespread.

In this work we explored some computational approaches to close the gap between current BRCA stratification methods and use of high-throughput next-generation sequencing data: we investigated the breast cancer subtype prediction task from the whole-transcriptome information present in TCGA and by comparing the performance of an L_1-regularized Logistic Regression, 2-hidden layer Feed

Forward Neural Networks and Variational Autoencoder based semi-supervised learners that also make use of pan-cancer TCGA data or breast tissue data from ARCHS4. With the Variational Autoencoder models, we aimed to learn a low-dimensional feature space in which we could project our original inputs, with the hypothesis of this latent space being more suitable for the classification task, as it tackles the issues due to the high dimensionality of genes in the used data.

We found the L_1-regularized Logistic Regression to perform at least as well as the deep and semi-supervised learning techniques, although learning curve analysis suggests that the latter may have been provided with insufficient unlabeled data to be effective and possibly overcome the results reached with simpler supervised model. Furthermore, the obtained results appear in good accordance with the biological literature on the subject. Therefore, the use of whole-transcriptome subtyping approaches could provide useful insights for additional biological investigations, aimed at increasing the awareness about the heterogeneity of breast cancer by assessing the role of many more genes than those hitherto considered significant for BRCA stratification.

Availability. The source code and the most relevant genes selected by means of the implemented analysis are available at: https://github.com/DEIB-GECO/brca_subtype.

Acknowledgments. Results shown here are in part based on data generated by the TCGA Research Network (https://www.cancer.gov/tcga/). This research is funded by the ERC Advanced Grant project 693174 GeCo (Data-Driven Genomic Computing), 2016–2021.

References

1. Bengio, Y., Courville, A., Vincent, P.: Representation learning: a review and new perspectives. IEEE Trans. Pattern Anal. Mach. Intell. **35**(8), 1798–1828 (2013)
2. Ciriello, G., Gaza, M., Beck, A., Wilkerson, M., Rhie, S., Pastore, A.: Comprehensive molecular portraits of invasive lobular breast cancer. Cell **163**(2), 506–519 (2015)
3. Dai, X., et al.: Breast cancer intrinsic subtype classification, clinical use and future trends. Am. J. Cancer Res. **5**(10), 2929–2943 (2015)
4. Gao, F., et al.: DeepCC: a novel deep learning-based framework for cancer molecular subtype classification. Oncogenesis **8**(9), 1–12 (2019)
5. Hu, Z., et al.: The molecular portraits of breast tumors are conserved across microarray platforms. BMC Genomics **7**(1), 96 (2006)
6. Kingma, D.P., Welling, M.: Auto-encoding variational Bayes. In: International Conference on Learning Representations, pp. 1–14 (2014)
7. Lachmann, A., et al.: Massive mining of publicly available RNA-seq data from human and mouse. Nature Commun. **9**(1), 506–519 (2018)
8. Motlagh, N.H., et al.: Breast cancer histopathological image classification: a deep learning approach. bioRxiv p. 242818 (2018)
9. NCI: National Cancer Institute. The Cancer Genome Atlas Program. https://www.cancer.gov/about-nci/organization/ccg/research/structural-genomics/tcga

10. Parker, J.S., et al.: Supervised risk predictor of breast cancer based on intrinsic subtypes. J. Clin. Oncol. **27**(8), 1160–1167 (2009)
11. Perou, C.M., et al.: Molecular portraits of human breast tumours. Nature **406**(6797), 747–752 (2000)
12. Sørlie, T., et al.: Gene expression patterns of breast carcinomas distinguish tumor subclasses with clinical implications. Proc. Natl. Acad. Sci. U.S.A **98**(19), 10869–10874 (2001)
13. Spanhol, F.A., Oliveira, L.S., Petitjean, C., Heutte, L.: Breast cancer histopathological image classification using convolutional neural networks. In: International Joint Conference on Neural Networks (IJCNN), pp. 2560–2567. IEEE (2016)
14. Stephens, Z.D., et al.: Big data: astronomical or genomical? Plos Biol. **13**(7), 1–11 (2015)

Learning Weighted Association Rules in Human Phenotype Ontology

Giuseppe Agapito, Mario Cannataro, Pietro H. Guzzi$^{(\boxtimes)}$ (iD),
and Marianna Milano

Magna Græcia University, 88100 Catanzaro, Italy
{agapito,cannataro,hguzzi.m.milano}@unicz.it

Abstract. Human Phenotype Ontology (HPO) provides information about medically relevant phenotypes and the association of disease and phenotype concepts to HPO terms through annotations. The specificity of each HPO terms is estimated by its Information Content (IC), which assess the specificity of a term. An important research area focuses on the analysis of annotated data to extract knowledge. Association Rules (AR) can be used to discover relevant associations from annotated data. Classical AR methods consider all annotation equally, do not take into account that the HPO terms have different Information Content, i.e., different relevance. This implies the generation of association rules with low IC. This paper presents HPO-Miner (Human Phenotype Ontology-based weighted association rules), a methodology for extracting Weighted Association Rules from the HPO Ontology considering the IC of terms. To assess our methods, we tested HPO-Miner on publicly available HPO annotation datasets. The results demonstrate that our method outperforms the current state of the art approaches. HPO-Miner is publicly available at https://github.com/hguzzi/HpoMiner.

Keywords: Human Phenotype Ontology · Gene Ontology · Weighted Association Rules · Annotation

1 Introduction

In computer science, biological knowledge is structured by using formal instruments such as controlled vocabularies and ontologies that offer a formal framework for data modelling and representation. In literature, there exist different ontologies that formalise different domains of knowledge. For instance, Gene Ontology points to provide a universal language to describe the biological aspects of genes and genes products [14]. Human Phenotype Ontology (HPO) [35] and Disease Ontology (DO) [37] are two novel ontologies focus on the description of relation among molecular biology and disease. In particular, HPO is a framework of phenotypic abnormalities encountered in human diseases. Human disease is encoded by using a unified identifier known as Online Mendelian Inheritance in Man (OMIM) [25]. DO is a repository describing the classification of human

© Springer Nature Switzerland AG 2020
P. Cazzaniga et al. (Eds.): CIBB 2019, LNBI 12313, pp. 245–256, 2020.
https://doi.org/10.1007/978-3-030-63061-4_22

diseases organized by etiology. The ontology terms are linked to biological and medical concepts by a process known as an annotation.

The annotation process is not a trivial task. The advancement of the knowledge about biological and medical phenomena implies a constant updating of annotation corpus and ontology terms [10]. The ontology update is regularly performed by expert curators who handle each change carefully to ensure the correct introduction of information. However, this process is slower compared to the rate of biological discoveries. In this contest, a challenge has regarded the development of computational methods to support manual curators and to the update and maintenance processes of ontologies to improve the speed and accuracy. In the past, different computational approaches have been developed to support the curators in the improvement of the GO annotations consistency [19] .

In contrast, the literature contains few computational methodologies to aid HPO curators. Recently, Faria et al. [19], Manda et al. [29], and Agapito et al. [2–5] have applied Association Rules (AR) to analyze the annotations, showing that they can be used to improve annotations consistency and uncover relationships among terms that do not seem explicitly related.

Starting from this consideration, we present HPO-Miner, an improved version of our previous work, GO-WAR [2]. HPO presents a single type of relationships, e.g., *is-a relationships*, whereas in GO are available several types of relationships, i.e., *is a*; *part of*; *has part*; *regulates, negatively*, and *positively regulates*. For that reason, it has been necessary to implement a different methodology in HPO-Miner. HPO-Miner is able to deal with this limited relationship data and it computes the weight of the HPO terms. Then, HPO-Miner uses that weighted terms to produce the weighted dataset, from which to mine weighted association rules (WAR). In [4,5,11] the effectiveness of WAR to extract relevant relationships between cross ontology terms is reported. In [3], a parallel AR mining methodology ables to deal with a huge amount of microarray data is described.

HPO-Miner tool learns WAR to inspect annotation consistency and to detect hidden relationships between two phenotype abnormalities from HPO. Classical association rule approaches are not able to distinguish between items; they are careless of the relevance of terms producing rules with low specificity. The specificity of each term is estimated by its information content (IC) [24,26].

The IC formulation of a term falls into two major classes, intrinsic and extrinsic. Intrinsic methods rely on the topology of the ontology graph, analyzing the positions of terms in a taxonomy (see [26] for a complete review). Instead, the extrinsic formulations involve the annotation data for a considered corpus.

The application of IC calculated for each HPO term provided the IC-weighted annotations as presented in the following: *OMIM100050: (HP:0001380, Joint laxity, 10.95), (HP:0009466, Radial deviation of finger, 9.85)*. HPO-Miner ensures the extraction of weighted association rules starting from an annotated dataset of diseases. The HPO-Miner methodology is based on two main steps: (i) Data preprocessing and transformation, necessary to remove corrupted data as well as, to convert OMIM term in transactional data. (ii) Weighted association rules mining, using a modified *FP-Tree* able to deal with the dimension of classical biological datasets. We use publicly available HPO annotation data to demonstrate our method.

The rest of the paper is organized as follows. Section 2 introduces the AR, describes the HPO ontology, as well as summarizes some of the available intrinsic IC calculation methods. Section 3 describes the HPO-Miner algorithm, and Sect. 4 describes the obtained results, as well as it discusses some of the mined weighted association rules. Finally, Sect. 5 concludes the paper.

2   Materials and Methods

2.1   The Human Phenotype Ontology

HPO is a structured and controlled vocabulary, available at the website: https://hpo.jax.org/app/, that includes more than 13,000 terms and more than 15,000 annotations describing the phenotypic abnormalities in human diseases. HPO comprises five independent sub-ontologies: *mode of inheritance, onset and clinical course, phenotypic abnormalities, clinical modifier,* and *frequency*.

HPO is structured as a direct acyclic graph (DAG). The relations among DAG's terms are modelled by means of *is_a* edges "relations", in order to distinguish between general or precise terms. A HPO class presents an individual identifier (e.g. *HP:0002604*), a label and a list of synonyms, describing a phenotypic abnormality i.e. *"Neoplasm"* as depicted in Fig. 1.

Fig. 1. HPO graph Example. The term *"Neoplasm"* is the root and the terms *"Neoplasm by anatomical site"* and *"Neoplasm by histology"* represent the leaves. Starting from the root node to the leaves nodes, the information content of the terms increases, from a general term to more specific ones.

Each HPO term is linked to diseases listed in the OMIM (Online Mendelian Inheritance in Man) database [25]. Diseases are annotated with terms of the HPO, meaning that HPO terms describe all the signs, symptoms, and other phenotypic manifestations that features the disease in question. The annotations of OMIM entries are both manual annotations, obtained by the HPO curators, and automated, obtained by matching the OMIM Clinical Synopsis to HPO term labels. Since HPO uses multiple terms to annotate a disease, the need for the definition of methodologies and tools to support HPO curators to improve annotation consistency and the structure of the ontology arises.

2.2   Association Rules

AR mining is a common task in data mining; it is used to discover hidden associations in market basket analysis and unknown relations among features in databases [6]. The association rules extraction problem may be stated as follows: let $I = \{i_1, i_2, \ldots, i_n\}$ be a set of items, and $D = \{t_1, \ldots, t_m\}$ a transactional database that contains a set of transactions, where a transaction t_j is a subset of items belonging to I. An association rule is a relationship of the form $A \rightarrow B$, where A and B are two disjoint sets. ARs may be ranked on the basis of two fundamental properties to define the relevance of the mined rules, *Support* and *Confidence*. Formally, the Support definition is:

$$S(A \rightarrow B) = \frac{\sigma(A \cup B)}{N} \tag{1}$$

Where N is the total number of transactions contained in D, and σ is called *support count*, i.e., the number of transactions that contain a particular item. The Confidence is defined as:

$$C(A \rightarrow B) = \frac{\sigma(A \cup B)}{\sigma(A)} \tag{2}$$

Where $\sigma(A)$ is the number of transactions in D containing A, and $\sigma(A \cup B)$ is the number of transactions in D that contains both items A and B.

2.3   Weighting HPO Term with Information Content

In literature, there exist different intrinsic methods proposed by Sanchez et al. [36], Harispe et al. [26], Resnick et al. [34], Seco et al. [27], Zhou et al. [40], with which to weight ontological terms.

The measure of Sanchez takes into account the number of leaves and the set of ancestors (subsumers) of a, including itself, *subsumers(a)*, and defines the root node as the number of leaves *max_leaves* in the IC assessment.

$$IC_{Sanchez\ et\ al.}(a) = -log\left(\frac{\frac{|leaves(a)|}{|subsumers(a)|} + 1}{max\_leaves + 1}\right) \tag{3}$$

Harispe et al., defines the specificity of leaves on the basis of the number of ancestors, considering *leaves(a)* = a concept when a is a root and evaluating *max_leaves* as the number of ancestors of a node.

$$IC_{Harispe\ et\ al.}(a) = -log\left(\frac{\frac{|leaves(a)|}{|subsumers(a)|}}{max\_leaves}\right) \tag{4}$$

Resnick et al. points out all the top-downs path from a concept a to the reachable leaves, $p(a)$, and then computes the log yielding the following formula:

$$IC_{Resnik}(a) = -log(p(a)). \tag{5}$$

Seco et al. considers the rate between the number of hyponyms in ontology, i.e. the number of descendant with respect to the whole number of ontological concepts.

$$IC_{Seco\ et\ al}(a) = \frac{log\left(\frac{hypo(a)+1}{max\_nodes}\right)}{log\left(\frac{1}{max\_nodes}\right)} \qquad (6)$$

The formulation provided from Zhou et al. employs the depth of a term in a taxonomy, $depth(a)$, the maximum depth of the taxonomy max_depth and k as factor that weights the contribution of the two evaluated features.

$$IC_{Zhou\ et\ al.}(a) = k * -\left(1 - \frac{log(hypo(a)+1)}{log(max\_nodes)}\right) + (1-k) * \left(\frac{log(depth(a))}{log(depth\_nodes)}\right) \quad (7)$$

3   The HPO-Miner Algorithm

In this Section, we briefly describe the *HPO-Miner* algorithm. HPO-Miner is able to extract weighted association rules form the HPO dataset.

WAR models the *importance* of a term by means of a *weight* (ω). A weight (ω) is a non-negative real number that reflects the relevance of an HPO term, for which high values represent essential items as reported in [22,23,39]. In our case, the relevance can be expressed by using the information content (IC).

HPO-Miner's core algorithm is developed, exploiting the WAR's concepts. In particular, we defined the $WeightedSupport(\omega S)$ of a generic item x_i as $\omega S(x_i) = w_i * \sigma(x_i)$ where ω_i is the information content of the i-th term and $\sigma(x_i)$ is the number of transaction containing x_i. Let $I = \{i_1 \dots i_m\}$ be a set of weighted items (HPO terms) and let WD be the weighted transactions database. We defined the *weighted minimum support* ($m\omega S$) as:

$$m\omega S = \left(\frac{\sum_{i=1}^{|WD|} \sigma(x_i) * \omega_i}{|WD|}\right) * p \qquad (8)$$

$|WD|$ is the number of transactions into the dataset, p is a threshold value given in input by the user, to define the relevance of items. Only the items for which the following constraint holds $\omega S(I) \geq m\omega S$, are significant and can be used as candidates to generate frequent item-sets and rules.

Algorithm 1 presents the main steps of *HPO-Miner* algorithm. The first step of the HPO-Miner algorithm concerns with the load and conversion of the input HPO dataset in a *weightedTable WT*, i.e., a data structure suitable to represent weighted transaction data (Algorithm 1, row 2). The WT is used to obtain the list of frequent weighted items (Algorithm 1, row 3). All the items into the frequent item list, for which is not verified, the following condition: $\omega S(I) \geq m\omega S$, are removed. Frequent weighted items are hence used to build a data structure based on $FP-Tree$. Finally, *HPO-Miner* iteratively analyzes the $FP-Tree$ to mine significant rules.

Algorithm 1. HPO Weighted Association Rules Miner (HPO-Miner)

Require: A table of HPO annotation as input dataset D
 1: *Data Structure initialization: WT, $\mathcal{FW}ItemsList$, **FPTree***
 2: $WT \leftarrow getTransactionalData(D)$
 3: $\mathcal{FW}ItemsList \leftarrow$ retrieve$\mathcal{FW}ItemsList(WT)$
 4: $FPTree$.create($\mathcal{FW}ItemsList$)
 5: $mineWeightedRules()$
 6: **end.**

---

4   Results

In this section, we present and discuss the meaning of the most significant rules extracted by using *HPO-Miner* that are: rules which have a high value of weight-edSupport and Confidence among all those extracted.

Five real weighted HPO datasets were produced to asses the performance of HPO-Miner. The five datasets have been obtained computing the IC for all the HPO terms, by using the five intrinsic methods presented in Sect. 2. The experiments were performed employing each produced datasets, for various combination of weightedSupport and Confidence values. Then we selected the values for the parameters able to ensure the best results in terms of reduced number of mined rules and at the same time with relevant values of weightedSupport and Confidence. The best combination of values was obtained for weightedSupport equal to 50% and Confidence greater than 80%.

We selected the first top 5 rules from each dataset, and we manually analyzed them in the literature to find claims that can prove the validity of the mined rules.

HPO-Miner can assist medical doctors in improving clinical findings that are not specific enough to allow a more accurate diagnosis. The doctors can use HPO-Miner to find a collection of relationships (rules) between HPO terms more specific for the individual diagnoses. For instance, Rule (R1) in Table 1 shows the relation between the terms *Giant cell hepatitis* and *Autosomal recessive inheritance* relationships not available explicitly in HPO. Thus, these relationship can contribute to help medical doctors to refine a diagnosis regarding the liver disorder, with a more detailed and accurate diagnosis.

4.1   Analysis of Mined Rules

In this Section, we evaluate and discuss the meaning and the biological relevance of the 5 top rules (due to space limitation), obtained analysing the five weighted HPO datasets generated using the different information content formulations presented above. The rules have been mined using HPO-Miner, and are ranked by their weightedSupport and Confidence values.

We begin to discuss the rules contained in Table 1 mined by HPO-Miner from the Resnik dataset. Rule 1R, *Giant cell hepatitis, Autosomal recessive inheritance*. As stated in [16,32] both terms could be related to defects in the biological mechanisms of the liver. Rule 2R, *Giant cell hepatitis, Elevated hepatic*

Table 1. The first five rules mined by HPO-Miner using Resnik Dataset with related Weighted Support (WS) and Confidence (C). (Rules are identified using IDs).

	Term 1	Term 2	WS	C	Function	Function
1R	HP:0200084	HP:0000007	1.00	1.00	Giant cell hepatitis	Autosomal recessive inheritance
2R	HP:0200084	HP:0002910	1.00	1.00	Giant cell hepatitis	Elevated hepatic transaminases
3R	HP:0200067	HP:0000006	1.00	1.00	Recurrent spontaneous abortion	Autosomal dominant inheritance
4R	HP:0100818	HP:0000774	1.00	1.00	Long thorax	Narrow chest
5R	HP:0100775	HP:0001537	1.00	1.00	Dural ectasia	Umbilical hernia

Table 2. The first five rules mined by HPO-Miner using Sanches Dataset with related Weighted Support (WS) and Confidence (C). (Rules are identified using IDs).

	Term 1	Term 2	WS	C	Function	Function
1S	HP:0100818	HP:0000774	0.88	1.00	Long thorax	Narrow chest
2S	HP:0030034	HP:0003774	0.88	1.00	Diffuse glomerular basement membrane lamellation	Stage 5 chronic kidney disease
3S	HP:0012743	HP:0001773	0.88	1.00	Abdominal obesity	Short foot
4S	HP:0012263	HP:0000007	0.88	1.00	Immotile cilia	Autosomal recessive inheritance
5S	HP:0012023	HP:0000007	0.88	1.00	Galactosuria	Autosomal recessive inheritance

Table 3. The first five rules mined by HPO-Miner using Harispe Dataset with related Weighted Support (WS) and Confidence (C). (Rules are identified using IDs).

	Term 1	Term 2	WS	C	Function	Function
1H	HP:0009577	HP:0004220	1.00	1.00	Short middle phalanx of the 2nd finger	Short middle phalanx of the 5th finger
2H	HP:0010105	HP:0010034	1.00	1.00	Short first metatarsal	Short 1st metacarpal
3H	HP:0000933	HP:0001305	1.00	1.00	Posterior fossa cyst at the fourth ventricle	Dandy-Walker malformation
4H	HP:0004704	HP:0004689	1.00	1.00	Short fifth metatarsal	Short fourth metatarsal
5H	HP:0001885	HP:0004209	1.00	0.99	Short 2nd toe	Clinodactyly of the 5th finger

Table 4. The first five rules mined by HPO-Miner using Seco Dataset with related Weighted Support (WS) and Confidence (C). (Rules are identified using IDs).

	Term 1	Term 2	WS	C	Function	Function
1Se	HP:0200084	HP:0000007	1.00	1.00	Giant cell hepatitis	Autosomal recessive inheritance
2Se	HP:0100818	HP:0000774	1.00	1.00	Long thorax	Narrow chest
3Se	HP:0100775	HP:0001537	1.00	1.00	Dural ectasia	humbilicalhernia
4Se	HP:0100775	HP:0000494	1.00	1.00	Dural ectasia	Downslanted palpebral fissures
5Se	HP:0100775	HP:0000316	1.00	1.00	Dural ectasia	Hypertelorism

Table 5. The first five rules mined by HPO-Miner using Zhou Dataset with related Weighted Support (WS) and Confidence (C). (Rules are identified using IDs).

	Term 1	Term 2	WS	C	Function	Function
1Z	HP:0002335	HP:0001305	0.97	1	Congenital absence of the vermis of cerebellum	Dandy Walker malformation
2Z	HP:0003031	HP:0002986	0.95	1	Bending of the diaphysis (shaft) of the ulna (Uknar bowing)	A bending or abnormal curvature of the radius (Radial bowing)
3Z	HP:0000176	HP:0000193	0.95	0.97	submucous clefts Hard-palate	Bifid uvula
4Z	HP:0001338	HP:0002007	0.95	0.94	Partial agenesis of the corpus callosum	Frontal Bossing
5Z	HP:0001338	HP:0000494	0.95	0.94	Partial agenesis of the corpus callosum	Downslated palpebral fissures

transaminases) consists of two terms involved in the hepatitis process. In [9] is highlighted that patients develop hepatic fibrosis or cirrhosis due to the presence of *Giant cell hepatitis*. Rule 3R, *Recurrent spontaneous abortion* and *Autosomal dominant inheritance*. In [15] the authors point out that the rare familial disorders are usually inherited as *Autosomal dominant inheritance*. Rule 4R, *Long thorax, Narrow chest* involved in the syndrome of Jeune and Ellis-Van Creveld syndrome as reported in literature in [7,17]. Browsing HPO Ontology with its on line browser did not reveal any information that allows the user to associate both abnormalities with the syndrome of Juene and Ellis-Van Creveld. Rule 5R, *Dural ectasia, Umbilical hernia* at first glance seems that there not exists a connection among the two terms. Analyzing the literature we found the work of Mizuguchi et al. [30] and Chen et al. [12].

Table 2 reports the rules mined by HPO-Miner from the Sanchez dataset. Rule 1S, *Long thorax, Narrow chest*) consists of two terms involved in the Asphyxiating Thoracic Dysplasia (Jeune Syndrome). In [18] is reported the role of both phenotypes in Jeune syndrome. Rule 2S, *Diffuse glomerular basement membrane lamellation, Stage 5 chronic kidney disease*. In [31] the glomerular basement membrane lamellation as a manifestation in patients after transplantation of kidneys, is described. Rule 3S, *Abdominal obesity, Short foot* we didn't find a correlation among *Abdominal obesity*(term 1) and *Short foot* (term 2) despite a depth research in literature has been conducted. Rule 4S, and Rule 5S, associate two pathologic phenotypes, *Immotile cilia* and (*Galactosuria* to *Autosomal recessive inheritance*). In fact, in [1] is reported that the immotile cilia syndrome seems due to an autosomal recessive disease; as well as galactosuria due to galactokinase deficiency in a newborn is inherited in an autosomal recessive manner [33].

Here we interpret the rule mined by HPO-Miner from the dataset Harispe and contained in Table 3. Rule 1H, *Short middle phalanx of the 2nd finger, Short middle phalanx of the 5th finger*. These abnormalities have been observed in the Adams-Oliver Syndrome, as reported in [28]. Rule 2H, contains the *Short first metatarsal, Short 1st metacarpal.* Although we conducted an in-depth analysis of stare of art, this rule is not confirmed in the literature. Rule 3H, *Posterior fossa cyst at the fourth ventricle Dandy-Walker malformation* involved in abnormality that affects brain development. Although we conducted an in-depth analysis of stare of art, this rule is not confirmed in the literature. Rule 4H, *Short fifth metatarsal, Short fourth metatarsalRule.* Analyzing the literature has not been possible to confirm this rule. Rule 5H, *Short 2nd toe, Clinodactyly of the 5th finger.* Searching the literature we found that both symptoms occurred in Carpenter Syndromeas states in the work of Gershoni et al. [21].

Table 4 contains the rules mined by HPO-Miner from the Seco dataset. Rule 1Se, *Giant cell hepatitis*and *Autosomal recessive inheritance.* The relation between two phenotypes in [13] is reported. Rule 2Se, *Long thorax, Narrow chest* is discussed above. Rule 3Se, *Dural ectasia, humbilical hernia* is not confirmed in literature. Rule 4Se, and the Rule 5Se, associate the phenotype*Dural ectasia* with*Downslanted palpebral fissures* and *Hypertelorism.* Carrying out a analysis in the state of art, we found a clinical case which report a patient with lateral meningocele syndrome (LMS) affected by both down slanting palpebral fissures and hyperteloris [12].

Finally, we discuss the rules contained in Table 5. Rule 1Z, *Congenital absence of the vermis of cerebellum* with *Dandy Walker malformation.* The link between the two phenotypes is confirmed in [8]. Regarding, rule 2Z, *Bending of the diaphysis (shaft) of the ulna (Ulnar bowing) A bending or abnormal curvature of the radius (Radial bowing)* and the Rule 3Z *submucous clefts Hard-palate Bifid uvula,* although we conducted an in-depth analysis of stare of art, these rules are not confirmed in literature. Rule 4Z, and Rule 5Z, associate the *Partial agenesis of the corpus callosum* to two abnormal phenotype: *Frontal Bossing* and *Downslated palpebral fissures* as confirmed in [38] and [20].

5   Conclusion

We presented a new methodology implemented in a software tool called HPO-Miner to mine weighted association rule from the HPO datasets. Weighted association rules are obtained, taking into account the relevance of terms its Information Content. The use of the information content as a weight allows to HPO-Miner to mine rare frequent itemsets that are items with low frequency but with high biological relevance, e.g., very informative items with a high value of IC as well as sufficient value of weighted support. These itemsets are used in turn to generate association rules that have high weighted support. The evidence proves the biological relevance of the mined rules by HPO-Miner found analyzing the literature. Finally, we are planning to make HPO-Miner available as a RESTFul or Cloud service, making it possible to use it through a standard web browser.

Acknowledgments. This work has been partially funded by the following research project funded by the Calabrian Region: "Smart Electronic Invoices Accounting-SELINA CUP:*J28C*1700016006".

References

1. Afzelius, B.A., Srurgess, J.: The immotile-cilia syndrome: a microtubule-associated defect. CRC Crit. Rev. Biochem. **19**(1), 63–87 (1985)
2. Agapito, G., Cannataro, M., Guzzi, P.H., Milano, M.: Using GO-WAR for mining cross-ontology weighted association rules. Comput. Methods Programs Biomed. **120**(2), 113–122 (2015)
3. Agapito, G., Guzzi, P.H., Cannataro, M.: Parallel and distributed association rule mining in life science: A novel parallel algorithm to mine genomics data. Inf. Sci. (2018)
4. Agapito, G., Milano, M., Guzzi, P.H., Cannataro, M.: Improving annotation quality in gene ontology by mining cross-ontology weighted association rules. In: IEEE International Conference on Bioinformatics and Biomedicine (BIBM), pp. 1–8. IEEE (2014)
5. Agapito, G., Milano, M., Guzzi, P.H., Cannataro, M.: Extracting cross-ontology weighted association rules from gene ontology annotations. IEEE/ACM Trans. Comput. Biol. Bioinform. **13**(2), 197–208 (2015)
6. Agrawal, R., Imieli, T., Swami, A.: Mining association rules between sets of items in large databases. SIGMOD Rec. **22**(2), 207–216, 1993. https://doi.org/10.1145/170036.170072
7. Baujat, G., Le Merrer, M.: Ellis-Van Creveld syndrome. Orphanet J. Rare Dis. **2**(6), 27 (2007)
8. Bordarier, C., Aicardi, J.: Dandy-Walker syndrome and agenesis of the cerebellar vermis: diagnostic problems and genetic counselling. Develop. Med. Child Neurol. **32**(4), 285–294 (1990)
9. Byrne, W.J., Kase, B.F., Bjorkhem, I., Haga, P., Pedersen, J.I.: Defective peroxisomal cleavage of the c27 steroid side chain in the cerebro. J. Pediatr. Gastroenterol. Nutr. **4**(4), 685 (1985)
10. Cannataro, M., Guzzi, P.H., Milano, M.: God: an R-package based on ontologies for prioritization of genes with respect to diseases. J. Comput. Sci. **9**, 7–13 (2015)
11. Cannataro, M., Guzzi, P.H., Veltri, P.: Impreco: distributed prediction of protein complexes. Future Gener. Comput. Syst. **26**(3), 434–440 (2010)
12. Chen, K.M., Bird, L., Barnes, P., Barth, R., Hudgins, L.: Lateral meningocele syndrome: vertical transmission and expansion of the phenotype. Am. J. Med. Genet. Part A **133**(2), 115–121 (2005)
13. Clayton, P., et al.: Familial giant cell hepatitis associated with synthesis of 3 beta, 7 alpha-dihydroxy-and 3 beta, 7 alpha, 12 alpha-trihydroxy-5-cholenoic acids. J. Clin. Invest. **79**(4), 1031 (1987)
14. Consortium, G.O., et al.: The gene ontology (GO) database and informatics resource. Nucleic Acids Res. **32**(Supp 1), D258–D261 (2004)
15. Coumans, A., et al.: Haemostatic and metabolic abnormalities in women with unexplained recurrent abortion. Human Reprod. **14**(1), 211–214 (1999)
16. Danks, D.M., Campbell, P.E., Jack, I., Rogers, J., Smith, A.L.: Studies of the aetiology of neonatal hepatitis and biliary atresia. Archiv. Dis. Childhood **52**(5), 360–367 (1977). https://doi.org/10.1136/adc.52.5.360
17. Elejalde, B.R., De Elejalde, M.M., Pansch, D., Opitz, J.M., Reynolds, J.F.: Prenatal diagnosis of Jeune syndrome. Am. J. Med. Genet. **21**(3), 433–438 (1985). https://doi.org/10.1002/ajmg.1320210304
18. Elejalde, B.R., De Elejalde, M.M., Pansch, D., Opitz, J.M., Reynolds, J.F.: Prenatal diagnosis of Jeune syndrome. Am. J. Med. Genet. **21**(3), 433–438 (1985)

19. Faria, D., et al.: Mining go annotations for improving annotation consistency. PLoS One **7**(7), e40519 (2012)
20. Gelman-Kohan, Z., Antonelli, J., Ankori-Cohen, H., Adar, H., Chemke, J.: Further delineation of the acrocallosal syndrome. Eur. J. Pediatr. **150**(11), 797–799 (1991)
21. Gershoni-Baruch, R.: Carpenter syndrome: marked variability of expression to include the Summitt and Goodman syndromes. Am. J. Med. Genet. **35**(2), 236–240 (1990)
22. Guzzi, P.H., Cannataro, M.: μ-cs: an extension of the TM4 platform to manage Affymetrix binary data. BMC Bioinform. **11**(1), 315 (2010)
23. Guzzi, P.H., Agapito, G., Cannataro, M.: coreSNP: parallel processing of microarray data. IEEE Trans. Comput. **63**(12), 2961–2974 (2013)
24. Guzzi, P.H., Milano, M., Cannataro, M.: Mining association rules from gene ontology and protein networks: promises and challenges. Procedia Comput. Sci. **29**, 1970–1980 (2014). 2014 International Conference on Computational Science. https://doi.org/10.1016/j.procs.2014.05.181. http://www.sciencedirect.com/science/article/pii/S1877050914003585.
25. Hamosh, A., Scott, A.F., Amberger, J.S., Bocchini, C.A., McKusick, V.A.: Online Mendelian inheritance in man (OMIM), a knowledgebase of human genes and genetic disorders. Nucleic Acids Res. **33**(suppl-1), D514–D517 (2005)
26. Harispe, S., Sanchez, D., Ranwez, S., Janaqi, S., Montmain, J.: A framework for unifying ontology-based semantic similarity measures: a study in the biomedical domain. J. Biomed. Inf. **48**, 38–53 (2013)
27. Hermjakob, H., et al.: The Hupo psi's molecular interaction format - a community standard for the representation of protein interaction data. Nat. Biotechnol. **22**, 177–183 (2004). https://doi.org/10.1038/nbt926
28. Kuster, W., Lenz, W., Kaariainen, H., Majewski, F., Opitz, J.M., Reynolds, J.F.: Congenital scalp defects with distal limb anomalies (Adams-Oliver syndrome): report of ten cases and review of the literature. Am. J. Med. Genet. **31**(1), 99–115 (1988)
29. Manda, P., McCarthy, F., Bridges, S.M.: Interestingness measures and strategies for mining multi-ontology multi-level association rules from gene ontology annotations for the discovery of new go relationships. J. Biomed. Inf. **46**(5), 849–856 (2013)
30. Mizuguchi, T., et al.: Heterozygous TGFBR2 mutations in Marfan syndrome. Nat. Genet. **36**(8), 855–860 (2004)
31. Nadasdy, T., Abdi, R., Pitha, J., Slakey, D., Racusen, L.: Diffuse glomerular basement membrane Lamellation in renal allografts from pediatric donors to adult recipients. Am. J. Surg. Pathol. **23**(4), 437–442 (1999)
32. Nassa, G., et al.: Comparative analysis of nuclear estrogen receptor alpha and beta interactomes in breast cancer cells. Mol. BioSyst. **7**(3), 667–676 (2011)
33. Pickering, W.R., Howell, R.R.: Galactokinase deficiency: clinical and biochemical findings in a new kindred. J. Pediatr. **81**(1), 50–55 (1972)
34. Resnik, P.: Using information content to evaluate semantic similarity in a taxonomy. In: IJCAI, pp. 448–453 (1995)
35. Robinson, P.N., Kohler, S., Bauer, S., Seelow, D., Horn, D., Mundlos, S.: The human phenotype ontology: a tool for annotating and analyzing human hereditary disease. Am. J. Hum. Genet. **83**(5), 610–615 (2008)
36. Sanchez, D., Batet, M., Isern, D.: Ontology-based information content computation. Knowl.-Based Syst. **24**(2), 297–303 (2011)
37. Schriml, L.M., et al.: Disease ontology: a backbone for disease semantic integration. Nucl. Acids Res. **40**(D1), D940–D946 (2012)

38. Taylor, W.B., Anderson, D.E., Howell, J., Thurston, C.S.: The nevoid basal cell carcinoma syndrome: autopsy findings. Arch. Dermatol. **98**(6), 612–614 (1968)
39. Wang, W., Yang, J., Yu, P.S.: Efficient mining of weighted association rules (WAR). In: Proceedings of the Sixth ACM SIGKDD International Conference on Knowledge Discovery and Data Mining. ACM, New York. https://doi.org/10. 1145/347090.347149
40. Zhou, Z., Wang, Y., Gu, J.: A new model of information content for semantic similarity in wordnet. In: Second International Conference on Future Generation Communication and Networking Symposia, FGCNS 2008, vol. 3, pp. 85–89. IEEE (2008)

Network Modeling and Analysis of Normal and Cancer Gene Expression Data

Gaia Ceddia$^{(\boxtimes)}$ (ID), Sara Pidò (ID), and Marco Masseroli (ID)

Politecnico di Milano - Dipartimento di Elettronica, Informazione e Bioingegneria,
Piazza Leonardo da Vinci 32, Milan, Italy
{Gaia.Ceddia,Sara.Pido,Marco.Masseroli}@polimi.it

Abstract. Network modelling is an important approach to understand cell behaviour. It has proven its effectiveness in understanding biological processes and finding novel biomarkers for severe diseases. In this study, using gene expression data and complex network techniques, we propose a computational framework for inferring relationships between RNA molecules. We focus on gene expression data of kidney renal clear cell carcinoma (KIRC) from the TCGA project, and we build RNA relationship networks for either normal or cancer condition using three different similarity measures (Pearson's correlation, Euclidean distance and inverse Covariance matrix). We analyze the networks individually and in comparison to each other, highlighting their differences. The analysis identified known cancer genes/miRNAs and other RNAs with interesting features in the networks, which may play an important role in kidney renal clear cell carcinoma.

Keywords: Gene networks · microRNA · Gene expression profiles · Complex networks · Similarity networks · Co-expression

1 Scientific Background

Network biology covers a wide range of scales, from molecular interactions in the cell to intercellular communications and connections between organisms. At the cell level, high-throughput next-generation sequencing technology is generating an enormous amount of genomic data from which qualitative and quantitative relationships between RNA molecules can be inferred [3]. In particular, gene expression data provide information about the synthesis of functional gene products, either proteins or not. Using mathematical and statistical techniques, from gene expression data we can generate biological networks, where genes are the network nodes and interactions between gene products are the edges in the network graph [3]. This process, named network inference or reverse engineering, has given important insights on complex biological processes and disease mechanisms within the cell [30]. Network inference has the advantage of being efficient

© Springer Nature Switzerland AG 2020
P. Cazzaniga et al. (Eds.): CIBB 2019, LNBI 12313, pp. 257–270, 2020.
https://doi.org/10.1007/978-3-030-63061-4_23

and inexpensive compared to experimental lab validation; thus, complex network techniques and algorithms have been increasingly deployed to understand inferred biological networks [3].

A complex network is a graph with non-trivial topological features [25], i.e., the patterns of connection between its elements are neither purely regular nor purely random. All biological processes can be modeled as networks, since they occur thanks to interactions among molecules. In biology, the most studied complex networks are gene networks, where typically genes encode for proteins; their interrelated activity determines protein abundance and related processes [25].

Most of the approaches used for inferring edges in gene networks are based on similarity (co-expression) measures. Co-expression measurement is based on the "guilt by association" definition, where genes with similar expression profiles are functionally associated due to their presumable co-regulation [30]. Thus, several different measures have been considered to assess co-expression, including Pearson's correlation and Euclidean distance. Pearson's correlation is the most common co-expression measure in the literature [30]. It has the benefit of being scalable, i.e., it can be efficiently computed for large numbers of genes, and it is not sensitive to linear transformations or different normalizations. However, its limitation lies on the fact that causality and direction of the gene interactions are ignored in the computation [22]. Zhang et al. [32] performed a *Weighted Gene Co-expression Network Analysis (WGCNA)* providing interesting communities of genes; nonetheless they carry several false positives. Some methods tried to handle the over-connectivity of co-expression networks by comparing the network structures among cancer types [5]. Other methods for the construction of gene networks include Bayesian network approaches, as well as regression and differential equation based models [3]. Bayesian networks are applied to represent conditional dependencies between genes given their expression levels, using a directed acyclic graph structure [3]. However, this procedure is applicable only to small networks, i.e., only a modest number of genes must be involved. Instead, regression and differential equation models are used for inferring gene regulatory networks, i.e., they assume that a particular subset of gene expression profiles is the most informing subset of all to predict expression profiles of target genes [3].

Here, we consider three different similarity measures for the construction of gene co-expression networks and we innovatively deal with the over-connectivity of similarity gene networks by using three statistical thresholding steps. In particular, we focus on co-expression networks built by computing Pearson's correlation, Euclidean distance and inverse Covariance metrics. The first similarity measure is calculated to capture the scale-free similarity of gene expression profiles, the second one to take into account the scale of different gene expression profiles, and the third one as a multivariate analysis representing conditional independence between variables. Using expression data from the TGCA project [31], we build two different gene co-expression networks for normal or cancer cells, respectively; normal and cancer gene networks are computed for each similarity measure, and comparison analyses are performed among them. To our knowledge, this study is a novel approach for comparing different similarity co-expression networks using human datasets; other attempts were done on *S. cerevisiae* and *S. pombe* organisms [6]. In addition, we integrate long RNA and miRNA expression data as done in Pian et al. [18], although we innovatively take advantage of three similarity measures to compare the overall differences of the gene networks in normal and cancer data by using the strength analysis. The novel use of strength comparisons lead us to find some relevant miRNAs by clearly displaying their dysregulation between normal and cancer Euclidean distance and the Pearson's correlation networks.

For the considered datasets, we integrate messenger RNA (mRNA), microRNA (miRNA) and long non-coding RNA (lncRNA) expression profiles, and we computed the co-expression networks among them; thus, our study is not limited to protein coding RNAs. MicroRNAs are small non-coding RNA molecules containing between 19 and 25 nucleotides, which work for RNA silencing and post-transcriptional regulation of gene expression [4]. The predominant function of miRNAs is to regulate protein translation by binding to complementary sequences in the 3' untranslated region (UTR) of target messenger RNAs, and thereby to negatively regulate mRNA translation [4]. A single miRNA can target hundreds of mRNAs, using base-pairing with complementary sequences within mRNA, and influence the expression of many genes often involved in a functional interaction pathway. However, miRNAs can also target lncRNAs, which are made of more than 200 nucleotides and are not translated into proteins. In this case, lncRNAs act as decoys for miRNAs silencing, allowing the translation of target mRNAs [17].

By focusing on whole gene co-expression networks in normal and cancer conditions we decide not to only select differentially expressed (DE) genes. DE genes are the ones showing statistically significant changes in read counts, or expression levels, between two experimental conditions. However, not significant DE genes, or genes with small changes in their expression levels, may play an important role due to the interaction of their products with other proteins and gene products; thus, our method is purely based on network comparison without any prior biological assumption on DE genes.

2   Materials and Methods

In this section, we explain our extraction and pre-processing pipeline for TCGA gene expression data and how we build pair networks for normal and cancer conditions, respectively, using three different similarity measures for each condition, resulting in a total of six networks. The whole process is represented in Fig. 1.

2.1   Data Extraction and Pre-processing

We consider both RNA-Seq and miRNA-Seq public data for the human GRCh38 assembly from the TCGA repository. GRCh38 miRNA-Seq data contains miRNA quantification (i.e., the calculated expression for all reads aligning to a particular miRNA) and is derived from the sequencing of microRNAs, whereas GRCh38 RNA-Seq data contains all gene expression quantification. For each

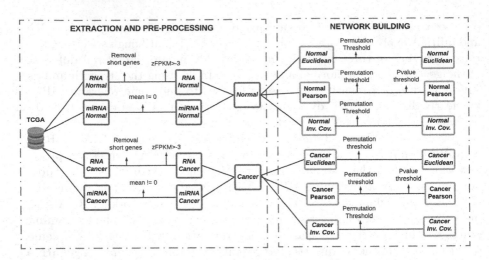

Fig. 1. Defined workflow that starts with the extraction and the pre-processing of TCGA data in order to build the gene co-expression networks.

miRNA-Seq and RNA-Seq dataset of each tumor type in TCGA, we compute the number of normal and cancer condition samples from patients. In our datasets, each patient corresponds to one sample, thus, in this study, the term patient and sample have the same meaning. Figure 2 shows the number of normal samples

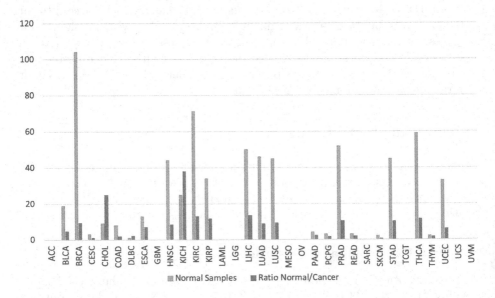

Fig. 2. Number of normal samples (blue bars) and ratio between the number of normal and cancer samples (red bars) for all tumor types in TCGA data. (Color figure online)

and the ratio between the number of normal and cancer samples for all tumor types in the TCGA repository. KIRC results as one of the tumor types with the highest ratio and number of normal samples, providing balanced normal and cancer datasets. Thus, we choose KIRC because it has the highest number of normal samples after BRCA, but BRCA has low ratio of normal/cancer samples. As shown in Table 1, KIRC RNA-Seq dataset resulted to have 72 and 534 samples for normal and cancer conditions, respectively, and KIRC miRNA-Seq dataset 71 and 541 samples for normal and cancer conditions, respectively. Thus, we use these KIRC data for our analysis. After the selection and extraction of KIRC RNA-Seq and miRNA-Seq datasets performed with GMQL [15], we have 60,483 RNAs and 1,881 miRNAs for each sample, as shown in Table 2.

Table 1. Number of samples in TCGA KIRC data.

	Normal	Cancer	Total
RNA-Seq	72	534	606
miRNA-Seq	71	541	612
Common samples	**71**	**487**	**558**

Table 2. Number of RNA molecules in each sample during the filtering steps.

	RNAs	miRNAs	Total
Extraction with GMQL	60,483	1,881	62,364
Removal of RNAs of short genes	27,144	1,881	29,025
Removal of RNAs with zero mean	26,706	1,397	28,103
Removal of RNAs with zFPKM< -3	**12,792**	**1,397**	**14,189**

Since RNA-Seq is designed for long gene sequencing, expression quantifications of short genes (i.e., shorter than 200 bp) can be considered as measure errors indeed. Thus, we remove them from the RNA-Seq dataset, and we select only data of protein coding and long non-coding genes (as reported in the second row of Table 2), which we integrate with the miRNA-Seq dataset ones, considering only common samples (as reported in the last row of Table 1).

We arrange these public gene expression data from the TCGA repository in the form of matrices; we assemble two RNA-Seq and two miRNA-Seq matrices (two for normal and two for cancer data) in which rows represent genes/miRNAs, columns represent samples and each matrix element represents an expression level. TCGA miRNA-Seq expression levels are available as reads per million miRNAs mapped (RPM); conversely, the expression levels in the TCGA RNA-Seq data are provided as fragments per kilobase per million mapped reads (FPKM). To integrate the two miRNA-Seq and RNA-Seq datasets, we transform miRNA expression data to be homogeneous with the RNA expression data; we convert RPM expression levels into FPKM ones by multiplying each element of the miRNA-Seq matrices by 1000 and dividing it by the double of the length of the corresponding miRNA [28].

After selecting the RNA molecules of interest for each dataset, i.e., protein coding, long non-coding genes and miRNAs, we delete miRNAs and RNAs with null expression in all normal and cancer samples, as reported in the third row of Table 2. Furthermore, to separate biologically relevant genes from low-expression noisy ones, on the RNA-Seq data we apply the zFPKM normalization method [8]. For normal and tumoral cases separately, we compute the mean and the standard deviation of the log-transformed expression distribution of each gene across all KIRC samples, and we normalize each logarithmic FPKM value of a gene by subtracting the gene computed mean and dividing the obtained value by the gene standard deviation (i.e., zFPKMs are Z-scores of log(FPKMs)). Then, we remove those genes with mean of their zFPKM distribution smaller than -3.0 in both normal and cancer conditions; this threshold separates expression levels of active genes from background genes as shown in [8].

Thus, we obtain two matrices, one for normal and one for cancer data, each with 12,792 long RNAs (either coding or non-coding) and 1,397 miRNAs, and regarding 71 normal and 487 samples with KIRC tumor, respectively (Table 2). These two matrices contain all the relevant FPKM values needed to build then the desired networks.

2.2 Building the Networks

To build adjacency matrices describing gene networks, we consider three different similarity measures: Euclidean distance, Pearson's correlation and inverse Covariance. As mentioned in Sect. 1, we use these three different similarity measures to find scale-free, scale-dependent and multivariate similarities, respectively.

The Euclidean distance between two points is the length of the path connecting them. If $p = (p_1, p_2, ..., p_n)$ and $q = (q_1, q_2, ..., q_n)$ are two points in an Euclidean n-space, then their distance d is given by the Pythagorean formula [1]:

$$d = \sqrt{\sum_{i=1}^{n}(q_i - p_i)^2} \tag{1}$$

We apply the Euclidean distance on each pair of genes/miRNAs in the datasets, considering the n samples in the datasets as the Euclidean n-dimensional space.

In statistics, the Pearson's correlation coefficient is a measure of the linear correlation between two variables X and Y (Eq. 2) [3]. Its values range between -1 and $+1$, where -1 indicates total negative correlation, 0 no linear correlation, and $+1$ total positive linear correlation. The Pearson's correlation between variable X and Y is defined as:

$$\rho_{X,Y} = \frac{\text{cov}(X, Y)}{\sigma_X \sigma_Y} = \frac{E[(X - \mu_X)(Y - \mu_Y)]}{\sigma_X \sigma_Y} \tag{2}$$

where $cov(X, Y)$ is the covariance of the two variables X and Y, i.e., the joint variability of X and Y, σ_X and σ_Y are the standard deviations of X and Y, respectively, and $cov(X, Y)$ can be expressed as the expected product of X and Y deviations from their individual expected values (i.e., their means μ_X and μ_Y, respectively). In our study we compute pairwise Pearson's correlation on each pair of genes/miRNAs in the datasets, and use the Pearson's coefficients to represent the weights of the edges connecting two nodes (i.e., genes/miRNAs) in the networks.

The inverse Covariance matrix, commonly referred to as *precision matrix*, displays information about the partial correlations of variables [27]. In the Covariance matrix, the (i,j)-th element represents the unconditional correlation between a variable i and a variable j [27]. The inverse Covariance matrix instead represents conditional dependence, such that its (i,j)-th element is equal to zero if i and j are conditionally independent [27]. In other words, it gives the co-variation of two variables while conditioning on the potential influence of the other variables involved in the analysis, i.e., it removes the effect of other variables. Thus, the precision matrix allows obtaining direct co-variation between two variables by capturing partial correlations. If X is the data matrix containing k variables and n observations, the Covariance matrix can be expressed as follows:

$$\mathbf{C} = \frac{1}{\mathbf{n-1}} \sum_{i=1}^{n} (\mathbf{X_i} - \mu)(\mathbf{X_i} - \mu)^\top \tag{3}$$

where $C \in \mathbb{R}^{k \times k}$, μ is the mean value of the variables, and $\top$ represents matrix transposition. In this study we consider genes/miRNAs as variables and samples as observations to compute the inverse of C, i.e., the precision matrix C^{-1}.

We build six different networks, three for the cancer and three for the normal conditions, based on the three similarity measures described. Networks are first built as fully connected graphs for all gene/miRNA pairs, where similarity coefficients are used as weights of the network node associations. Then, we randomize the expression data and compute again the similarity measures to obtain a reference null distribution [3]; we do so by computing the average null distribution on 10 permuted repetitions of the gene/miRNA expression dataset. From the comparison between real and average permuted distributions of each similarity measure, we derive relevant associations in the networks [3]. In other words, we identify the limit values of each permuted distribution and use them as thresholds in the correspondent real distribution. Table 3 shows how the number of edges changes after each filtering step. E.g., Fig. 3(a) shows that the average permuted distribution for the normal Pearson's correlation has values ranging from -0.2 to 0.4; thus, values of the real normal distribution greater than 0.4 and smaller than -0.2 are considered as representing the only relevant associations, and links whose values range from -0.2 to 0.4 are deleted.

Furthermore, since the networks created with Pearson's correlation are very dense, we use the computed p-value of the Pearson's statistic to further threshold them. We sort the computed Pearson's p-values and we only consider the network edges associated with the 99th percentile of the first ten percent of these p-values (i.e., the 0.1% of the edges of the fully-connected network). The third column of Table 3 shows the number of edges after the p-value threshold.

Table 3. Number of edges in the networks. Step 1 represents the creation of networks phase, Step 2 is the filtering phase by the permutation method and Step 3 filters non-relevant edges from Pearson's networks by p-value analysis.

	Step 1	Step 2	Step 3
Normal Pearson	194,140,866	39,392,104	3,959,308
Cancer Pearson	200,789,224	22,728,618	2,249,150
Normal Euclidean	201,247,218	141,846	141,846
Cancer Euclidean	201,313,190	113,492	113,492
Normal Inverse Covariance	194,073,019	14,300	14,300
Cancer Inverse Covariance	200,788,919	14,230	14,230

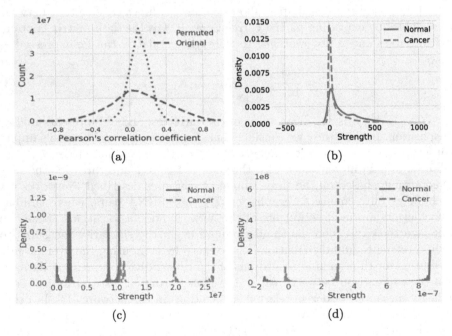

Fig. 3. (a) Red dashed line represents the distribution of Pearson's correlation coefficients for the gene/miRNA expression dataset in normal condition. Dotted green line represents the distribution of the average Pearson's correlation coefficients on 10 permuted repetitions of the gene/miRNA expression dataset in normal condition; (b)–(d) Strength distributions in normal and cancer networks are shown in blue full line and red dashed line, respectively, for the networks built with each of the similarity measures considered, i.e., Pearson's correlation (b), Euclidean distance (c) and inverse Covariance (d), respectively. Density is the proportion of network nodes having certain strengths. (Color figure online)

3   Results

The six constructed networks have same nodes and different edges/weights, depending on the similarity measure used for each network construction (as shown in the fourth row of Table 2 and in the third column of Table 3). We focused our unsupervised analysis on the computation of each node *strength*, i.e., the sum of the total weighted connections of each gene/miRNA, in each of the six networks.

3.1   Pearson's Correlation Networks

Strength distributions of Pearson's correlation networks for normal and cancer condition are shown in Fig. 3(b), where the x-axis represents the strength values and the y-axis is the proportion of network nodes having certain strengths. Interestingly, the proportion of nodes with strength around 0 gets higher in cancer condition (red dashed line), meaning that in cancer many genes/miRNAs have lost, or relevantly lowered, their correlation with other genes/miRNAs. We perform a gene set enrichment analysis on the set of genes whose strength changes from high/low in the normal network to almost 0 in the cancer network (180 genes out of 12,792). We find this gene set significantly enriched for several KEGG pathways related to cancer, particularly for *metabolic pathways*, as shown in Table 4; indeed, KIRC is known as a metabolic disease [12].

Table 4. Results of KEGG gene set enrichment analysis: first column contains the term name of KEGG pathways, second column reports the term ID, and the third column contains the adjusted p-values, which is the correction of p-values performed by [20]

KEGG pathway	ID	Adj. p-value
Metabolic pathways	**01100**	5.697×10^{-27}
Gastric cancer	05226	1.128×10^{-5}
Pathways in cancer	05200	7.181×10^{-5}
Proteoglycans in cancer	05205	3.244×10^{-4}
Transcriptional misregulation in cancer	05202	7.710×10^{-4}
Hepatocellular carcinoma	05225	1.743×10^{-3}

MiRNAs having high/low strength in normal condition and almost 0 strength in cancer are 9 (out of 1,397), including *hsa-mir-192*, *hsa-mir-194-1* and *hsa-mir-194-2*, which are well known miRNAs involved in cancer [26]. Out of the other 6, *hsa-mir-1266* is associated with epithelial tissue diseases, *hsa-mir-210*, *hsa-mir-218-1* and *hsa-mir-218-2* are known to be involved in breast cancer, *hsa-mir-934* is up-regulated in papillary renal cell carcinoma, and *hsa-mir-22* acts as an oncogenic mirna in renal cell carcinoma [7,11,14,16,23].

3.2   Euclidean Networks

Figure 3(c) shows the strength distribution for the nodes of the Euclidean networks, i.e., the networks built using the Euclidean distance as similarity measure between each pair of genes/miRNAs in cancer (red dashed line) or normal (blue full line) condition, respectively. Figure 3(c) shows higher values of strength in cancer compared to the strengths in the normal network, i.e., $[4.0 \times 10^3, 2.65 \times 10^7]$ vs. $[1.5 \times 10^3, 1.0 \times 10^7]$, respectively. The y-axis scale permits the identification of a set of outlier nodes having high values of strength in both normal and cancer conditions, i.e., *hsa-mir-10b*, *hsa-mir-30a*, *hsa-mir-22* and *hsa-mir-143*; these miRNAs maintain high Euclidean distances with all the other genes/miRNAs in the dataset from normal to cancer condition. In the literature *hsa-mir-10b* is known to be associated with *Non-Alcoholic Fatty Liver Disease* and *Bladder Cancer* [26]. Also *hsa-mir-30a* has been studied for its involvement in cancer development, in particular for its potential role as a diagnostic or prognostic marker of gliomas [29]. *Hsa-mir-143* has been associated with *Burkitt Lymphoma* and *Diffuse Large B-Cell Lymphoma* [26]. All three miRNAs are related to *MicroRNAs in cancer* pathway [26]. Moreover, as mentioned in Sect. 3.1, *hsa-mir-22* has been studied for its ability to repress cancer progression in *clear cell renal cell carcinoma* [7]. Instead, *hsa-mir-10a* has one of the highest strength in the normal network and low strength in cancer, with FPKM values overexpressed but not significantly in normal condition compared to cancer, where its regulatory activity could be disrupted. Biologically, *hsa-mir-10a* is associated with several diseases, including *renal cell carcinoma* [2]. Moreover, it is involved in two relevant pathways: *Proteoglycans in cancer* and *MicroRNAs in cancer* [26].

3.3   Inverse Covariance Networks

The inverse Covariance networks show different strength distributions in normal and cancer conditions, as presented in Fig. 3(d). The dependencies between pairs of genes/miRNAs conditioned for all the other genes/miRNAs, here used as edge weights of the inverse Covariance networks, are lower in cancer than in normal network. However, Fig. 3(d) shows that inverse Covariance values in both normal and cancer networks are very close to 0; this means that, even if inverse Covariance coefficients have greater values in normal than in cancer, they do not represent a real dependency between genes/miRNAs in either condition.

3.4   Network Comparison

The strength analysis performed allows us to identify relevant RNAs to be further investigated. For example, hsa-mir-22 has an interesting behaviour in both Pearson's correlation networks and Euclidean distance networks. It has high values of Pearson's correlation coefficients with all the other genes/miRNAs in normal condition, however it does not maintain these high correlations in cancer. It also has one of the highest value of strength in both Euclidean distance networks, i.e., it has very distant FPKM expression values from any other gene/miRNA in the network, both in cancer and normal condition; furthermore, these Euclidean distances get wider in cancer, where hsa-mir-22 doubles its strength compared to the one in the normal network, with its FPKM mean value increasing in cancer (to 396,490 from 332,072 in the normal condition). Gong et al. [7] found that hsa-mir-22 targets directly PTEN in *renal cell carcinoma*; thus, the increase of expression levels in hsa-mir-22 leads to the downregulation of the PTEN protein, indicating an oncogenic effect of the miRNA. However, this miRNA has 15 targets out of 250 significantly related to the KEGG *Pathways in cancer*; thus, its loss of correlations in the Pearson's correlation cancer network may cause a cascade of dysregulation in cancer. These features together make hsa-mir-22 a miRNA of interest for the analysis of gene/miRNA interactions in KIRC.

Another interesting miRNA is hsa-mir-10a; it is one of the outliers with high value of strength in the normal Euclidean distance network, and it has very low strength in the cancer Euclidean distance network. Moreover, its strength values in Pearson's correlation networks are very different from normal to cancer condition (1,128 vs. 380, respectively). Thus, in normal condition this miRNA has FPKM expression values distant from those of the other genes/miRNAs, but highly correlated with them, whereas in cancer they get closer to the ones of the other genes/miRNAs and their correlation to them decreases. Hsa-mir-10a has 290,026 and 140,536 mean FPKM values in normal and cancer condition, respectively; thus, it is over-expressed in normal condition, but not statistically significant. The antitumor role of hsa-mir-10a has been studied in Arai et al. [2] for its interaction with the SKA1 oncogene, explaining the computed downregulation in normal condition. Moreover, 21 targets out of 463 of hsa-mir-10a are significantly involved in the KEGG *Pathways in cancer*; thus, the reported change in correlations between normal and cancer condition may represent abnormal co-regulations of the miRNA-RNAs interaction network.

4 Conclusions

In this study we propose an unsupervised data-driven framework based on complex networks to better represent and understand gene/miRNA relationships and interactions based on gene expression data. We implement a novel pipeline to compute the gene co-expression networks that comprises the pre-processing and the construction phases[1]. Normally, these steps are taken for granted; indeed, it is very difficult to find a complete and efficient workflow.

To this aim, we preprocess the public gene expression data of kidney renal clear cell carcinoma from the TCGA project, and we compute three different similarity measures between genes/miRNAs to get different normal and cancer network representations. Comparative analysis of the six networks obtained lead us to identify two interesting miRNAs: hsa-mir-22 and hsa-mir-10a. They are not differentially expressed; yet, they display important features in both Euclidean and Pearson's correlation networks. According to Euclidean distance

---

[1] https://github.com/DEIB-GECO/GeneNetFusion/blob/master/preprocessing.py.

networks, hsa-mir-22 has highly different expression from other genes/miRNAs in both normal and cancer conditions, and hsa-mir-10a only in normal condition; however, based on Pearson's correlation networks, from normal to cancer condition both miRNAs lose many correlations with other genes/miRNAs, i.e., they co-regulate with a lower number of genes/miRNAs. Interestingly, in miRNet[2] hsa-mir-10a and hsa-mir-22 share an interaction network of 12 genes enriched in a particular KEGG pathway called *Pathways in cancer*. Among them ERBB2, PIK3CG, PTEN and XIAP are known in the literature to have a relevant role in KIRC development [9,13,19,21]. Dysregulated miRNAs play an important role in cancer initiation and progression involving their targets [24]; they have also shown great potential as novel diagnostic/prognostic biomarkers of cancer [10]. Our findings support this assumption and stress the importance of understanding the function of miRNAs as gene suppressors. Future work will further explore the created networks with ad hoc network algorithms, and will deeper investigate the role of miRNAs in the networks.

Acknowledgments. This research is funded by the ERC Advanced Grant project 693174 "GeCo" (Data-Driven Genomic Computing), 2016–2021.

References

1. Anton, H., Rorres, C.: Elementary Linear Algebra. Wiley, Hoboken (1994)
2. Arai, T., et al.: Regulation of spindle and kinetochore-associated protein 1 by antitumor miR-10a-5p in renal cell carcinoma. Cancer Sci. **108**(10), 2088–2101 (2017)
3. Banf, M., Rhee, S.Y.: Computational inference of gene regulatory networks: approaches, limitations and opportunities. Biochim. Biophys. Acta Gene Regul. Mech. **1860**(1), 41–52 (2017)
4. Bartel, D.P.: MicroRNAs: target recognition and regulatory functions. Cell **136**(2), 215–233 (2009)
5. Care, M.A., Westhead, D.R., Tooze, R.M.: Parsimonious gene correlation network analysis (PGCNA): a tool to define modular gene co-expression for refined molecular stratification in cancer. NPJ Syst. Biol. Appl. **5**(1), 1–17 (2019)
6. Deshpande, R., VanderSluis, B., Myers, C.L.: Comparison of profile similarity measures for genetic interaction networks. PloS One **8**(7), e68664 (2013)
7. Gong, X., Zhao, H., Saar, M., Peehl, D.M., Brooks, M., James, D.: miR-22 regulates invasion, gene expression and predicts overall survival in patients with clear cell renal cell carcinoma. Kidney Cancer **3**(2), 119–132 (2019)
8. Hart, T., Komori, H.K., LaMere, S., Podshivalova, K., Salomon, D.R.: Finding the active genes in deep RNA-seq gene expression studies. BMC Genomics **14**(1), 778 (2013). https://doi.org/10.1186/1471-2164-14-778
9. Kedzierska, H., et al.: Decreased expression of SRSF2 splicing factor inhibits apoptotic pathways in renal cancer. Int. J. Mol. Sci. **17**(10), 1598 (2016)
10. Lan, H., Lu, H., Wang, X., Jin, H.: MicroRNAs as potential biomarkers in cancer: opportunities and challenges. Biomed. Res. Int. **2015**, 15–31 (2015)

---

[2] https://www.mirnet.ca/miRNet.

11. Li, Q., Zhu, F., Chen, P.: miR-7 and miR-218 epigenetically control tumor suppressor genes RASSF1A and Claudin-6 by targeting HoxB3 in breast cancer. Biochem. Biophys. Res. Commun. **424**(1), 28–33 (2012)

12. Linehan, W.M., Srinivasan, R., Schmidt, L.S.: The genetic basis of kidney cancer: a metabolic disease. Nat. Rev. Urol. **7**(5), 277–285 (2010)

13. Liontos, M., et al.: Expression and prognostic significance of VEGF and mTOR pathway proteins in metastatic renal cell carcinoma patients: a prognostic immunohistochemical profile for kidney cancer patients. World J. Urol. **35**(3), 411–419 (2016). https://doi.org/10.1007/s00345-016-1890-7

14. Luo, W., et al.: hsa-mir-3199-2 and hsa-mir-1293 as novel prognostic biomarkers of papillary renal cell carcinoma by COX ratio risk regression model screening. J. Cell. Biochem. **118**(10), 3488–3494 (2017)

15. Masseroli, M., et al.: Processing of big heterogeneous genomic datasets for tertiary analysis of next generation sequencing data. Bioinformatics **35**(5), 729–736 (2019)

16. Pasculli, B., et al.: Hsa-miR-210-3p expression in breast cancer and its putative association with worse outcome in patients treated with Docetaxel. Sci. Rep. **9**(14913), 1–9 (2019)

17. Perkel, J.M.: Visiting "noncodarnia" (2013)

18. Pian, C., Zhang, G., Wu, S., Li, F.: Discovering the 'Dark matters' in expression data of miRNA based on the miRNA-mRNA and miRNA-lncRNA networks. BMC Bioinf. **19**(1), 379 (2018). https://doi.org/10.1186/s12859-018-2410-0

19. Que, W.C., Qiu, H.Q., Cheng, Y., Liu, M.B., Wu, C.Y.: PTEN in kidney cancer: a review and meta-analysis. Clin. Chim. Acta **480**, 92–98 (2018)

20. Reimand, J., Kull, M., Peterson, H., Hansen, J., Vilo, J.: g:profiler – a web-based toolset for functional profiling of gene lists from large-scale. Nucleic Acids Res. **35**(suppl_2), W193–W200 (2007)

21. Reuter, S., et al.: Thiocolchicoside exhibits anticancer effects through downregulation of NF-κB pathway and its regulated gene products linked to inflammation and cancer. Cancer Prev. Res. **3**(11), 1462–1472 (2010)

22. Saint-Antoine, M.M., Singh, A.: Network inference in systems biology: recent developments, challenges, and applications. Curr. Opin. Biotechnol. **63**, 89–98 (2020)

23. Seifeldin, N.S., El Sayed, S.B., Asaad, M.K.: Increased Micro RNA-1266 levels as a biomarker for disease activity in psoriasis vulgaris. Int. J. Dermatol. **55**(11), 1242–1247 (2016)

24. Sharma, G., Dua, P., Mohan Agarwal, S.: A comprehensive review of dysregulated miRNAs involved in cervical cancer. Curr. Genomics **15**(4), 310–323 (2014)

25. de Silva, E., Stumpf, M.P.: Complex networks and simple models in biology. J. R. Soc. Interface **2**(5), 419–430 (2005)

26. Stelzer, G., et al.: The GeneCards suite: from gene data mining to disease genome sequence analyses. Curr. Protoc. Bioinf. **54**(1), 1–30 (2016)

27. Van Kampen, N.G.: Stochastic Processes in Physics and Chemistry, vol. 1. Elsevier, Amsterdam (1992)

28. Wagner, G.P., Kin, K., Lynch, V.J.: Measurement of mRNA abundance using RNA-seq data: RPKM measure is inconsistent among samples. Theory Biosci. **131**(4), 281–285 (2012). https://doi.org/10.1007/s12064-012-0162-3

29. Wang, K., et al.: Analysis of hsa-miR-30a-5p expression in human gliomas. Pathol. Oncol. Res. **19**(3), 405–411 (2013). https://doi.org/10.1007/s12253-012-9593-x

30. Wang, Y.R., Huang, H.: Review on statistical methods for gene network reconstruction using expression data. J. Theor. Biol. **362**, 53–61 (2014)
31. Weinstein, J.N., et al.: The cancer genome atlas pan-cancer analysis project. Nat. Genet. **45**(10), 1113–1120 (2013)
32. Zhang, B., Horvath, S.: A general framework for weighted gene co-expression network analysis. Stat. Appl. Genet. Mol. Biol. **4**(1) (2005). Article 17

Regularization Techniques in Radiomics: A Case Study on the Prediction of pCR in Breast Tumours and the Axilla

Eunice Carrasquinha[1], João Santinha[1], Alexander Mongolin[2], Maria Lisitskiya[1], Joana Ribeiro[3], Fátima Cardoso[3], Celso Matos[4], Leonardo Vanneschi[2], and Nickolas Papanikolaou[1(✉)]

[1] Center for the Unknown, Champalimaud Foundation, Computational Clinical Imaging Group, Av. Brasilia, 1400-038 Lisbon, Portugal
nickolas.papanikolaou@research.fchampalimaud.org
[2] Nova Information Management School (NOVA IMS), Universidade Nova de Lisboa, Campus de Campolide, 1070-312 Lisbon, Portugal
lvanneschi@novaims.unl.pt
[3] Champalimaud Clinical Center, Breast Unit, Av. Brasilia, 1400-038 Lisbon, Portugal
joana.ribeiro@fundacao.champalimaud.org
[4] Radiology Department, Centre for the Unknown, Champalimaud Foundation, Av. Brasília, 1400-038 Lisbon, Portugal

Abstract. Clinicians have shown an increasing interest in quantitative imaging for precision medicine. Imaging features can extract distinct phenotypic differences of tumours, potentially they can be used as a non-invasive prognostic tool and contribute for a better prediction of pathological Complete Response (pCR). However, the high-dimensional nature of the data brings many constraints, for which several approaches have been considered, with regularization techniques in the cutting-edge research front. In this work, classic lasso, ridge and the recently proposed priority-lasso are applied to high-dimensional imaging data, regarding a binary outcome. A breast cancer dataset, with radiomics, clinical and pathological information as features, was used. The application of sparsity techniques to the dataset enabled the selection of relevant features extracted in MRI of breast cancer patients, in order to identify the accuracy of those features and predict the pCR in the breast and the axilla.

Keywords: Radiomic features · High-dimensional data · Regularization techniques · Breast cancer

1 Scientific Background

High throughput experiments provide large amounts of data, holding the potential to improve our knowledge about biological processes, and eventually provide more insights regarding the treatment of many diseases, particularly cancer. In the last few years, the use of *omics* data has contributed to the identification of

© Springer Nature Switzerland AG 2020
P. Cazzaniga et al. (Eds.): CIBB 2019, LNBI 12313, pp. 271–281, 2020.
https://doi.org/10.1007/978-3-030-63061-4_24

biomarkers, and improving diagnosis, prognostication and treatment of cancer patients. However, most of the studies are making use of molecular biomarkers including genomics and proteomics, which require biopsies or invasive surgeries to extract and analyze tumour samples.

More recently, [1] proposed the use of radiomics, which is based on the extraction of a large number of imaging features, that could contain complementary information and provide a more comprehensive view of the entire tumour. In this sense, radiomics is an emerging field that converts imaging data into a high-dimensional mineable feature space, where the number of features greatly exceeds the number of observations. Due to the high-dimensionality of this type of data, classic statistical approaches, such as generalized linear models (glm) [13], are no longer suitable, and according to [2] a multivariate regression model is likely to be reliable if the number of features (p) is less than 1/10 the number of sample (n).

Several solutions to cope with this dimensionality problem can be found in the literature. The most common choice is the use of machine learning algorithms to extract knowledge from data, possibly without any assumption on the shape and characteristics of the model [3]. Some machine learning approaches have been considered so far. However, due to the nature of the data, regularized optimization techniques have revealed their importance [4] and [5]. Regularization techniques usually work by adding constraints to the cost function, with the objective of improving the generalization ability of the model. Ridge, lasso, and elastic net penalties are the most common examples of regularization techniques. The difference between them lies in the particular L_p norm used. For $p = 1$, the regularization technique is called lasso (Least Absolute Shrinkage and Selection Operator) regression, and for $p = 2$ it is called ridge regression. The result of combining the L_1 (lasso) and L_2 (ridge) norms is called elastic net [5]. In this work, the lasso and ridge regularization techniques are used.

In the context of cancer diseases, heterogeneous types of information (genomic, radiomic and clinical) are very common, and the inclusion of all information may improve the prediction accuracy of the model. Nevertheless, each type of data has generally a different structure. There are in the literature methodologies that can handle different structures/groups of data. Group lasso [6,14], integrative lasso with penalty factors [7] and priority-lasso [8] are some of the methodologies used to integrate multiple omics data to select a subset of features for prediction. In this work, we focused on the priority-lasso [8] a hierarchical lasso extension penalization framework to predict a clinical outcome using different types of omics data. This method is similar to the lasso regularization method, but it takes into account groups of variables, called blocks. The idea is to prioritize blocks that explain the largest possible part of the variability in the outcome.

A breast cancer dataset provided by the Breast Cancer Unit of the *Fundação Champalimaud* (FC) was used to evaluate the performance of those techniques. The radiomic features were extracted from Magnetic Resonance (MR) Mammography before receiving neoadjuvant chemotherapy (NAC) and the pathological features were obtained at pre-treatment biopsy. Two different outcomes were considered: breast tumor pathological complete response (pCR) and axilla pCR. Notice that for the axilla pCR data, the patients without lymph node involvement were removed.

There are in the literature several studies regarding radiomics analysis in breast cancer. [9], presented a review of the recent literature on this topic. [10], studied the use of quantitative imaging features (radiomics) extracted from the tumor and tumor environment on breast MR imaging characterizing tumor biological features relevant to outcome of targeted therapy. [11] and [12], both used lasso to obtain the optimal radiomic features. However, there is a gap of studies that integrates different types of information (such as demographic, clinical/histopathological) with radiomics features.

The main goal of this work is to evaluate the predictive performance of logistic regression on the breast tumor pCR and axilla pCR, using different regularization techniques, namely the ridge, lasso and model extensions based on priorities, by integrating not only radiomic features but also demographic, clinical/histopathological features, in order to provide an accurate alternative to surgical resection for assessment of pCR.

2 Materials and Methods

2.1 Regularization Methods

In this section, the regularization techniques used for a logistic regression model are briefly presented. Logistic regression is a popular classification method [15] that describes the relationship between one or more independent variables and a binary outcome variable Y, given by the logistic function:

$$p_i = P(y_i = 1 | \mathbf{x}_i) = \frac{\exp(\mathbf{x}_i^T \boldsymbol{\beta})}{1 + \exp(\mathbf{x}_i^T \boldsymbol{\beta})}, \tag{1}$$

where x_i corresponds to the feature value for observation i (p is the number of features and n is the number of observations), p_i is the probability of success for observation i and $\boldsymbol{\beta} = (\beta_1, \beta_2, \ldots \beta_p)$ are the unknown regression coefficients. The $\boldsymbol{\beta}$ parameters are estimated by maximizing the log likelihood function of the logistic model given by:

$$l(\boldsymbol{\beta}) = \sum_{i=1}^{n} \left\{ y_i \log p_i + (1 - y_i) \log[1 - p_i] \right\} + F(\boldsymbol{\beta}), \tag{2}$$

with $F(\boldsymbol{\beta})$ denoting the regularization term, which for the elastic net penalty takes the form:

$$F(\boldsymbol{\beta}) = \lambda \{ \alpha \|\boldsymbol{\beta}\|_1 + (1 - \alpha) \|\boldsymbol{\beta}\|_2^2 \}, \tag{3}$$

with $\lambda > 0$, which controls the penalization of the weights and $0 \leq \alpha \leq 1$ gives the balance between L_1 and L_2 norms, with the L_1 part being responsible for achieving sparsity. For $\alpha = 0$, leads to the ridge regression, for $\alpha = 1$ it corresponds to the lasso regression.

As previously mentioned, multiple types of data can improve the prediction accuracy. In order to incorporate all different types of data, different groups of features (*blocks*) should be considered. The extension of lasso, the priority lasso [8] builds a predictive model, based on different blocks of features.

Let $\boldsymbol{\pi} = (\pi_1, ..., \pi_M)$ be the permutation of $(1, ..., M)$ indicating the priority order, where M is the number of blocks. At a first step, the features from π_1 are used to fit the lasso regression model with the highest priority. Following the notation of Eq. (2), the regression coefficients $\boldsymbol{\beta}$ can be estimated by minimizing:

$$l(\boldsymbol{\beta}) = \sum_{i=1}^{n} \left\{ y_i \log p_i + (1 - y_i) \log[1 - p_i] \right\} + \lambda^{\pi_1} \| \boldsymbol{\beta}^{\pi_1} \|_1. \qquad (4)$$

Secondly, lasso is applied to the block with second highest priority. In this step, a linear score obtained from the previous step is used to force the model with coefficient fixed to 1 (*offset*). Finally, in the third step, lasso is applied to the block with third highest priority, using the linear score of the previous step as an *offset*. This process is repeated until all blocks have been considered. The priority-lasso is a hierarchical approach, since blocks with higher priority tend to explain the largest part of the variability in the outcome. For more details regarding this methodology, the reader is referred to [8].

2.2   Breast Cancer Data

The breast cancer dataset that we have used was based on a retrospective, single institution study comprising 131 consecutive patients diagnosed with early breast cancer, that underwent MR Mammography examination at the Breast Unit of the Champalimaud Clinical Center before receiving NAC. Following the completion, all patients underwent to surgery, and 58 patients presented with pCR while 73 with no-pCR as far as the breast tumor concerned. Consequently, a subgroup of patients with axillary lymph nodes involvement was assessed, where 31 patients presented with pCR and 36 with no-pCR for the axilla lymph nodes.

Each dataset is composed by two groups of features: demographic, clinical/histopathological and radiomics features. The clinical variables were nodal clinical status, tumor stage and menopausal status. All histopathological features were obtained from pretreatment (NAC) biopsy including: tumor morphology, histological grade and biological subtype; age was the only demographic feature considered. The categorical subtypes variable is a combination between positive or negative of estrogen receptors (ER), progesterone receptor (PR) and HER, having in total six categories. On the other hand, the biological subtypes variable is a derivation from the previous, with three categories: HER2, Luminal and triple negative (TNBC).

The dynamic contrast enhanced (DCE) sequence of the breast MR Mammography examinations were reviewed by two radiologists (M.L. with 11 years of experience and A.U., a last year radiology resident). The arterial phase images of the DCE-MRI sequence were selected for the analysis and the radiologists manually delineated the entire tumor in multiple consecutive slices, resulting in a volume of interest (VOI), using ITK-SNAP [16]. A second VOI comprising peritumoral tissue was created automatically by creating a zone with 5 mm of thickness around the external borders of the tumor. An example of tumoral and peritumoral VOIs are presented in Fig. 1.

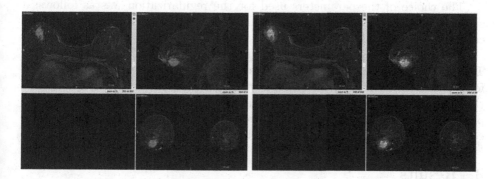

Fig. 1. Example of a tumoral (left) and peritumoral (right) VOI.

Both tumoral and peritumoral VOIs were used to extract a total of 4300 (2150 extracted from the tumor and 2150 extracted from a 5 mm peritumoral zone) radiomic features (first order, shape, gray-level co-occurrence matrix (GLCM), gray-level run length matrix (GLRLM), gray-level size zone matrix (GLSZM), neighbouring gray-tone difference matrix (NGTDM), and gray-level dependence matrix (GLDM) of both original and filtered images - Laplacian of Gaussian, LoG, with sigma 1 mm, 2 mm, 3 mm; wavelets - level 1 and 2) using Pyradiomics [17].

2.3   Statistical Analysis

Before performing the multivariate analysis, a segmentation stability analysis was conducted to obtain only stable radiomic features, regarding their variability arising from differences in segmentations of different observers. The intraclass correlation coefficient (ICC), with a threshold of 0.81, selecting 3023 out of 4300 radiomic features, was used. By using the ICC, a generalization of the radiomic features to different segmentations can be obtained. Notice that the radiomic features used for modeling were extracted from the segmentations performed by the senior radiologist.

In addition, the feature correlation matrix was computed and the absolute values of pair-wise correlations were obtained. Only features with low correlation were considered, using a threshold of 0.95. All radiomic features were normalized based on the z-score, and zero-variance and near-zero-variance features were removed from the analysis.

Since classic lasso and ridge do not work with feature groups, RFE was used to find 10 of the most important features. This number was chosen based on feature heatmaps and resulted in a more stable features selection across repetitions.

The lasso and ridge regularization techniques were applied only for the radiomic features, and the priority-lasso was used first by considering only the radiomic features and then for all the features involved in the study.

The choice of the parameters used for the regularization was as follows: a cross-validation procedure was used to optimize λ, considering a training set composed of 75% of samples and 5 folds.

The open-source software R [18] was used to perform the statistical analysis. The main libraries used for the analysis were: `glmnet`, for lasso regularization for the radiomics features, and `prioritylasso`, to conduct the priority lasso for the different types of features. The open-source library `scikit-learn` [19] and `python-glmnet` were used to perform recursive feature elimination and score collection for the lasso and ridge methods.

3   Results

3.1   Breast Tumor pCR

Table 1 reports the results for the AUC (Area Under the Curve) ROC (Receiver Operating Characteristics) and the AUC Precision-Recall (PR) values and corresponding 95% confidence intervals, obtained using logistic regression for breast tumor, without grouping features. The first column shows the results achieved with recursive feature elimination (RFE). Note that, for this analysis, the AUC-PR was used, due to the fact that the dataset is unbalanced (58 pCR Vs 73 no-pCR). To obtain the results, the following methodology was employed: 100 independent executions of each one of the algorithms were performed. For each one of these executions, 75% of the observations were randomly selected (with uniform probability) to form the training/validation set for stratified 5-fold cross validation, and the remaining 25% were used as a test set (hold out). Feature selection was executed inside of each fold, so the testing/validation samples are not exposed to the RFE algorithm.

In Table 1, "train" stands for the mean score obtained on the validation set, while "test" is the score obtained on the test set, both averaged over the 100 executions.

From Table 1, at least two facts can be observed: first, lasso, when compared to ridge and no regularization techniques, is the method that provides the slightly higher performance, regardless of the use or not of RFE, for both AUC-ROC and AUC-PR. Secondly, in practically all the presented experiments, the results obtained on the validation set are very similar to the ones obtained on the test set (hold-out), which indicates that the obtained models are not overfitting the training data.

Considering the lasso method with RFE five tumoral features (wavelet LLH first order Skewness, wavelet2 HLH glcm Correlation, LoG sigma 3 mm 3D first

Table 1. AUC scores on breast tumor pCR for lasso, ridge and no regularization of logistic regression, for the Receiver Operating Characteristics (ROC) and the Precision Recall (PR).

	With RFE	No RFE
Lasso	ROC train: 0.59 CI (0.53; 0.65) ROC test: 0.59 CI (0.50; 0.68) PR train: 0.57 CI (0.51; 0.63) PR test: 0.55 CI (0.49; 0.61)	ROC train: 0.59 CI (0.53; 0.65) ROC test: 0.60 CI (0.52; 0.68) PR train: 0.56 CI (0.50; 0.62) PR test: 0.56 CI (0.45; 0.67)
Ridge	ROC train: 0.56 CI (0.50; 0.62) ROC test: 0.53 CI (0.42; 0.65) PR train: 0.54 CI (0.48; 0.60) PR test: 0.50 CI (0.38; 0.62)	ROC train: 0.55 CI (0.48; 0.62) ROC test: 0.52 CI (0.42; 0.62) PR train: 0.52 CI (0.45; 0.59) PR test: 0.52 CI (0.39; 0.65)
No regularization	ROC train: 0.55 CI (0.49; 0.61) ROC test: 0.54 CI (0.44; 0.64) PR train: 0.52 CI (0.46; 0.58) PR test: 0.51 CI (0.39; 0.63)	ROC train: 0.54 CI (0.47; 0.61) ROC test: 0.54 CI (0.45; 0.63) PR train: 0.51 CI (0.45; 0.57) PR test: 0.50 CI (0.39; 0.61)

order Kurtosis, logarithm ngtdm Busyness, square root first order Kurtosis) and five peritumoral (wavelet LLH firstorder Kurtosis,wavelet2 HHH first order Median, wavelet HHL first order Skewness, wavelet2 LHH first order Skewness, exponential glszm Size Zone Non Uniformity Normalized) were selected.

To compare the performance of the techniques above, the priority lasso was used first by considering only the radiomic features. Two blocks were considered: tumoral radiomic features and peritumoral radiomic features. From the analysis the block with tumoral radiomic features should be prioritized first. The AUC-ROC values for the training was 76% with a confidence interval (CI) (0.67; 0.86) and for the test set was 57% with CI (0.34; 0.80). For the AUC-PR values obtained were, 79% with a CI (0.68; 0.89) and 50% with CI (0.31; 0.76), for the training and test set respectively. By using the priority lasso the features selected were five tumoral (LoG sigma 2 mm 3D glcm Idn, LoG sigma 2 mm 3D gldm Small Dependence Low Gray Level Emphasis, wavelet LHL glcm Imc1, wavelet2 LHH first order 90 Percentile, wavelet2 LHL glszm Zone Variance) and one peritumoral (exponential gldm Large Dependence High Gray Level Emphasis).

When considering different types of features (age, clinical/histopathological, radiomic), using priority-lasso, the results show that the AUC-ROC values for train and test slight depend on the block that is prioritized. For this analysis, two blocks were considered: one block for the age, clinical/histopathological features and a second block for the radiomic features. If demographic, clinical/histopathological features were prioritized, the AUC-ROC values for training and test were 89% and 79%, respectively, and the AUC-PR values for training and test were 89% and 66%, respectively. When the block of radiomic features had the highest priority, the AUC-ROC values were 88% and 75%, for the training and test cohorts, respectively. Regarding the AUC-PR values were 89% and 63%, for the training and test cohorts, respectively. For more details on the results, the reader is referred to Table 3.

Notice that although the AUC-ROC and AUC-PR for the training obtained higher values, when compared to the lasso (with or without RFE), however the difference between training and test values increased, showing problems regarding overffiting of the data.

Regarding the features selected by this approach, when the highest priority block considered was age and clinical/histopathological features, four features of this block (i.e. age, tumor stage, subtypes and biological subtypes), and nine radiomic features were selected. Concerning these nine radiomic features, two are tumoral (LoG sigma 2 mm 3D glcm Imc2 and wavelet2 LHL glszm ZoneVariance) and seven peritumoral (exponential gldm Large Dependence High Gray Level Emphasis, square gldm Small Dependence Low Gray Level Emphasis, square ngtdm Contrast, wavelet LHL glcm Imc1, wavelet LHL ngtdm Strength, wavelet2 LHH gldm Dependence Variance and wavelet2 LLH ngtdm Contrast).

On the other hand, when the highest priority block was the one of the radiomic features, two features from the age and clinical/histopathological block were selected (subtypes and biological subtypes) and 11 from the radiomic features, among which three tumoral (LoG sigma 2 mm 3D glcm Idn, LoG sigma 2 mm 3D gldm Small Dependence Low Gray Level Emphasis and wavelet2 LHL glszm Zone Variance) and eight peritumoral (exponential gldm Large Dependence High Gray Level Emphasis, original glcm Idmn, square ngtdm Contrast, wavelet HLH glcm Imc2, wavelet LHL glcm Imc1, wavelet LHL ngtdm Strength, wavelet2 LHH gldm Dependence Variance, wavelet2 LHH glszm Zone Variance).

3.2 Axilla pCR

Table 2 reports the results obtained for the AUC-ROC values for the axilla pCR. This table shows better performance with Ridge regularization with RFE, but it is still very low. AUC-PR is not added to the table, since the dataset is balanced, and thus the AUC-ROC is more representative.

Table 2. AUC scores on axilla pCR for lasso, ridge and no regularization of logistic regression, for the ROC.

	With RFE	No RFE
Lasso	ROC train: 0.55 CI (0.45; 0.65) ROC test: 0.52 CI (0.40; 0.64)	ROC train: 0.54 CI (0.44; 0.64) ROC test: 0.51 CI (0.40; 0.62)
Ridge	ROC train: 0.59 CI (0.50; 0.68) ROC test: 0.55 CI (0.40; 0.70)	ROC train: 0.51 CI (0.42; 0.60) ROC test: 0.50 CI (0.37; 0.63)
No regularization	ROC train: 0.56 CI (0.46; 0.66) ROC test: 0.57 CI (0.44; 0.70)	ROC train: 0.51 CI (0.42; 0.60) ROC test: 0.55 CI (0.41; 0.69)

From the ten selected features obtained from lasso regularization using RFE, six tumoral (square gldm Large Dependence Low Gray Level Emphasis, wavelet2 HHL first order Skewness, square glrlm Long Run Low Gray Level Emphasis, square glszm Large Area Low Gray Level Emphasis, LoG sigma 2 mm 3D first order Skewness and exponential glszm Large Area Low Gray Level Emphasis) features, and four peritumoral (wavelet2 LHH firstorder Skewness, original glcm Idmn, wavelet LLH first order Kurtosis and wavelet LHH glszm Large Area High Gray Level Emphasis) features were selected.

By using priority lasso, with only the radiomic features, the AUC-ROC values were for the training set 83% with a CI (0.71; 0.95) and for the test set 42% with a CI (0.10; 0.73), and seven tumoral features were selected (exponential first order Median, LoG sigma 2 mm 3D first order Skewness, LoG sigma 2 mm 3D gldm Large Dependence High Gray Level Emphasis, LoG sigma 3 mm 3D glcm Correlation, LoG sigma 3 mm 3D glszm Zone Variance, logarithm glszm Large Area Low Gray Level Emphasis, wavelet LHL gldm Low Gray Level Emphasis).

For the priority-lasso, with all the features included, the blocks considered were the same as the ones mentioned in the previous section. The results show that if the block with the highest priority was the one of the demographic and clinical/histopathological features, the AUC-ROC values for training and test cohorts were 87% and 88%, respectively. On the other hand, when the radiomic features block had the highest priority, AUC-ROC values were 94% and 86%, for training and test cohorts, respectively. For more details on the results, the reader is referred to Table 3.

Regarding the features selected, if the demographic and clinical/ histopathological block was considered with the highest priority, the selected features were age, tumor stage, tumor morphology, histologic grade, subtypes and biological subtypes. On the other hand, when the radiomic block was considered as priority, five features were selected from the demographic and clinical/histopathological block (tumor stage, tumor morphology, histologic grade, subtypes and biological subtypes) and four features were selected from radiomic block, among which three tumoral (exponential first order Median, LoG sigma 2 mm 3D first order Skewness and wavelet LHL gldm Low Gray Level Emphasis) and one peritumoral (original glcm Idn).

Table 3. AUC scores and confidence intervals (CI) on breast tumor pCR and axilla pCR for priority-lasso for the ROC and PR.

	1st priority: demographic, clinical and histopathological features	1st priority: radiomics features
Breast tumor pCR	ROC train: 0.89 CI (0.82; 0.96) ROC test: 0.79 CI (0.63; 0.94) PR train: 0.89 CI (0.83; 0.95) PR test: 0.66 CI (0.54; 0.78)	ROC train: 0.88 CI (0.80; 0.95) ROC test: 0.75 CI (0.56; 0.94) PR train: 0.89 CI (0.82; 0.97) PR test: 0.63 CI (0.49; 0.77)
Axilla pCR	ROC train: 0.87 CI (0.77; 0.97) ROC test: 0.88 CI (0.71; 1)	ROC train: 0.94 CI (0.87; 1) ROC test: 0.86 CI (0.69; 1)

4   Conclusion

In this work, we evaluated the predictive power of logistic regression using the ridge, lasso and priority-lasso regularization techniques. The presented results showed that a better prediction accuracy of pCR after NAC for breast cancer can be obtained using priority-lasso for both cases: breast and axilla pCR. Although, for the breast tumor dataset, the results shown some overfitting using priority-lasso. classic models could not achieve same high level of performance irrespectively of whether RFE was used or not. The presented results indicated that data of different nature should be considered partitioning them into different groups, and the used regularization method should support prioritization of the different groups. There are many application studies in the literature were the use of regularization techniques (lasso or grouped lasso) are very common, especially for genomics data, however the introduction of priorities applied to radiomics data, it is a novelty. This work may pave the way for further research, aimed at defining other regularization methods to support different blocks of features. Also, considering the high scores achieved in this work, similar data collection and classification methods could be used by clinicians in the future, and may provide a better understanding of the disease contributing for a more precise medicine by using less invasive surgeries or biopsied, contributing to the well-being of the patient. In the context of the breast cancer dataset, this could provide an accurate alternative to surgical resection for assessment of pCR before NAC, thus decreasing the extent and morbidity of surgical resection.

Acknowledgments. The authors acknowledge partially supported by national Portuguese funds through Portuguese Foundation for Science & Technology (FCT) under project PTDC/CCI-INF/29168/2017 (BINDER).

References

1. Lambin, P., et al.: Radiomics: extracting more information from medical images using advanced feature analysis. Eur. J. Cancer **48**(4), 441–446 (2012)
2. Harrell, F.: Regression modeling strategies: with applications to linear, logistic and ordinal regression, and survival analysis. Springer, (2015)
3. Boulesteix, A.-L., Wright, M.N., Hoffmann, S., König, I.R.: Statistical learning approaches in the genetic epidemiology of complex diseases. Human Genet. **139**(1), 73–84 (2019). https://doi.org/10.1007/s00439-019-01996-9
4. Tibshirani, R.: Regression shrinkage and selection via the lasso. J. Royal Stat. Soc. Series B (Methodol.) **58**(1), 267–288 (1996)
5. Zou, H., Hastie, T.: Regularization and variable selection via the elastic net. J. Royal Stat. Soc. Series B (Stat. Methodol.) **67**(2), 301–320 (2005)
6. Yuan, M., Lin, Y.: Model selection and estimation in regression with grouped variables. J. Royal Stat. Soc. Ser. B Stat. Methodol. **68**(1), 49–67 (2006)
7. Boulesteix, A.-L., De Bin, R., Jiang, X., Fuchs, M.: Computational and mathematical methods in medicine, IPF-LASSO: integrative -penalized regression with penalty factors for prediction based on multi-omics data. J. Royal Stat. Soc. Ser. B Stat. Methodol. (2017). https://doi.org/10.1155/2017/7691937
8. Klau, S., Jurinovic, V., Hornung, R., Herold, T., Boulesteix, A.-L.: Priority-Lasso: a simple hierarchical approach to the prediction of clinical outcome using multi-omics data. BMC Bioinform. **19**(1), 1–14 (2018)

9. Valdora, F., Houssami, N., Rossi, F., Calabrese, M., Tagliafico, A.S.: Rapid review: radiomics and breast cancer. Breast Cancer Res. Treat. **169**(2), 217–229 (2018). https://doi.org/10.1007/s10549-018-4675-4
10. Braman, N., et al.: Association of peritumoral radiomics with tumor biology and pathologic response to preoperative targeted therapy for HER2 (ERBB2)-positive breast cancer. JAMA Network Open **2**(4), e192561–e192561 (2019). https://doi.org/10.1001/jamanetworkopen.2019.2561
11. Liu, J., et al.: Radiomics analysis of dynamic contrast-enhanced magnetic resonance imaging for the prediction of sentinel lymph node metastasis in breast cancer. Front. Oncolo. **9**, 980 (2019).https://www.frontiersin.org/article/10.3389/fonc.2019.00980
12. Tagliafico, A.S., et al.: Breast cancer Ki-67 expression prediction by digital breast tomosynthesis radiomics features. Eur. Radiol. Exp. **3**(1), 1–6 (2019). https://doi.org/10.1186/s41747-019-0117-2
13. Nelder, J.A., Wedderburn, R.W.M.: Generalized linear models. Royal Stat. Soc. Wiley **135**(3), 370–384 (1972)
14. Meier, L., van de Geer, S., Buhlmann, P.: The group lasso for logistic regression. J. Royal Stat. Soc. Ser. B **70**, 53–71 (2008)
15. Tang, J., Alelyani, S., Liu, H.: Feature selection for classification: a review. In: Data Classification: Algorithms and Applications, p. 34 (2014)
16. Yushkevich, P.A., et al.: User-guided 3D active contour segmentation of anatomical structures: significantly improved efficiency and reliability. Neuroimage **31**(3), 1116–1128 (2006)
17. van Griethuysen, J.J., et al.: Computational radiomics system to decode the radiographic phenotype. Cancer Res. **77**(21), e104–e107 (2017)
18. R Core Team. R: A Language and Environment for Statistical Computing. R Foundation for Statistical Computing, Vienna, Austria (2012)
19. Pedregosa, F.: Scikit-learn: machine Learning in Python. J. Mach. Learn. Res. **12**, 2825–2830 (2011)

Modeling and Simulation Methods for Computational Biology and Systems Medicine

In Silico Evaluation of Daclizumab and Vitamin D Effects in Multiple Sclerosis Using Agent Based Models

Marzio Pennisi[1](✉) , Giulia Russo[3] , Giuseppe Sgroi[2] ,
Giuseppe Alessandro Parasiliti Palumbo[2] , and Francesco Pappalardo[3]

[1] Computer Science Institute, DiSIT, University of Eastern Piedmont,
Alessandria, Italy
marzio.pennisi@uniupo.it
[2] Department of Mathematics and Computer Science,
University of Catania, Catania, Italy
giuseppe.sgroi@unict.it, giuseppe.parasilitipalumbo@phd.unict.it
[3] Department of Drug Science, University of Catania, Catania, Italy
{giulia.russo,francesco.pappalardo}@unict.it

Abstract. We present an improved version of an agent-based model developed to reproduce the typical oscillating behavior of relapsing remitting multiple sclerosis, a demyelinating autoimmune disease of the central nervous system. The model now includes the effects of vitamin D, a possible immune-modulator that can potentially influence the disease course, as well as the mechanisms of action of daclizumab, a monoclonal antibody that was previously reported as the unique third line treatment for MS, i.e., to be used only in patients who had an inadequate response to the other therapies, but then retired from market due to the arising of severe side effects. The use of this computational approach, capable of qualitatively reproducing the main effects of daclizumab, is used to grasp some useful insights to delineate the possible causes that led to the withdrawal of the drug. Furthermore, we explore the possibility to combine vitamin D administration with a reduced dosage of daclizumab, in order to qualitatively delineate if a combined treatment can lead to similar efficacy, thus entitling a reduced risk of adverse effects.

Keywords: Agent based models · Multiple sclerosis · Computational modeling · In silico · Daclizumab · Vitamin D

1 Introduction

1.1 Background

Multiple Sclerosis (MS) is a chronic progressive autoimmune inflammatory disease that affects the Central Nervous System (CNS). It is mediated by CD4+ helper T cells action [16] and it is characterized by demyelination of nerve fibers.

© Springer Nature Switzerland AG 2020
P. Cazzaniga et al. (Eds.): CIBB 2019, LNBI 12313, pp. 285–298, 2020.
https://doi.org/10.1007/978-3-030-63061-4_25

Even if the etiology is not yet fully understood, both clinical and experimental scientific evidences suggest that at the base of MS development an autoimmune reaction is triggered by some combination of external, environmental and genetic factors. Such factors lead to a reaction against the sheath that covers the axons, i.e., the myelin. As a result, an inflammatory process is established, and circumscribed areas of the central nervous system are involved in the destruction of myelin and oligodendrocytes.

Common symptoms include fatigue, vision problems, numbness and tingling, muscle spasms, stiffness and weakness, mobility problems, problems with thinking, learning and planning, depression and anxiety, speech and swallowing difficulties. The most common form of MS is represented by Relapsing-Remitting MS (RRMS), in which periods of worsening of the disease, characterized by the reactivation of previously encountered symptoms or the appearing of new symptoms, are followed by periods of remission of the disease, represented by partial or total recovery of the previously encountered symptoms. However, MS progression cannot be arrested, and often evolves into a more severe form, Secondary-Progressive MS (SPMS) [22]. Besides genetic factors, environmental factors are also considered to have a significant role i.e., Epstein–Barr viral infection [1,21,23] and some dietary factors. Vitamin D and turmeric play a protective role in MS and neurodegeneration. Turmeric protects the brain from neurodegeneration and vitamin D is considered one of the most important factors to prevent MS [11].

Up to now, there is no cure for MS. The main available treatments are disease modifying therapies (DMTs), whose goal is mainly represented by the reduction of the incidence of relapses, reduction of their severity, prevention, delay of the progression of the disease. These drugs are generally classified as: first line drugs, characterized by a balance between efficacy and safety, and thus suitable for dealing with the disease in its early stages or in its less aggressive forms; second line drugs, which often own a greater power of action, but potentially able to expose patients to a higher risk of side effects.

Daclizumab (DAC) (commercial name Zynbrita) was the only drug that until February 2018 was considered as a third line treatment, i.e., used only in patients who had an inadequate response to at least two DMTs and for which the treatment with any other drug was contraindicated or otherwise unsuitable. However, the treatment has been retired due to the arising of some severe side effects, including 12 cases of serious inflammatory brain disorders, of which 4 of them with fatal outcome [15]. In this work, we extend a previously developed Agent-Based Model (ABM) for RRMS to reproduce the main effects of DAC treatment, trying also to grasp some useful information to delineate the possible causes that led daclizumab to its withdrawal. Furthermore, we try to combine vitamin D with a reduced dosage of daclizumab, in order to check if a combined strategy may lead to a similar efficacy, with a reduced risk of adverse effects.

1.2   Daclizumab Treatment

DAC is a humanized IgG1 monoclonal antibody directed to the alpha subunit of the high-affinity interlukin-2 receptor (IL2-R), also called (CD25), which is expressed at high levels on activated T cells and CD4$^+$, CD25$^+$, FoxP3 regulatory T cells (Treg). DAC therapy has been associated with a relatively mild decline of circulating T cells, but a significant expansion of CD56^{bright} natural killer (NK) cells [6]. This unexpected expansion of NK cells has been shown to be driven by the increased availability of IL-2 with intermediate activity on CD56^{bright} NK cells, as a result of CD25 blockade.

The CD25 blockade leads to several changes [7]:

(1) decrease of the stimulation of autoreactive T cells by autocrine IL-2.
(2) increase of the availability of soluble IL-2 for cells that exhibit intermediate affinity on IL-2R, such as CD56^{bright} NK cells, leading to an expansion of this cell population.
(3) blockage of the expression of CD25 on dendritic cells (which show on the surface only the α chain of IL-2R) preventing a "presentation" of the α chain to the IL-2R with intermediate affinity on T cells and therefore the formation of high-affinity IL-2R, which would involve stimulation of T cells [26].
(4) CD25 blockade also leads to a reduction of lymphoid tissue inducers in the central nervous system [10].

2   Materials and Methods

2.1   The Multi-agent System Based Approach

To model treatment effects over the course of the pathology, we started from an ABM that we previously developed using the NetLogo software [25]. Firstly, the model demonstrated to be able to mimic the typical oscillatory behavior associated with the most common form of the pathology [19], and then capable of both mimicking the possible beneficial effects of vitamin D over the disease course [17] and the effects of possible treatments tailored to regulate the opening of the Blood Brain Barrier (BBB) [20]. NetLogo is a software that offers a development environment for the realization of simulation models with agents, networks and dynamic systems. It is simple to learn and to manage, and has demonstrated to be a useful tool to model many diseases [8].

ABMs represent a computational simulation methodology whose goal is the representation of a global behavior that arises from the description of an individual behavior. In ABMs, particles called agents are described and followed individually; agents can move and interact each others in the environment; to them, we can associate different states and behaviors, that can be modified by the interactions with other agents. Space is an intrinsic feature of such models, so ABMs can be used to describe both spatial and temporal evolution of a given phenomenon. ABMs permit to use a very accurate description of the phenomena we are dealing with, without adding complexity to the solution of the model as it may happen with differential equations based models. Overall, they represent an excellent modeling and simulation tool if computational resources are not an issue. When the complexity of biological systems starts to become an issue, strategies based on the use of GPU-powered simulations can be now adopted, as in the work by Beccuti et al. [3] for ODE based models, or in the one by Chimeh et al. [9] for ABMs.

3   An Agent Based Model for Multiple Sclerosis

The complete description of the logic implemented into the model can be found in our previous works [17,19,20]. We will present here only a brief description in order to give to the reader an idea of the large picture.

The fundamental hypothesis is given by the presence, due to some genetic predisposition, of both self-reactive effector T cells (Teff) and regulatory T cells (Treg). Teff cells can potentially react against myelin based protein. However they are usually counterbalanced by Treg suppression mechanisms in healthy individuals. As shown in [19], the presence of such self-reactive cells may be not sufficient alone to justify the occurrence of the disease. The presence of a breakdown of the Teff-Treg cross-regulation mechanisms is a mandatory condition for the clinical developing of the disease. We supposed that this breakdown can be reproduced by an immune system fault at peripheral level i.e., a lower duplication rate for Treg, as supposed in [24]. In this case, due to the lack of control, Teff are able to spread and cause neural damage. However, self-reactive Teff must be activated in order to become harmful for the host. This could happen through a mechanism called antigenic mimicry, in which a third actor that acts as an external triggering mechanism leads to the erroneous activation of self-reactive Teff. It has been shown that such trigger could be elicited by Epstein-Barr Virus (EBV) [23].

The NetLogo model we developed uses a grid of 51×51 cells (namely patches) that are used to represent the myelin that covers the axons in the brain. This zone is colored in light gray to represent a small portion of white matter. Both resting Teff and Treg can normally move in the compartment. However, after random spreads of the external trigger that is introduced into the simulation using stochastic pulse trails, they can be activated and will release cytokines such as IL-2. Then, activated Teff cells can destroy the myelin, receiving as a result the stimulus to duplicate. In this case, two types of damage are represented: a recoverable and an unrecoverable damage. The first one, represented by dark gray zones, is a kind of damage that can be recovered if the action of self-reactive Teff is promptly stopped by activated Treg actions (that in this case will be also stimulated to duplicate). The second one, represented by black zones, appears if the activated Teff will continue to attack that portion of space.

The time-step of the simulation is $\Delta(t) = 2.4$ h. Such a choice allowed to obtain a good trade-off between a sufficient degree of granularity, necessary to reproduce the relapses that usually can go from days to weeks or months, and a simulation in a reasonable time frame of 5 years (18,250 time steps).

3.1 Modeling of Vitamin D Effects over Disease Course

As previously stated, vitamin D (VD) represents one of the factors that may potentially lead to better disease course. To this end, we introduced vitamin D as a new entity inside the model. Its administration can be done at any time step. Vitamin D will be captured and consumed by both Teff and Treg that, as a consequence of such, will modify their behavior. In particular, two major induced immunoregulatory capabilities of VD have been taken into account:

- T cell proliferation suppression: VD can potentially suppress the proliferation of Teff [4]. Such a behavior has been modeled as follows: if an active Teff finds a VD in the same patch, the Teff will consume it and behave as previously described except for the fact that it will not duplicate;
- promotion of T regulatory cells: VD can potentially improve the Treg actions through a series of mechanisms described in [2,12,13,18] that we reassumed with the following statement: if an active Treg finds in the same patch a VD, it will consume it and will behave as previously described, except for the fact that its lower duplication rate (i.e., ill patients) will be temporary restored to the same level of Teff duplication rate for that time-step.

3.2 Extension of the Model with Daclizumab

In order to introduce the Mechanisms of Action (MoA) of DAC into the NetLogo model, it was necessary to intervene on the model both modifying the behavior of some agents and introducing others. The high-affinity CD25 receptor for IL-2 present in T cells is the main therapeutic target of DAC, which therefore blocks the self-stimulation mechanisms of these lymphocytes promoted by this interleukin. The role of CD56^{bright} NK cells should also be emphasized, as in vivo studies their number results increased during the administration of the treatment.

These two entities, in addition to DAC, were then introduced into the model. IL-2 is a cytokine that is released from the T cells during their activation. The latter, once activated, will be able to duplicate only in the presence of IL-2, that will be consumed during the duplication phase. IL-2 is also fundamental for NK cells, which will always be introduced in the simulation as resting cells (not activated). NKs are able to benefit from the IL-2 action becoming activated. Active NKs can also be further stimulated by IL-2, and this will result in a slight increase in their half-life within the model.

These benefits will in any case have a probability of occurring two orders of magnitude lower than in the case of T cells. This is due to the fact that NK cells have an IL-2 receptor that has a much lower affinity (about 1 in 100) with respect to the one on T lymphocytes. An active NK that holds the same position of an active Teff will be able to eliminate the Teff, as observed in the paper by Bielekova and Becker [5]. The homeostasis of NK cells is ensured by a function that stochastically introduces new resting NK cells into the simulation only when the average level of NK resting falls below the mean value.

The administration of DAC takes place once every 300 time-steps (about once a month), as described in the guidelines. At each administration, 1500 agents representing the drug molecules are randomly distributed within the simulation space. The MoA of DAC is reproduced in the following way: if an active T cell (Teff or Treg) is in the same position (patch) of a unit of DAC, this cell will consume the drug and become inhibited, disappearing at the end of its life cycle. It is important to note that an inhibited entity will not be able to be stimulated by the IL-2 and therefore duplicate.

During the modeling phase we decided to mainly target the active T lymphocytes, as these tend to overexpress the CD25 receptor compared to the resting T lymphocytes, thus having a greater chance of being tailored by the drug effects. It should also be noted that, in line with current literature, the action of DAC influences both Teff and Treg cells. The conceptual representation of the logic implemented inside the model is presented in Fig. 1. The model is available at: https://www.combine-group.org/assets/MS_viruses_DACLIZUMAB_VD.html.

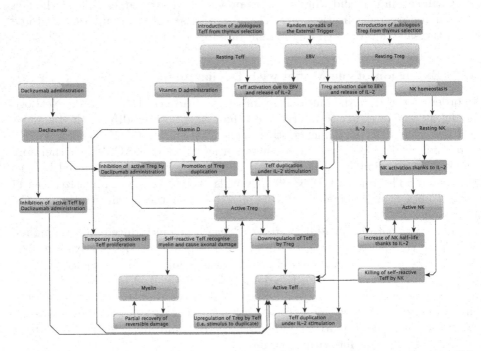

Fig. 1. Conceptual representation of the logic implemented inside the NetLogo model. Both autologous Teff and Treg cells are introduced into the system by means of stochastic pulse trails. Random spreads of EBV may elicit the activation of such cells that release IL-2, an immunostimulatory cytokine that promotes their actions and duplication. Active Teff may attack and destroy the myelin causing recoverable first, and unrecoverable then, neural damage. Treg cells may stop Teff actions receiving as feedback the stimulus to duplicate. IL-2 may be also responsible of the activation of NK cells that can kill activated Teff. VD can promote and limit Treg and Teff actions, respectively. Daclizumab inhibits both activated Treg and Teff duplication.

4   Results

4.1   Analysis of Daclizumab Administration

To qualitatively analyze the behavior of the model described so far and the effects of daclizumab treatment, we performed 200 distinct simulations (100 in the presence of treatment and 100 in the absence) using 100 different random seeds. In other words, we used the same seed for each simulation without the treatment and the corresponding simulation with the treatment in order to start with the same initial conditions. As different random seeds will entitle completely different pseudo-random number sequences, these will influence the initial arrangement of all the agents, the movements within the simulation space, the probabilities of interaction, the introduction times and positions of viruses, autologous lymphocytes and other entities, leading in general to different behaviors. We can therefore imagine to associate a distinct patient with a different clinical history to each seed, and then to summarize the average trend by analyzing the behavior on different distinct individuals.

DAC treatment, when present, is administered once a month, as per real administration protocol. In particular, 1500 DAC entities are randomly placed inside the simulation space every 300 time-steps (30 days). We remind here that underlying hypothesis in both the two simulation groups is the presence of a breakdown in the peripheral tolerance mechanisms, modeled as an impaired capability of Treg cells to duplicate in respect to Teff cells. Comparison between ill and healthy patients (i.e., with the same replication probabilities) has been already presented in [19].

We did not simulate the behavior of the model when the starting of the treatment is delayed (i.e., after 2 years of disease) because, as a disease modifying therapy, DAC has no clinical impact on past disease history, in the sense that it cannot resolve the disease and/or restore lost functionality, but it can only be used to control the present and future disease course. An already highly damaged portion of brain tissue (i.e., after 2 years of simulation) will therefore not benefit of DAC administration. Conversely, a healthy portion of tissue will benefit of DAC protective effects.

From the simulation results it is possible to see how the model shows a reduction in the number of active Teff and Treg when DAC is administered (Fig. 2), coherently with the MoA of the drug and the experimental results. It is worth to note here that, even if the initial number of resting cells has been set to 100 for both the Teff and Treg populations, we can observe in general a lower mean number of activated Treg cells in respect to the number of activated Teff cells. This is probably due to the lower duplication probability (used to represent the breakdown of local tolerance) of Treg cells in respect to Teff cells.

The model also shows a reduction in the total damage (Fig. 3), and the oscillatory behavior in the treated case mostly suggests that only recoverable damage occurs in general. The treatment is in fact able to control the disease progression in most, but not in all instances, as it is suggested by the change in the behavior for the treated case observed in Fig. 3 around time step 500.

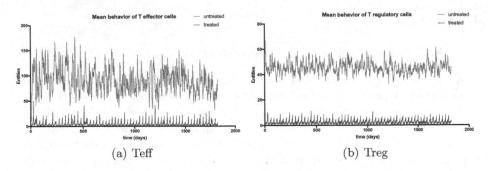

(a) Teff   (b) Treg

Fig. 2. Mean behavior of Teff (left) and Treg (right) over 100 different simulations, in presence (blue lines) and absence (red lines) of DAC treatment. Simulation time has been set to 5 years (18250 time-steps). Treated graphs show an overall lower number of entities ($p < 0.0001$). (Color figure online)

In this case, the mean value settles to a non zero value, suggesting that (at least) in one of the simulations some unrecoverable damage occurs even in presence of DAC.

A slight increase in the number of activated NK and in the quantity of IL-2 can be also observed (Fig. 4), showing a behavior that can be considered, from a qualitative point of view, in line with clinical observations. Note here that the mean number of activated NK is always very low. This is due to the fact that the mean is calculated over several simulations where very few simulations have activated NK at a given time point, while the remaining simulations have no activated NK cells at that time, and can be justified by taking into account the following facts: 1) the random virus spreads that activate the immune response happen at different times for different simulations, entitling also different time-windows for the activation of such cells; 2) differently from Treg and Teff, NK cells do not duplicate and do no last for long (no immunological "memory"), remaining low in number and disappearing rapidly if no infection is present. It must also be underlined that even if the difference between the untreated and treated groups for both NK and IL-2 may seem very limited, the Wilcoxon Rank Sum test for matched pairs that was used to compare the temporal mean behaviors of all the entities within the treated and untreated groups ensured that the difference was always statistically significant ($p < 0.0001$ for all the tested scenarios).

4.2   Combining Vitamin D with a Reduced Dosage of Daclizumab

Despite its undoubted efficacy, in 2018 daclizumab was retired from market due to the arising of severe side effects, including 12 cases of serious inflammatory brain disorders. While the real causes that led to the arising of such side effects have to be yet investigated, thanks to the use of this ABM, we draw in the conclusions some considerations about the motivations that may be responsible of the adverse effects. It is worth to note that in some cases, the arising of side

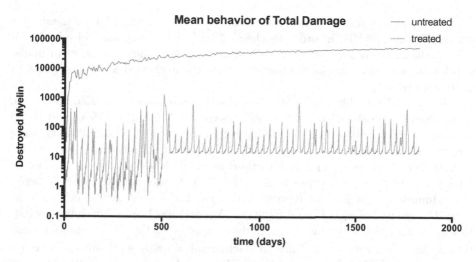

Fig. 3. Mean behavior of total damage (recoverable + unrecoverable) calculated over 100 different simulations, in presence (green lines) and absence (purple lines) of DAC treatment. Simulation time has been set to 5 years (18250 time-steps). The administration of the treatment leads to an overall lower total damage ($p < 0.0001$). Moreover, in the treated graph the oscillatory behavior suggests only the presence of recoverable damage.

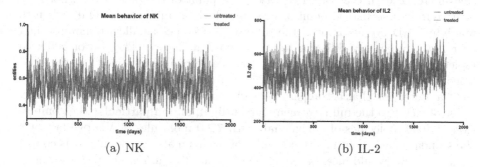

(a) NK   (b) IL-2

Fig. 4. Mean behavior of activated NK (left) and IL-2 (right) over 100 different simulations, in presence (green lines) and absence (purple lines) of DAC treatment. Simulation time has been set to 5 years (18250 time-steps). Treated graphs show slight, but statistically different ($p < 0.0001$), higher numbers for both NK and IL-2 when the treatment is administered, resulting in good agreement from a qualitative point of view with literature observations.

effects may be due to the excessive dosage of the treatment. If this happens, it would be advisable to reduce the dosage, but this would also entitle the risk of loosing the minimal required efficacy to maintain protection. In order to deal with this issue, a possible approach would be to combine a reduced dosage of the treatment with other less harmful substances that may have a positive role in the disease course.

Bearing this in mind, we decided to combine the daclizumab treatment with various dosages of VD, in order to check if such a strategy may be viable. Of course, thanks to the use of in silico models, this approach can be in principle applied to any combination of substances and diseases with a reduced effort in time and costs.

We tested 5 possible combinations of daclizumab dosages (0, 250, 500, 1000, and 1500 molecules) with 5 possible VD dosages (0, 2500, 5000, 7500, and 10000 molecules). For each possible combination we executed 100 simulations using different random seeds, for a total of 2500 simulations. Daclizumab was administered every 30 days as per standard therapy, while VD was administered every 10 days, as we found in our previous work [17] that lager time windows between VD administrations lead to unsatisfactory results. As marker of efficacy, the total damage at the end of the simulation (5 years) has been taken into account. In Fig. 5, we present the mean $\pm$ standard error $(\frac{\sigma}{\sqrt{n}})$ for all the combinations tested. Each line refers to a different daclizumab dosage, while on the x and y axis we report the VD dosages and the final neural damage, respectively. We used a log-scale for the y axis as it allows to better discern among small differences, that would be invisible otherwise. The difference in efficacy of all combinations has been assessed with a one-way Anova with multiple t-tests comparisons.

From Fig. 5 it appears clear that the treatments that involve 0 and 250 daclizumab molecules (blue and red lines) are in general less effective, and indeed the difference with the untreated case (first point of the blue line) is not significative from a statistical point of view for all the combinations. Only when the maximum VD dosage is used the difference becomes smaller, remaining however not significantly different from the untreated case. The use of 500 DAC molecules (yellow line) leads to a better disease course in almost all the situations, but only the use of 1000 molecules of daclizumab (purple line) is able to elicit a protection that is comparable (difference not significant from a statistical point of view) to the full treatment (green line), and in particular the combination of 1000 molecules of daclizumab with 7500 molecules of VD overlaps with the standard treatment (1500 molecules of daclizumab), even when it is coupled with VD. This would mean a reduction of about 33% in respect to the standard treatment. We note here for completeness, that the addition of VD to the standard treatment (1500 DAC molecules, green line) does not lead to a difference that is significant from a statistical point of view for all the VD dosages.

It is worth to note that for 10000 VD molecules we observe in some cases a small (but not significant) worsening of the disease outcome. This could be due to some sort of competition between the administered drugs that are basically directed towards the same targets, and thus it will deserve further investigation.

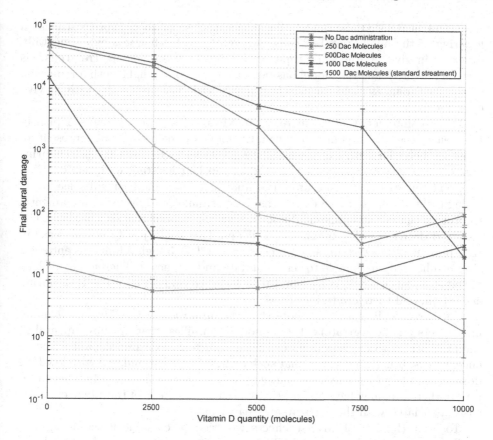

Fig. 5. Different daclizumab dosages combined with different VD dosages. Mean values $\pm$ standard error $(\frac{\sigma}{\sqrt{n}})$ are reported for all the combinations tested. Each line refers to a different daclizumab dosage. On the x and y axis the VD dosages and the final neural damage are reported, respectively. The y axis is on a log-scale to better discern among small differences.

5   Conclusions

We presented an ABM to describe the MoA of daclizumab, a monoclonal antibody previously defined as the only third line treatment for MS and then retired from market due to the insurgence of severe side effects, like the onset of encephalitis in 12 patients treated with it. The therapy with DAC has been associated with a relatively mild decline of circulating T cells, but also with a significant expansion of CD56^{bright} NK cells. Since then, it has been shown that this unexpected expansion of NK cells is driven by the increased availability of IL-2 to intermediate activity on CD56^{bright} NK cells, as a result of CD25 blockade.

Through the use of an agent model created using NetLogo, it was possible to model and simulate the mechanism of action of DAC, in particular in the treatment of RRMS. The model allowed us to observe, in case of treatment, a lowering of the number of Teff cells, coherently with what was experimentally observed in vivo. At the same time, an inhibition of autologous Treg cells has been highlighted, and the model has been also able to highlight both a reduction of the total damage, and an increase of NK and IL-2, in accord with in vivo experiments.

It is worth to note that the increases in NK and IL-2 levels were not explicitly foreseen by the rules and interactions introduced by the model, but represent an emerging behavior typical of the use of agent models, which in this case confirms how the mechanics envisaged by the model can be consistent with reality.

Final remarks that arose after the analysis of the model results led to us to highlight the importance of IL-2 as Treg stimulator. Tregs are fundamental to protect against autoimmune diseases, such as autoimmune encephalitis. The observed decrease in the number of Teff and Treg could justify the increase in IL-2 present in the system, as such cells, whose CD25 high-affinity receptor for IL-2 is inhibited, are unable to self-stimulate and duplicate. Therefore, this lack of self-stimulation could correspond to a lower consumption of the IL-2 during activation, and therefore to its excess.

Furthermore, it is worth to note that the inhibition of CD25 by daclizumab also affects all the regulatory T lymphocytic families, that are fundamental in the prevention of most autoimmune diseases such as, for example, autoimmune encephalitis. It has been shown that such cells, if further stimulated with low IL-2 dosages, can be protective in the development of this pathology [14]. Their inhibition, therefore, may have had a primary role in the onset of the 12 encephalitis cases associated with the daclizumab treatment.

To limit the risk of adverse effects we tried to combine lower dosages of daclizumab with the less harmful VD, in order to see if it was possible to achieve a similar protection with less risks for the health. In particular we found that constant use of VD may allow a reduction of up 33% in the required daclizumab dosage, and thus a reduced risk of severe side effects.

Even if such considerations need to be verified in vivo, a qualitative analysis through an in silico simulation platform, like the one we used so far, could have been, a priori, capable of suggesting potential mechanisms responsible of the development of adverse effects such as those that were then observed in the case of administration of daclizumab. In particular, the decrease in the number of active T regulators seen in the simulations could have represented a possible red flag way before the starting of the experimentation on real individuals. Finally, the use of In Silico platforms like the one used so far could be also useful to suggest possible countermeasures, as the use of adjuvants like VD to decrease the risk of possible adverse effects.

References

1. Ascherio, A., et al.: Epstein-Barr virus antibodies and risk of multiple sclerosis: a prospective study. JAMA **286**(24), 3083–3088 (2001)

2. Barrat, F., et al.: In vitro generation of interleukin 10-producing regulatory CD4(+) T cells is induced by immunosuppressive drugs and inhibited by T helper type 1 (Th1)- and Th2-inducing cytokines. J. Exp. Med. **195**(5), 603–613 (2002)
3. Beccuti, M., et al.: GPU accelerated analysis of Treg-Teff cross regulation in relapsing-remitting multiple sclerosis. In: Mencagli, G., et al. (eds.) Euro-Par 2018. LNCS, vol. 11339, pp. 626–637. Springer, Cham (2019). https://doi.org/10.1007/978-3-030-10549-5_49
4. Bhalla, A., Amento, E., Glimcher, B.S.L.: 1,25-Dihydroxyvitamin D3 inhibits antigen-induced T cell activation. J. Immunol. **133**(4), 1748–1754 (1984)
5. Bielekova, B., Becker, B.L.: Monoclonal antibodies in MS: mechanisms of action. Neurology **74**(Suppl 1), 31–40 (2010)
6. Bielekova, B., et al.: Regulatory CD56(bright) natural killer cells mediate immunomodulatory effects of IL-2Rα-targeted therapy (daclizumab) in multiple sclerosis. Proc. Natl. Acad. Sci. USA **103**(15), 5941–5946 (2006)
7. Bielekova, B., et al.: Intrathecal effects of daclizumab treatment of multiple sclerosis. Neurology **77**(21), 1877–1886 (2011)
8. Chiacchio, F., Pennisi, M., Russo, G., Motta, S., Pappalardo, F.: Agent-based modeling of the immune system: NetLogo, a promising framework. BioMed Res. Int. **2014**, 907171 (2014)
9. Chimeh, M.K., Richmond, P., Heywood, P., Pennisi, M., Pappalardo, F.: Parallelisation strategies for agent based simulation of immune systems. BMC Bioinform. **20**(Suppl 6), 1–14 (2019). https://doi.org/10.1186/s12859-019-3181-y
10. Gold, R., et al.: Daclizumab high-yield process in relapsing-remitting multiple sclerosis (SELECT): a randomised, double-blind, placebo-controlled trial. Lancet **381**, 2167–75 (2013)
11. Godin, D.: The causal cascade to multiple sclerosis: a model for MS pathogenesis. PLoS One **4**(2), e4565 (2009)
12. Gorman, S., et al.: Topically applied 1,25-Dihydroxyvitamin D3 enhances the suppressive activity of CD4+CD25+ cells in the draining lymph nodes. J. Immunol. **179**(9), 6273–6283 (2007)
13. Gregori, S., Casorati, M., Amuchastegui, S., Smiroldo, S., Davalli, A., Adorini, L.: Regulatory T cells induced by 1 alpha, 25-Dihydroxyvitamin D3 and mycophenolate mofetil treatment mediate transplantation tolerance. J. Immunol. **167**(4), 1945–1953 (2001)
14. Lim, J.A., et al.: New feasible treatment for refractory autoimmune encephalitis: low-dose interleukin-2. J. Neuroimmunol. **299**, 107–111 (2016)
15. Luessi, F., Engel, S., Spreer, A., Bittner, S., Zipp, F.: GFAPα IgG-associated encephalitis upon daclizumab treatment of MS. Neurol. Neuroimmunol. Neuroinflamm. **5**(5), e481 (2018)
16. Martin, R., McFarland, H.F., McFarlin, D.E.: Immunological aspects of demyelinating diseases. Ann. Rev. Immunol. **10**, 153–187 (2003)
17. Pappalardo, F., Pennisi, M., Rajput, A.M., Chiacchio, F., Motta, S.: Relapsing-remitting multiple scleroris and the role of vitamin D : an agent based model. In: ACM-BCB, pp. 744–748 (2014)
18. Penna, G., et al.: Expression of the inhibitory receptor ILT3 on dendritic cells is dispensable for induction of CD4+Foxp3+ regulatory t cells by 1,25-dihydroxyvitamin D3. Blood **106**(10), 3490–3497 (2005)
19. Pennisi, M., Rajput, A.M., Toldo, L., Pappalardo, F.: Agent based modeling of Treg-Teff cross regulation in relapsing-remitting multiple sclerosis. BMC Bioinf. **14**(Suppl 16), S9 (2013)

20. Pennisi, M., Russo, G., Motta, S., Pappalardo, F.: Agent based modeling of the effects of potential treatments over the blood-brain barrier in multiple sclerosis. J. Immunol. Methods **427**, 6–12 (2015)
21. Ponsonby, A., et al.: Exposure to infant siblings during early life and risk of multiple sclerosis. JAMA **293**(4), 463–469 (2005)
22. Sospedra, M., Martin, R.: Immunology of multiple sclerosis. Annu. Rev. Immunol. **23**, 683–747 (2005)
23. Sundström, P., et al.: An altered immune response to Epstein-Barr virus in multiple sclerosis: a prospective study. Neurology **62**, 2277–2282 (2004)
24. Vélez de Mendizábal, N., et al.: Modeling the effector - regulatory T cell cross-regulation reveals the intrinsic character of relapses in multiple sclerosis. BMC Syst. Biol. **5**(1), 1–15 (2011)
25. Wilensky, U.: NetLogo, Center for connected learning and computer-based modeling, Northwestern University, Evanston, IL (1999). http://ccl.northwestern.edu/netlogo/
26. Wuest, S.C., et al.: A role for interleukin-2 transpresentation in dendritic cellmediated T cell activation in humans, as revealed by daclizumab therapy. Nat. Med. **17**(5), 604–609 (2011)

Multiple Sclerosis Disease: A Computational Approach for Investigating Its Drug Interactions

Simone Pernice[1]([✉]), Marco Beccuti[1], Greta Romano[1], Marzio Pennisi[2],
Alessandro Maglione[3], Santina Cutrupi[3], Francesco Pappalardo[4],
Lorenzo Capra[5], Giuliana Franceschinis[2], Massimiliano De Pierro[1],
Gianfranco Balbo[1], Francesca Cordero[1], and Raffaele Calogero[6]

[1] Department of Computer Science, University of Turin, Turin, Italy
pernice@di.unito.it
[2] Computer Science Institute, DiSIT, University of Eastern Piedmont,
Alessandria, Italy
[3] Department of Clinical and Biological Sciences, University of Turin,
Orbassano, Italy
[4] Department of Drug Sciences, University of Catania, Catania, Italy
[5] Department of Computer Science, University of Milan, Milan, Italy
[6] Department of Molecular Biotechnology and Health Sciences, University of Turin,
Turin, Italy

Abstract. Multiple Sclerosis (MS) is a chronic and potentially highly disabling disease that can cause permanent damage and deterioration of the central nervous system. In Europe it is the leading cause of non-traumatic disabilities in young adults, since more than 700,000 EU people suffer from MS. Although recent studies on MS pathophysiology have been performed, providing interesting results, MS remains a challenging disease. In this context, thanks to recent advances in software and hardware technologies, computational models and computer simulations are becoming appealing research tools to support scientists in the study of such disease. Motivated by this consideration, we propose in this paper a new model to study the evolution of MS in silico, and the effects of the administration of the daclizumab drug, taking into account also spatiality and temporality of the involved phenomena. Moreover, we show how the intrinsic symmetries of the model we have developed can be exploited to drastically reduce the complexity of its analysis.

Keywords: Multiple Sclerosis · Computational model · Colored Petri Nets

1 Introduction

Multiple Sclerosis (MS) is a long-term and autoimmune disease of the Central Nervous System (CNS). During the progression of the disease, cells of immune

© Springer Nature Switzerland AG 2020
P. Cazzaniga et al. (Eds.): CIBB 2019, LNBI 12313, pp. 299–308, 2020.
https://doi.org/10.1007/978-3-030-63061-4_26

system attack the principal components of the CNS, the neurons, removing the enveloping myelin and preventing the efficient transmission of the nervous signals. Relapsing-Remitting MS (RRMS) is the predominant type of MS since it is diagnosed in about 85%–90% of MS cases [12]. In RRMS, the disease alternates two phases: (1) relapse phase is characterized by a disease worsening due to the active inflammation damaging the neurons; (2) in the remission phase there is a complete or partial lack of the symptoms [7]. Recently, many treatments were proposed and studied to contrast the RRMS progression. Among these drugs, daclizumab [4] (commercial name Zynbrita), an antibody tailored against the Interleukin-2 receptor (IL2R) of T cells, exhibited promising results. Unfortunately, its efficacy was accompanied by an increased frequency of serious adverse events as infections, encephalitis, and liver damages. For these reasons daclizumab has been withdrawn from the market worldwide.

In [11] we proposed a model to investigate the effect of the daclizumab administration in RRMS. It involves the following seven main actors of MS: Epstain-Barr virus (EBV), Effector T lymphocytes cells (Teff), Regulatory T lymphocytes cells (Treg), Natural Killer cells (NK), Oligodentrocytes cells (ODC), Interleukin-2 (IL2) and daclizumab (DAC). In details, the EBV was considered since several studies [13] commonly agree on the hypothesis that viruses may play a role in RRMS pathogenesis acting as environmental triggers, and in particular the presence of this virus represents a well established risk factor in MS [8]. Effector T cells (Teff) are instead immune cells with a protective role against pathogens in healthy people. However, in RRMS a hypothesis is that the reactivation of EBV latent infection could bring to the activation of autologous Teff lymphocytes against myelin, due to a structure similarity between one viral protein and myelin protein (molecular mimicry). Regulatory T cells (Treg) are immune cells acting as balancing of the immune response since they contribute to suppress and modulate the Teff cells activity when no longer needed, or when there is a high risk of inflammation that can cause injuries to the tissues of the host. Other important actors within this context are the natural killer (NK) cells, a family of immune cells that acts as host-rejection of infected cells. Oligodendrocytes (ODC) are instead cells supporting the neurons since they produce and are able to partially restore the myelin around the neurons whenever a not excessive damage occurs. IL2 is an immunomodulatory cytokine released by Teff in order to self-stimulate to duplicate and to propagate their immune actions. Finally, we included in the model the drug daclizumab, a humanized monoclonal antibody used in MS as drug against the Interleukin-2 receptor (IL2R) that is able to break the autoimmune reaction by suppressing the immune cells proliferation [4].

Thus, to help scientists in improving their knowledge of these phenomena, in this work we extend the RRMS models presented in [11] considering the cells movement into a three-dimensional grid. In details, in this paper we show how the use of a graphical formalism, i.e. the Extended Stochastic Symmetric Net (ESSN) formalism [10,11], allows to easily deal with this complex three-dimensional model whose direct definition in terms of ODE system becomes clearly unfeasible even for a small three-dimensional grid.

Indeed, for instance considering three-dimensional grid with dimension 3 × 3 × 3 the ESSN model is a bipartite graph with only 38 nodes (i.e. 13 places and 25 transitions) and approximately 90 arcs, while its underlying deterministic process comprises 433 ODEs.

Moreover, the high level of parametrization and flexibility provided in the model through this graphical formalism enables to study different grid dimensions in a fairly easy manner and without the need of redrawing the whole model. Similarly, in the analysis phase the ESSN model provides a powerful methodology that automatically exploits the system symmetries to reduce the complexity (in terms of number of equations) of the underlying deterministic process. Indeed in [2] we proposed an algorithm that directly derives a *compact* ODE system from a ESSN model in a symbolic way, through algebraic manipulation of ESSN annotations.

2   Scientific Background

In this section we introduce the Petri Nets (PNs) formalism used to describe our model. PNs and their extensions are effective formalisms to model biological systems thanks to their capability of representing in a simple and clear manner the system features and to provide efficient techniques to derive system qualitative and quantitative properties. In details, PNs are bipartite directed graphs with two types of nodes called *places* and *transitions*. Places, graphically represented as circles, correspond to the state variables of the system, while transitions, graphically represented as boxes, correspond to the events that can induce a state change. The arcs connecting places to transitions (and vice versa) express the relations between states and event occurrences. Places can contain tokens, drawn as black dots. The state of a PN, namely a *marking*, is defined by the number of tokens in each place. The system evolution is provided by the firing of an enabled transition, where a transition is enabled if and only if each input place contains a number of tokens greater than or equal to a given threshold defined by the cardinality of the corresponding input arc. The firing of an enabled transition removes a fixed number of tokens from its input places and adds a fixed number of tokens into its output places (according to the cardinality of its input/output arcs).

In this work we focus on Stochastic Symmetric Nets (SSNs) a high level formalism that extends PNs with *colors* and *stochastic firing delays* [6]. Colors provide a more compact, readable and parametric representation of the system thanks to the possibility of having tokens with different characteristics.

More specifically, the *color domain* associated with place p, denoted $cd(p)$, specifies the color of the tokens contained in this place, whereas the color domain of a transition defines the different ways of firing it (i.e. the possible *transition instances*). In order to specify these firings, a color function is attached to each arc which, given a color of the transition connected to the arc, determines the number of colored tokens that will be added to or removed from the corresponding place. A color domain is defined as Cartesian product of *color classes* which may be viewed as primitive domains. A color class can be partitioned into *static subclasses*. The colors of a class have the same nature (e.g. T cells), whereas the colors inside a static subclass have the same potential behavior (e.g. Teff). Stochastic firing delays, sampled from negative exponential distributions, allow to automatically derive the underlying Continuous Time Markov Chain (CTMC) that can be studied to quantitatively evaluate the system behaviour. In the literature, different techniques are proposed to solve the underlying CTMC; in particular, in case of very complex models, the so-called deterministic approach [9] can be efficiently exploited. According to this, in [3] we proposed a method to derive a deterministic process, described through a system of Ordinary Differential Equations (ODEs), which well approximates the stochastic behavior of an SSN model assuming all reactions follow the Mass Action (MA) law. In the same paper we also described an efficient translation method based on the SSN formalism, which is able to reduce the size (in terms of equations number) of the underlying ODE system through the automatic exploitation of system symmetries. Practically, the complete set of ODEs, which can be derived from an SSN model is partitioned into equivalence classes of ODEs which have same solution so that a representative equation, called *symbolic equation*, can be pointed out for each equivalence class. Then, a reduced ODE system may be derived including only these symbolic equations whose solution mimics the behavior of the original model. Recently this result was further improved in [2] where a new algorithm is discussed to generate the symbolic equation for each equivalence class of ODEs without deriving the complete ODE system. This is achieved thanks to a recent extension of a symbolic calculus for the computation of SSN structural properties [5].

Furthermore, in [11] we introduced the Extended SSNs (ESSNs) to deal with more complex biological laws splitting the set T of all the transitions of the model into two subsets: T_{ma} and T_g. The former subset contains transitions (that are called *standard*) whose rates are specified as MA laws. The latter includes instead all the transitions (that are called *general*) whose random firing times have rates that are defined by means of general real functions. In our definition, we assumed that the general function associated with a transition $t \in T_g$ is a real function which depends only on time and on the input places of t. So, if $x_{p,c}(\nu)$ represents the average number of tokens of color c in the place p at time ν, then the rate at which the instance $\langle t, c, c' \rangle$, $t \in T_g$ will move tokens with color c in place $x_{p,c'}(\nu)$ is given by $f_{\langle t,c,c' \rangle}(\hat{x}(\nu), \nu)$, where $\hat{x}(\nu)$ is the vector characterized by the average number of tokens of the input places of transition t.

3   Materials and Methods

In this section, we report our extension of the Relapsing-Remitting Multiple Sclerosis (RRMS) model defined with the ESSN formalism and originally presented in [11] assuming that the cells may move within a cubic grid. In Fig. 1a) is depicted a portion of the CNS, showing: the neuron with its myelin sheath, and the 7 elements characterizing the MS disease distributed within a 3D cubic grid. The respective

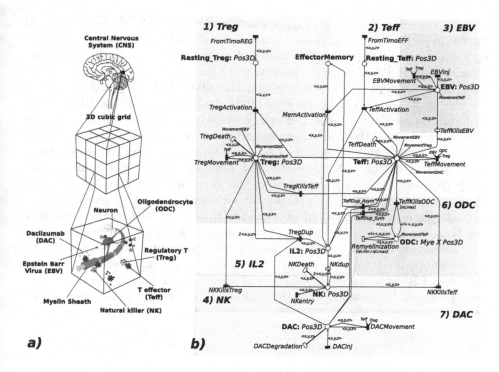

Fig. 1. a) Representation of the three-dimensional model. b) The ESSN model.

ESSN model is shown in Fig. 1b), consisting of 13 places and 25 transitions. For the sake of clarity, the white transitions are standard transitions, while the black ones are general transitions. This model is organized in seven modules corresponding to the biological entities characterizing RRMS. Briefly, the EBV module simulates the virus reactivation by means of a series of injections of virus particles in the system at given times, while the Treg and Teff modules encode the activation of the T cells, the annihilation of the virus by the Teff action, the control mechanism of the Treg over the Teff. The NK module describes the killing of self-reactive Teff and Treg cells respectively, due to NK cells. The IL2 module is focused on the IL2 role. IL2 is consumed by the Treg, Teff and NK functions and it is produced by the Teff activation. The ODC module describes instead the ODC behaviour, characterized particularly by the damage caused by Teff cells on ODC cells. Indeed, when the myelin level reaches the lowest value, an irreversible damage occurs and a remyelination of the neurons is no more possible. Finally, the DAC module encodes the drug administration and its pharmacokinetics inhibition of the expansion of Treg and Teff.

The model is characterized by four color classes: $PosX$, $PosY$, and $PosZ$ representing the coordinates of the position of a molecule in a 3D cubic grid; Mye encoding the myelination levels of ODC. Mye is divided into five static subclasses ranging from $Lmin$ (no myelination) to $Lmax$ (full myelination). Then, all the places except the ODC and EffectorMemory are characterized by the color domain defined as $Pos3D = PosX \times PosY \times PosZ$, i.e. the three-dimensional Cartesian product of the three coordinates color classes. Instead, the ODC place is characterized by the three coordinates plus the myelination levels, so that its color domain is $Pos3D \times Mye$. Finally, the EffectorMemory place has neutral color domain. Moreover, we assume that the EBV, Teff, Treg and DAC cells are able to move in all the cubic cells of the grid. Practically, the EBVs move uniformly in all the cells, the Teff cells move with higher probability towards a location in which there is higher concentration of EBV, and Treg and DAC cells move with higher probability towards a location in which there is higher concentration of Teff cells. Hereafter, the notation of the color combinations $\langle p_x, p_y, p_z \rangle$ and $\langle q_x, q_y, q_z \rangle$, representing the location coordinates, is simplified to $\langle \mathbf{p} \rangle$ and $\langle \mathbf{q} \rangle$, respectively. In particular, we define $x_{CellType_{\langle \mathbf{p} \rangle}}$ as the number of $CellType$ in the location $\langle \mathbf{p} \rangle$ at a specific time point. Hence, the movement functions can be defined as follows. The transition **EBVMovement** simulates the movements of EBV cells from point (with coordinates represented by the color combination) $\langle \mathbf{p} \rangle$ to point $\langle \mathbf{q} \rangle$. The speed of this movement (the rate of transition EBVMovement) is uniform in all directions and is captured in the following formula by assuming that the probability to move is equally distributed among all the grid cells.

$$f_{\langle EBV Movement, \mathbf{p}, \mathbf{q} \rangle}(\hat{x}(\nu), \nu) = r_{moves} * p_{\langle \mathbf{q} \rangle}^{EBV} * x_{EBV_{\langle \mathbf{p} \rangle}}$$

where r_{moves} is a coefficient that we set equal to 0.1 in our numerical experiments. Differently, the transition **TeffMovement** simulates the movement of Teff cells from point $\langle \mathbf{p} \rangle$ to point $\langle \mathbf{q} \rangle$, and its speed is inversely related to the number of EBV cells in $\langle \mathbf{p} \rangle$ (since more the virus in $\langle \mathbf{p} \rangle$ less the Teff cells are tempted to leave the position) and depends on the number of EBV in $\langle \mathbf{q} \rangle$ (a greater number of EBV cells leads to a higher probability to move into that location). This is captured by the following formula

$$f_{\langle Teff Movement, \mathbf{p}, \mathbf{q} \rangle}(\hat{x}(\nu), \nu) = r_{moves} * (exp(-\frac{x_{EBV_{\langle \mathbf{p} \rangle}}}{C_{EBV}})) * p_{\langle \mathbf{q} \rangle}^{Teff} * x_{Teff_{\langle \mathbf{p} \rangle}}$$

where r_{moves} is again set equal to 0.1; the second term of the function, defined as $exp(-\frac{x_{EBV_{\langle \mathbf{p} \rangle}}}{C_{EBV}})$, varies in the interval $[1, 0)$, simulating the decreasing of the movement velocity with respect to the number of EBV cells present in the starting point; $p_{\langle \mathbf{q} \rangle}^{Teff} = \frac{x_{EBV_{\langle \mathbf{q} \rangle}}}{EBV_{tot}}$ represents the probability to move in the cell with coordinates $\langle \mathbf{q} \rangle$ where EBV_{tot} is the total number of EBV in the grid at time ν; and C_{EBV} is an experimental constant that we set equal to 1000. All these quantities are functions of the time ν which is omitted in the formula to keep the notation simpler.

Transitions **TregMovement** and **DACMovement** represent the movements of the Treg and DAC cells (respectively) from point $\langle \mathbf{p} \rangle$ to point $\langle \mathbf{q} \rangle$. Similarly to what explained for transition TeffMovement, their speeds are inversely related to the number of Teff and T (= Treg+Teff) cells in $\langle \mathbf{p} \rangle$ and depend on the number of Teffs and Ts in $\langle q \rangle$. The detailed expressions of the formulas that encode the firing rate dependencies for these two transitions, with the information regarding all the transitions and the files exploited thorough this study, are freely available at https://github.com/qBioTurin/ESSNandRRMS/tree/master/DeterministicModel/Multidimensional.

We exploited the GreatSPN tool [1] to develop the ESSN model Fig. 1b), in particular the corresponding system of ODEs is automatically generated from the ESSN model using the C/C++ module PN2ODE embedded in GreatSPN. In details, the ODEs system is defined by one equation for each pair $\langle p, c \rangle$, where p is a place and $c \in cd(p)$ is a color which encodes a coordinate in the 3D grid, plus the myelination level (the latter only in place ODC). Differently, the system of SODE has been generated with the SNespression tool (http://di.unito.it/~depierro/SNexpression) and integrated with the definition of the functions for the *general* transitions and the initial marking. In this case there is one equation for each pair $\langle p, \hat{c} \rangle$, where p is a place and $\hat{c}$ corresponds to a subset of $cd(p)$ with equivalent behavior. The SODE for a given $\hat{c}$ is representative of all ODEs in the *unfolded* system for all $c \in \hat{c}$. The identification of the equivalence classes in $cd(p)$ and of the generation of the representative SODE is completely automatized.

4 Results

In this work we studied the RRMS considering a tissue portion explicitly modeled through a cubic grid consisting of 27 cubic cells (Fig. 1a)). To achieve this, we defined the color classes $PosX = \{x_1, x_2, x_3\}$, $PosY = \{y_1, y_2, y_3\}$ and $PosZ = \{z_1, z_2, z_3\}$. For all the simulations, we assumed 500 ODC with level L_{max} of neuronal myelinization, 1687 resting Teff cells, 63 resting Treg cells, 375 NK cells and 1000 IL2 molecules, and zero cells in the other places (for more details see [11]).

This model is equivalent to a system of 433 ODEs, but with few assumptions it is possible to derive the corresponding reduced ODEs system including only the symbolic equations. In details, let us define the set of all the 27 location coordinates as $\mathbf{P} = \{\langle p_x, p_y, p_z \rangle, p_x \in PosX, p_y \in PosY, p_z \in PosZ\}$. Then, we consider three disjoint subsets of $\mathbf{P}$, namely $P1$, $P2$, $P3$; the first two correspond to grouping the EBV and DAC injection locations, respectively, while the third $P3$ groups all the remaining locations. For simplicity, and to maintain the symmetries into the system as well, the EBV and DAC injection locations do not change over the simulation time and do not overlap. Given this, it is possible to derive the symbolic ODEs (SODEs) system characterized by 49 equations only. Indeed, the 433 original equations can be partitioned into 49 groups of similar equations. Each group is expressed in the reduced model by one representative equation. The grouping derives from the observation that the behaviors of the modeled elements do not depend on their actual positions, but only on the presence of the EBV and/or DAC cells. When the grid size grows the number of groups does not change, as long as the number of locations where different quantities of EBV and/or DAC cells appear is fixed; instead the size of each group of equivalent ODEs increases with the grid size. A further reduction is represented by the number of terms in each SODE, representative of each group of ODEs, with respect to the number of terms appearing in the ODEs in the equivalence class. This reduction is due to the factorization obtained by exploiting symmetries. Other examples are reported in Table 1, where the R file dimension and the number of differential equations of the complete and reduced models are compared considering different cubic grid dimensions, from $3 \times 3 \times 3$ to $5 \times 5 \times 5$. It is easy to see that an increasing number of locations is associated with an increase in the number of ODEs and of the R file containing them, while the SODE system does not change. Note that when the $5 \times 5 \times 5$ grid is considered, the ODEs generation procedure fails because it exceeds the available memory.

The advantage can also be observed from the point of view of the simulation time, we obtained a speed up from 8.927205 h to 12.76043 s on an Intel Xeon processor @ 2 GHz, by using one core. Note that the simulation was performed considering $3 \times 3 \times 3$ cubic grid, one year interval and assuming EBV injections at regular times (every two months), and each injection introduces into the system 10000 EBV copies.

Table 1. Comparing the ODE and SODE system, varying the cubic grid dimension.

Number of locations	R File dimension ODEs/SODEs	Number of ODEs/SODEs
27 ($3 \times 3 \times 3$)	0.43 MiB/0.023 MiB	433/49
64 ($4 \times 4 \times 4$)	5.0 MiB/0.023 MiB	1025/49
125 ($5 \times 5 \times 5$)	Out of memory/0.023 MiB	2001/49

A possible evolution of the system is shown in Fig. 2, where the red circles represent the location in which EBV is injected. For each plot, the three rows represent the z-planes and the columns refer to the time points in which the injections are done. Fixing the time point and the z-plane, the corresponding 3×3 square reports the number of ODCs damaged into the nine grid cells obtained varying the x and y coordinates. As expected, the panel A of Fig. 2 shows the progressive accumulation of ODC irreversibly damaged until day 365. Instead, in panel B of Fig. 2 is reported the results of the simulation of the DAC effect. In details, every month after two months of simulation, two injections are simulated (green squares) introducing 300 DAC copies for each administration. These results agree with those proposed in [11] since the number of irreversibly damaged ODCs decreases in the case with DAC administration with respect to the case in which no drug is injected. With DAC the percentage of irreversibly damaged ODCs ranges from 28% to 45%, while with no DAC the number of irreversibly damaged ODCs is between 70% and 85%.

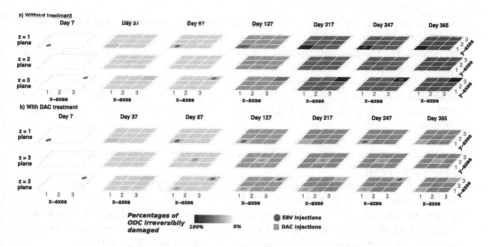

Fig. 2. Percentages of ODCs irreversibly damaged a) without and b) with DAC treatment. (Color figure online)

5 Conclusion

In this work we extended the model presented in [11] including the spatial coordinates of all entities in a cubic tissue portion. This gives the opportunity to model more realistic scenarios, where different quantities of virus enter into the system from different directions.

Moreover, we described how the intrinsic symmetries of the derived ESSN model may be automatically exploited to reduce the complexity of the analysis step. This allows us to study models which are independent from the grid size, while with the classical approach it is hard to generate the ODEs system corresponding to the model with a $5 \times 5 \times 5$ grid.

As further work, we will focus our experiments on the dosage of DAC and on the representation of the DAC pharmacokinetics in order to simulate the up taking of DAC by the body, its biotransformation and the distribution of DAC and its metabolites in the tissues.

References

1. Amparore, E.G., Balbo, G., Beccuti, M., Donatelli, S., Franceschinis, G.: 30 years of GreatSPN. In: Fiondella, L., Puliafito, A. (eds.) Principles of Performance and Reliability Modeling and Evaluation. SSRE, pp. 227–254. Springer, Cham (2016). https://doi.org/10.1007/978-3-319-30599-8_9
2. Beccuti, M., Capra, L., De Pierro, M., Franceschinis, G., Follia, L., Pernice, S.: A tool for the automatic derivation of symbolic ODE from symmetric net models. In: Proceedings of 27th IEEE International Symposium on Modeling, Analysis, and Simulation of Computer and Telecommunication Systems, MASCOTS 2019, Rennes, France, 21–25 October 2019, pp. 36–48 (2019)

3. Beccuti, M., et al.: From symmetric nets to differential equations exploiting model symmetries. Comput. J. **58**(1), 23–39 (2015)
4. Bielekova, B.: Daclizumab therapy for multiple sclerosis. Cold Spring Harb. Perspect. Med. **9**(5), a034470 (2019)
5. Capra, L., De Pierro, M., Franceschinis, G.: Computing structural properties of symmetric nets. In: Campos, J., Haverkort, B.R. (eds.) QEST 2015. LNCS, vol. 9259, pp. 125–140. Springer, Cham (2015). https://doi.org/10.1007/978-3-319-22264-6_9
6. Chiola, G., Dutheillet, C., Franceschinis, G., Haddad, S.: Stochastic well-formed coloured nets for symmetric modelling applications. IEEE Trans. Comput. **42**(11), 1343–1360 (1993)
7. Dutta, R., Trapp, B.: Mechanisms of neuronal dysfunction and degeneration in multiple sclerosis. Prog. Neurobiol. **93**(1), 1–12 (2011)
8. Guan, Y., Jakimovski, D., Ramanathan, M., Weinstock-Guttman, B., Zivadinov, R.: The role of Epstein-Barr virus in multiple sclerosis: from molecular pathophysiology to in vivo imaging. Neural Regen. Res. **14**(3), 373–386 (2019)
9. Kurtz, T.G.: Solutions of ordinary differential equations as limits of pure jump Markov processes. J. Appl. Probab. **1**(7), 49–58 (1970)
10. Pernice, S., et al.: Integrating Petri nets and flux balance methods in computational biology models: a methodological and computational practice. Fundamenta Informaticae **171**(1–4), 367–392 (2019)
11. Pernice, S., et al.: A computational approach based on the colored Petri net formalism for studying multiple sclerosis. BMC Bioinform. **20**, 1–17 (2019)
12. Sospedra, M., Martin, R.: Immunology of multiple sclerosis. Semin. Neurol. **36**(2), 115–127 (2016)
13. Virtanen, J., Jacobson, S.: Viruses and multiple sclerosis. CNS Neurol. Disord. Drug Targets **11**(5), 528–544 (2012)

Observability of Bacterial Growth Models in Bubble Column Bioreactors

Paola Lecca[1,2](✉) and Angela Re[3]

[1] Faculty of Computer Science, Free University of Bozen-Bolzano,
NOI Techpark, via Volta 13/A, 39100 Bolzano, Italy
Paola.Lecca@unibz.it
[2] Member of Gruppo Nazionale per l'Analisi Matematica,
la Probabilità e le loro Applicazioni, Istituto Nazionale di Alta Matematica
"Francesco Severi", Città Universitaria La Sapienza di Roma,
P.le A. Moro 5, 00185 Rome, Italy
[3] Centre for Sustainable Future Technologies, Fondazione Istituto Italiano
di Tecnologia, Environment Park - Parco Scientifico Tecnologico per l'Ambiente,
Via Livorno 60, 10144 Turin, Italy
Angela.Re@iit.it

Abstract. Observability being the dual of controllability is an advisable property for any dynamic model of bio-based chemical production processes encompassing substrate consumption, bacterial growth and products formation. In this study we show a mathematical model of these processes and present a novel observability analysis. The invertibility properties of the observability mapping of this model in a space-time domain are analysed independently of the discretization of such domains and indicate the existence of subdomains where the measurement of the output may not be sufficiently accurate to allow reconstructing the states of the system.

Keywords: Bioreactor · Bacterial growth · Advection-diffusion equation · Observability

1 Introduction

Valorization of highly diversified gaseous streams such as synthesis gas into new-to-nature, added-value compounds by biocatalysts is a cornerstone to set the course for the transition towards sustainable business models. Even though some long-sought transfer of bench-scale processes into industrially relevant platforms aimed at the gaseous, inorganic carbon fixation into valuable compounds have been attained, a great deal of improvement is still needed to achieve cost-effective process design and control technology. The adoption of a mechanistic modelling framework is expected to help in process design only when it does not limit itself to describe the phenomenon of interest faithfully but also shows the key characteristic of observability. Indeed, the fulfilment of this property ensures the

© Springer Nature Switzerland AG 2020
P. Cazzaniga et al. (Eds.): CIBB 2019, LNBI 12313, pp. 309–322, 2020.
https://doi.org/10.1007/978-3-030-63061-4_27

possibility to determine uniquely the state of the system from its measurable outputs. The determination of the observability property, already quite complex for systems with non-linear dynamics, becomes even more so in the case in which the non-linear dynamics has a spatial as well as a temporal domain. For such a dynamics, the observability map, i.e the vector function relating the internal states of a system to its external outputs, has a tensorial structure, i.e. it is a multidimensional (spatial and temporal) matrix. On such a structure the results of the application of classical systems control theory gives results depending on the pace size of the discretization of space domain [18]. Classical methods [20, 36] to assess the observability of non-linear systems calculate the derivatives of the output (by using the Lie derivative), build with them the columns of the observability matrix and require that the rank of this matrix is full (i.e. equal to the dimension of the state-space of the system). In this study, we present a method to establish whether a model is observable or not from given outputs, and apply this method to a model of biomass growth, substrate consumption, and products formation. Our method studies the differentiability property of the observability map and establishes the conditions for its invertibility, without resorting to the numerical calculation of the rank of the observability map, which is cumbersome to implement as the map matrix has a tensor structure, and not reliable given its dependency on the choice of the spatial discretization. As an illustration we show how this method is applied to a model of a bubble column reactor for converting carbon monoxide (CO) gas into ethanol through a *C. autoethanogenum* biocatalyst. We note the model allow to compute an agile closed-form solution of the spatio-temporal dynamics of the species of interest.

In Sect. 2 we describe the reactor and the fermentation processes taking place in it, in Sect. 3 we describe our bacterial growth and fermentation model, and in Sect. 4 we present the application of our observability analysis approach to this case study. In each of these sections we provide the reader with the appropriate literature references to afford a timely comparison of all our analyses and results with the current state of the art. In Sect. 5, we finally draw the conclusions.

2   Bubble Column Reactor

Gas fermentation is a technologically appealing option to valorize single-carbon (C1) gas resulting from manifold industrial processes into chemicals and fuels [31, 33, 40]. It can employ anaerobic autotrophic acetogenic bacteria, aerobic chemolithoautotrophic bacteria [4], and methanotrophs [6]. Acetogenic bacteria - which rely on the energy-conserving Wood-Ljungdahl metabolic pathway for C1 fixation into cell biomass - stand out among the most highly pursued cell factories in syngas fermentation bioprocesses [35]. The development of producer strains requires a fundamental understanding of acetogens carbon, energy and redox metabolism [25, 26, 41]. *Clostridium autoethanogenum* and *Clostridium Ljung-dahlii*, owing to the burgeoning number of published studies, are common model acetogens [21,32]. The bioreactor configuration is a crucial choice to overcome mass transfer limitations and favour high cell density and thus obtain elevated syngas conversion [5]. Bubble column reactor is one of the bets for industrially relevant configurations because of low capital and operational costs along with suitable heat and mass transfer rates. Importantly, this kind of reactor handles gaseous substrates mixing by gas bubbling without the need for power-consuming mechanical agitation [39]. Directly accessing to the behaviour of chemical species within syngas fermentation reactors is strongly conditioned by the technological challenges of settling sensor configurations compatible with process control. The growing interest into syngas fermentation process development has stirred the construction of computational spatio-temporal models of syngas fermenting reactors [13,14]. The assessment of model observability can be advantageously adopted to identify particularly effective modelling choices. This study presents an innovative observability analysis of a mathematical model inspired, at least in part, to the mass balance equations of gas substrates and biomass by-products previously adopted for a *C. autoethanogenum* biocatalyst grown in a bubble column reactor. This study presents a model of a bioreactor consisting of a bubble column for converting carbon monoxide (CO) gas into ethanol through a *C. autoethanogenum* biocatalyst. CO gas flows from the bottom of the bubble column. A pure culture of *C. autoethanogenum* is subsequently injected at the bottom of the column; therein cells are dispersed in the liquids and consume the dissolved gas and release by-products such as ethanol and acetic acid. A spatial gradient establishes in the column since gas concentration decreases as gas flows up due to cellular consumption. Consequently, cellular growth and by-product secretion are affected by spatially varying dissolved gas concentrations. The model accounts for four species representing the biomass, the CO substrate in the liquid phase, and two by-products - ethanol and acetic acid. While the equations of the ethanol and acetate dynamics are taken from Chen's work [13], in our model the dynamics of the substrate is described by an advection-diffusion equation with no sink/source terms depending on the biomass. This allows the calculation of closed-form solution for the substrate dynamics, that is used in the calculation of the bacterial growth rate. The sink/source term in the equation for the biomass is just governed by this growth rate. The model embeds parameters estimated in experimental studies [12,13]. Our modelling choices yield a closed-form solution of the spatio-temporal dynamics of the species of interest, and consequently an analytical treatment of the model observability, where previous studies rely on numerical algorithms for the solution of systems of partial differential equation and/or numerical algorithm for dynamic flux balance analysis (FBA) [17]. These algorithms are notoriously sensitive to numerical instabilities and also to small errors in the specification of reaction networks. Small errors could be an inaccurate value of a parameter or missing a variable in a reaction or missing a single reaction. These errors are often due to an incomplete *a priori* knowledge of the physical processes that govern the interactions between molecular species. On the other hand during the modelling phase it is not possible to have the complete picture of the system. In FBA methods, errors in a single reaction could result in a entirely wrong flux distribution [27,34].

The model accounts for the following assumptions: (a) the uptake rates of CO mimics Monod kinetics with inhibition by products; (b) products-induced inhibition is only activated after a threshold concentration is achieved; (c) biomass growth is a function of substrate uptake rate; (d) products formation is defined by the substrate-to-products conversion rates of the biomass; (e) substrate, products and biomass are controlled by the diffusion/advection equation.

3   Mathematical Model

The syngas feed CO (hereafter "the substrate") is fed into the bottom of the bubble column of the reactor. Gas concentration decreases as gas flows up due to cellular consumption. Therefore, the column has spatially varying dissolved gas concentrations that affect cellular growth and products synthesis.

3.1   Modelling Diffusion-Advection of Substrates

We consider that the substrate concentration S is only diffusion-advection controlled and that there is no source or sink term in its dynamics. Our model reflects two transport mechanisms: (i) advective transport with the mean flow of the liquid in the reactor, and (ii) the diffusive transport due to concentrations gradients. We do not consider the source and/or sink term, since we model the case in which the substance is completely mixed over the cross-section, implying that a source/sink term is considered to mix *instantaneously* over the cross-section.

In cylindrical coordinates (t, r, θ, z) (where $0 \leq r \leq R$, $0 \leq z \leq L$, and $t \in [t_{\text{in}}, t_{\text{fin}}]$), the diffusion-advection equation describing the transport of the particles of the species S has the following form

$$\frac{\partial S}{\partial t} = \boldsymbol{\nabla} \cdot (D_S \boldsymbol{\nabla} S) - \boldsymbol{\nabla} \cdot (\boldsymbol{v} S)$$

$$= D_S \left[\frac{1}{r} \frac{\partial S}{\partial r} + \frac{\partial^2 S}{\partial r^2} + \frac{1}{r^2} \frac{\partial^2 S}{\partial \theta^2} + \frac{\partial^2 S}{\partial z^2} \right] - \boldsymbol{\nabla} \cdot (\boldsymbol{v} S) + R \qquad (1)$$

where D_S is the diffusion coefficient of S (here assumed to be space and time independent), $\boldsymbol{v}$ is the velocity field that the quantity S is moving with, and R R describes sources or sinks of the quantity S.

The geometry of the bioreactor considered in this study is such that $R \ll L$. Moreover we assume there are neither sources nor sinks, and the velocity field describes an incompressible flow, so that the Eq. (1) becomes the well known linear diffusion-advection equation

$$\frac{\partial S}{\partial t} + c \frac{\partial S}{\partial z} = \nu \frac{\partial^2 S}{\partial t^2} \qquad (2)$$

where c is the advection velocity (here assumed to be constant) and ν is the fluid velocity. As in the procedure of Mojtabi et al. [30], we impose homogeneous Dirichlet boundary conditions $S(0,t) = S(L,t) = 0$ and the initial condition $S(z,0) = \sin \pi z$.

To obtain a closed form solution, we apply the change of variables suggested in [30], i.e.

$$S(z,t) = Y(z,t)e^{\alpha z + \beta t}, \tag{3}$$

Substituting the Eq. (3) in Eq. (2) and simplifying by the exponential, we obtain the governing equation for $Y(z,t)$

$$\frac{\partial Y}{\partial t} = -[\beta + \alpha(c - \alpha\nu)]Y - (c - 2\alpha\nu)\frac{\partial Y}{\partial x} - (c - 2\alpha\nu)\frac{\partial Y}{\partial z} + \nu\frac{\partial^2 Y}{\partial z^2} \tag{4}$$

where α and β are free parameters. In order to obtain the standard heat equation from the Eq. (4), we set

$$\beta + \alpha(c - \alpha\nu) = 0$$
$$c - 2\alpha\nu = 0$$

that gives $\alpha = \frac{c}{2\nu}$ and $\beta = -\frac{c^2}{4\nu}$. Therefore, the equation for Y becomes

$$\frac{\partial Y}{\partial t} = \nu\frac{\partial^2 Y}{\partial z^2} \tag{5}$$

subject to the homogeneous conditions $Y(0,t) = Y(L,t) = 0$ and the initial condition

$$Y(z,0) = -\sin \pi x e^{-\alpha z}$$

The Eq. (5) can be solved with the method of separation of variables, i.e. we set

$$Y(z,t) = P(z)Q(t) \tag{6}$$

and, omitting the algebraic passages, this leads to the solution

$$Y(z,t) = \sum_{k=0}^{\infty}\left(A_k \sin\frac{k\pi z}{2} + B_k \cos\frac{k\pi z}{2}\right)e^{-\frac{\nu k^2 \pi^2}{4}t} \tag{7}$$

where the boundary conditions impose that $A_{2p+1} = B_{2p} = 0$ for $p = 0, 1, 2, \ldots$. Therefore the Eq. (7) becomes

$$Y(z,t) = \sum_{p=0}^{\infty}\left(A_{2p}\sin(p\pi z)e^{-\nu p^2 \pi^2 t} + B_{2p+1}\cos\left(\frac{2p+1}{2}\pi z\right)e^{-\nu\frac{(2p+1)^2}{4}\pi^2 t}\right). \tag{8}$$

Mojtabi et al. [30] showed that the coefficients A_{2p} and B_{2p+1} can be calculated using the orthogonality properties of the Fourier polynomials, and the standard

trigonometrical relations, and that, when the viscosity goes to zero the final
solution becomes

$$S(z,t) = 8\pi^2 \left(\frac{\nu}{c}\right)^3 \exp\left[\frac{c}{4\nu}\left(z+1-\frac{c}{2}t\right)\right]$$

$$\times \left[\sum_{h=0}^{\infty}(-1)^h\left(2\,h\sin(h\pi z) + (2h+1)\cos\left(\frac{2h+1}{2}\pi z\right)\right)\right]. \quad (9)$$

For very small values of the kinematic viscosity the calculation of the solution
according the Eq. (9) produces a numerical overflow due to the presence of
the exponential term. In literature, we find a number of method to treat this
ill-behaviour (see for example [1,3,7–9,11,15,16]). Here we adopt the solution
of Mojtabi et al. [30] that is the most computationally efficient for our case
study. Assuming periodic boundary conditions instead of Dirichlet boundary
conditions, the periodic solution of the diffusion-advection equation is the Fourier
solution, that in case of inviscid liquid is approximated as follows

$$S(z,t) = -\sin(\pi(z-ct))e^{\nu\pi^2 t} \approx -\sin(\pi(z-ct))(1-\pi^2 t + \mathcal{O}(\nu^2)). \quad (10)$$

The solution given in Eq. (10) will be used in the equations for the dynamics
of biomass, as we will see in the next section.

3.2 Biomass and Product Equations

We model the evolution of the biomass concentration X with the following
equation

$$\frac{\partial X}{\partial t} = \nu\frac{\partial^2 X}{\partial z^2} - c\frac{\partial X}{\partial z} + \mu(\tilde{S})X \quad (11)$$

where

$$\mu(\tilde{S}) = \frac{\mu_{\max}\tilde{S}}{K_S + \tilde{S}} \quad (12)$$

is the bacterial growth rate and $\tilde{S} = \nu_{\text{uptake}}S$. The term ν_{uptake} is the rate of
uptake of the substrate, therefore $\mu(\tilde{S})$ is the growth rate expressed as a function
of the actual portion of substrate consumed. Assuming a n-product inhibited
Monod kinetics, the uptake rate has the following expression

$$\nu_{\text{uptake}} = \frac{\nu_{\text{uptake\_max}}\,S}{\left(K_M + S\right)\left(1 + \frac{\sum_{j=1}^{n} P_j}{K_I}\right)} \quad (13)$$

where $\nu_{\text{uptake\_max}}$ is the maximum uptake rate, K_M the saturation constant and
K_I the inhibition constant. From the Eq. (13) in Eq. (12), we obtain that

$$\mu(\tilde{S}) = \frac{\mu_{\max}\,\nu_{\text{uptake\_max}}\,S^2}{\nu_{\text{uptake\_max}}\,S^2 + K_S(K_M + S)\left(1 + \frac{\sum_{j=1}^{n} P_j}{K_I}\right)}. \quad (14)$$

From the Eq. (14) we see that the inhibitory effect of the products becomes preponderant as $\sum_{j=1}^{n} P_j \geq S$, because the infinity order of the denominator reaches the infinity order of the numerator. Based on this observation, in Eq. (14), we set $\frac{\sum_{j=1}^{n} P_j}{K_I} \approx \frac{nS}{K_I}$. This choice is justified by the fact that the closed form of the spatio-temporal dynamics of the products is the same as that of the substrates when the dynamics of the substrate is controlled mainly by advection and diffusion, and no source/sink term is present.

The equations for the product P_i, templated according to Chen et al. [13] are:

$$\frac{\partial P_i}{\partial t} = M_i \nu_i X - c\frac{\partial P_i}{\partial z} + \nu\frac{\partial^2 P_i}{\partial z^2}, \qquad (15)$$

where ν_i is the flux of the product and M_i is its molecular weight, and i = Ethanol, Acetate. The values of the parameters and their short description and literature reference are reported in Table 1. We refer the reader to the link in [23], where the Python code (constantly updated) and the .mpeg4 movie showing the numerical simulations of the time evolution of the spatial distribution of $E(z,t)$, $A(z,t)$, and $X(z,t)$, and $S(z,t)$. Four screenshots of this movie at time at t = 72 h are shown in Fig. 1. Data reported in scientific literature are not apt to validate the spatially resolved predictions due to the technical challenge represented by acquiring stepwise directly comparable measurements inside syngas fermentation reactors. When referring to alternative modelling studies of bio-catalysed conversions of gaseous substrates into end-products, caution is required due to peculiarities in fermentation process configuration and modelling strategies of both metabolic and physical processes occurring in the reactor. In this sense, we set a tentative comparison with Chen et al. [13], which presents a spatio-temporal metabolic model for bubble column reactor with a syngas fermentation bacterium, finding the trends of modelled variables to appear in line with ours.

Table 1. The table reports the parameters pertaining to the reactor geometry, the advection-diffusion transport mechanisms and to bacterial growth characteristics. CDW stands for "Cell Dry Weight".

Symbol	Description	Measure	Measure unit	Reference
L	Reactor height	25	m	In this study
α	Advective constant	0.5	m^{-1}	In this study
β	Diffusion coefficient	0.25	h^{-1}	In this study
ν	Water kinematic viscosity	0.6959×10^{-6}	mm^2/s	[19]
c	Advection velocity	0.25	m/h	In this study
μ_{max}	Maximal specific growth rate	0.195	h^{-1}	[29]
K_I	Inhibition constant	10	g/L	[13]
K_S	Half velocity constant	0.033	h^{-1}	[29]
ν_{max}	Maximum specific CO uptake rate	34.364	$g_{CDW}^{-1}\ h^{-1}$	[29]
K_M	Monod half saturation constant (CO)	0.02	mmol/L	[13]
M_A	Acetate molecular weight	0.06	g/mmol	[2]
ν_A	Acetate flux	3.9	mmol $g_{CDW}^{-1}\ h^{-1}$	[41]
M_E	Ethanol molecular weight	0.046	g/mmol	[2]
ν_E	Ethanol flux	3.8	mmol $g_{CDW}^{-1}\ h^{-1}$	[41]

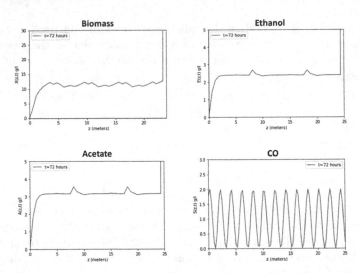

Fig. 1. Numerical simulation of the spatial distribution of biomass, ethanol, acetate and CO, after 72 h from the substrate injection. Except for $z = 0$ and $z = L$, the presence of edges in the curve is due to the spatial resolution chosen in these simulations ($= 0.65$ m). Biomass, ethanol and acetate curve are instead non-differentiable in $z = 0$ and $z = L$.

4 Observability

The equations for the substrate, biomass and products describe a non-linear input-free system Σ. The system is input-free because of the absence of source/sink term in the equation of substrate. The dynamics of such systems can be described by the following general equations

$$\Sigma : \quad \dot{x}(z,t) = f(z,t,x), \quad f : [z_0, L] \times [t_0, T] \times E \subset \mathbb{R}^n \longrightarrow \mathbb{R}^n \tag{16}$$

whose experimentally observed output is

$$y(z,t) = h(z,t,x), \quad h : [z_0, L] \times [t_0, T] \times E \subset \mathbb{R}^n \longrightarrow \mathbb{R}^n \tag{17}$$

In this system $x(z,t)$ is not available for direct measurement. The initial state $x(z,t_0) \equiv x_0$ is unknown and belongs to the set $E_0 \subset E$. The observability problem consists of determining whether there exist relations binding the state-variables $x(z,t)$ to outputs $y(z,t,x)$ and in defining them uniquely in terms of measurable quantities [10,38,43]. If such relations do not exist, the initial state of the system cannot be deduced from observing its output behaviour. In such a case, the lack of observability implies the lack of identifiability, and no chance to control the system, because infinitely many parameter sets produce exactly the same output for different values of x_0 and thus the model parameters cannot be estimated from any experimental measurements.

We suppose that the k-th order derivatives of f and h exist for every $x \in E$, every $t \in [t_0, T]$, every $z \in [z_0, L]$, and that $y(z, t)$ is smooth, so that we can approximate $y(z^*, t) \equiv y(z, t)|_{z=z^*}$ by a truncated Taylor series as follows

$$y(z^*, t) = y(z^*, t_0) + \dot{y}(t_0)(t - t_0) + \frac{\ddot{y}(z^*, t_0)}{2!}(t - t_0)^2 + \cdots + \frac{y^{(k)}(z^*, t^*)}{k!}(t - t_0)^k$$

(18)

where $t^* \in (t_0, T)$, and

$$y(z^*, t_0) = h(z^*, t_0, x(t_0)) \equiv h_0(z^*, t_0, x(t_0))$$

$$\dot{y}(z^*, t_0) = \frac{\partial h_0}{\partial t}(z^*, t_0, x(t_0)) + \left(\frac{\partial h_0}{\partial x}(z^*, t_0, x(t_0)) \right) f(z^*, t_0, x(t_0))$$

$$\equiv h_1(t_0, x(t_0))$$

$$\vdots$$

$$y^{(k-1)}(z^*, t_0) = \frac{\partial h_{k-2}}{\partial t}(z^*, t_0, x(t_0)) + \left(\frac{\partial h_{k-2}}{\partial x}(z^*, t_0, x(t_0)) \right) f(z^*, t_0, x(t_0))$$

$$\equiv h_{k-1}(z^*, t_0, x(t_0)).$$

These equations can be written as a non-linear map [22], called the *observability mapping* of Σ, that is

$$\mathbf{Q}(z^*, x(t_0)) = \begin{bmatrix} y(z^*, t_0) \\ \dot{y}(z^*, t_0) \\ \vdots \\ y^{(n-1)}(z^*, t_0) \end{bmatrix} = \begin{bmatrix} h_0(z^*, x(t_0), t_0) \\ h_1(z^*, x(t_0), t_0) \\ \vdots \\ h_{k-1}(z^*, x(t_0), t_0) \end{bmatrix}.$$

(19)

The system Σ is completely observable if $\mathbf{Q}$ is a one-to-one mapping [22] [1]. In fact, by definition, a system described by Eqs. (16) is completely observable in set E_0 of initial states on time interval $[t_0, T]$ if there exist a one-to-one correspondence between the set E_0 and the set of observed outputs $y(t)$ for $t \in [t_0, T]$. Therefore, if $\mathbf{Q}$ is a one-to-one mapping from E_0 to $\mathbf{Q}(E_0)$, $x(t_0)$ can be uniquely determined from $y(z, t)$, and consequently $x(z, t)$ can be determined by recursion at any t.

While for linear system it is immediate that, if the Jacobi matrix of $\mathbf{Q}$ is full rank, the system is observable, for non-linear systems it can be shown that, even though the Jacobi matrix of $\mathbf{Q}$ is full rank $\mathbf{Q}$ needs not to be one to one. This fact has put the conditions of Kostyukovskii in error, and most of the subsequent analyses on the observability of non-linear systems have relegated it to local theories mostly of which developed by invoking the standard Implicit Function Theorem. New analyses always relying on the properties of the Jacobi matrix have been recently made in a different direction to determine the necessary and sufficient conditions for observability for non-linear systems (see for example [24, 28, 42, 43]). However, for systems which show both temporal and spatial dimensions, the numerical analyses of the invertibility of Jacobian

---

[1] Here $\mathbf{Q}$ is a tensor, as its entries are indexed with respect to £z£ and t.

matrix, and more in general the analyses using the classical control systems theory, give results that are dependent on the step size of the spatially discretised system [18].

A possible way to avoid running into approximation problems, and in order to provide necessary and sufficient conditions for observability, is suggested by the invertibility theorems of the differentiable mapping and diffeomorphism of class C^k. A mapping $f : U \longrightarrow V$, where U and V are Banach spaces, is a C^1-mapping if f is differentiable and if the natural mapping $Df : U \longrightarrow L(U, V)$, where $L(U, V)$ is the Banach space of linear continuous operators $A : U \longrightarrow V$ with the operator norm, is continuous. Similarly, the mapping f is a C^{k+1}-mapping if $D^k f$ is differentiable at any point $u \in U$ and the mapping $D^{k+1} f : U \longrightarrow L^{k+1}(U, V)$ is continuous. The map f is called C^k-diffeomorphism if f homeomorphically maps U onto V and f and f^{-1} are C^k-mappings. Finally, a local C^k diffeomorphism at the point $u \in U$ is defined as a mapping $f : U \longrightarrow V$ for which there exists a neighbourhood $U' \subset U$ of the point u such that the restriction $f|_{U'}$ of f to U' is a C^k-diffeomorphism between U and an open subset of V.

The following theorems [37] give the necessary and sufficient condition for a map f to be a *local* C^k-diffeomorphism (Theorem 1), and a C^k-diffeomorphism (Theorem 2).

Theorem 1. *Let A and B Banach spaces, let $U \subset A$ be an open set, and let $k \in \mathbb{N}$. The C^k-mapping $f : U \longrightarrow B$ is a local C^k-diffeomorphism at the point $u_0 \in U$ if and only if the derivative*

$$(Df)_{u_0} : A \longrightarrow B$$

is a continuously invertible operator.

Theorem 2. *Let A and B Banach spaces, let $U \subset A$ be an open set, and let $k \in \mathbb{N}$. The C^k-mapping $f : U \longrightarrow B$ is a C^k-diffeomorphism if and only if*

1. *f is injective*
2. *f is surjective*
3. *$(Df)_{u_0} : A \longrightarrow B$ is a continuously invertible operator.*

Let the ethanol concentration be the measurable output of the system, i.e. $y = E(z, t)$. The components of the observability mapping $\mathbf{Q}$ are therefore the derivatives with respect to the time of $E(z, t)$. From the Eq. (15), and since X is a differentiable curve, it can be easily proved that the closed form solution for $E(z, t)$ is differentiable too.

Consequently, $\mathbf{Q}$ is a differentiable and hence continuous mapping. Furthermore, $\mathbf{Q}$ is open, i.e. the image by $\mathbf{Q}$ of any open set in $\Omega = [z_0, L] \times [t_0, T] \times E$ is open, because each component $E^{(k)}(z, t)$ is analytic and different from zero. In order to use Theorem 2, we have to verify that $\mathbf{Q}$ is also injective and surjective. We note that $\mathbf{Q}$ satisfies the first sufficient condition for injectivity and surjectivity of Theorem 3 [37].

Theorem 3. *Let A and B Banach spaces, let $U \subset A$ be an open set, and Q : $A \longrightarrow B$ a continuous mapping. If the following conditions are satisfied*

1.

$$\inf_{q_1, q_2 \in \Omega;\ q_1 \neq q_2} \frac{\|Q(q_1) - Q(q_2)\|_B}{\|q_1 - q_2\|_A} > 0, \quad \text{almost everywhere}$$

2. Q is open,

then Q is injective and surjective.

The condition 1 of Theorem 3 has a simple geometrical interpretation. Q maps a segment l_{q_1, q_2} into a curve $Q(l_{q_1, q_2})$, which does not intersect itself (Fig. 2). The mapping **Q** in our case study satisfies the first condition of Theorem 3 because it is differentiable and its domain Ω is convex. Consequently, since **Q** is also an open map, **Q** is injective and surjective (by Theorem 3). We finally note that, since $E(z, t)$ is a monotonic function, the derivative of **Q** is a continuous invertible operator, and then we conclude that **Q** is C^k-diffeomorphism, and thus, by Theorem 2, the model of bioreactor is observable except at $z = 0$ and $z = L$ where the boundary conditions prevent the functions to have finite derivatives.

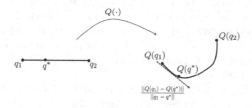

Fig. 2. Mapping of a line segment by a map Q satisfying the condition 1 of Theorem 3.

5   Conclusions

Observability is a key property of any modelling approach aiming to help in process design since it guarantees the possibility to determine uniquely the state of the system from measurable outputs. We verified the system observability property of equations used in the literature to model bacterial fermentation processes in the presence of advection-diffusion. Our analysis, based on the study of invertibility properties of differentiable mappings and diffeomorphisms, does not require numerical computations of the invertibility mapping, and thanks to this, it prevents the possibility that assessment of the observability depends on the discretization of the spatial and temporal domains. The system of our case study is formed by equations having the same structure for biomass and for the two products. In this system it is possible to obtain the closed-form solution of the equation for the substrate which, inserted in the biomass equation, is responsible for the oscillating and globally monotonous increasing trend which can be differentiated everywhere except at the top and bottom boundaries of the reactor. The absence of differentiability of the outputs in these regions of the reactor suggests that the measurements of the outputs carried out in these points do not allow an univocal reconstruction of the states of the entire system.

Finally, we wish to point out, that the observability analysis that we presented in this study is not specific to the case study considered here. It applies to any dynamic system whose observability is to be assessed given the experimental acquisition of the measurements of one or more variables (outputs). Our method studies the invertibility properties of the observability mapping and if these are incompatible with the necessary and sufficient conditions posed by Theorem 3, it asserts that the system is not observable.

References

1. Chetverushkin, B.N., Fitzgibbon, W., Kuznetsov, Y.A., Neittaanmäki, P., Periaux, J., Pironneau, O. (eds.): Contributions to Partial Differential Equations and Applications. CMAS, vol. 47. Springer, Cham (2019). https://doi.org/10.1007/978-3-319-78325-3. https://www.xarg.org/ref/a/3030086755/
2. PubChem (2020). https://pubchem.ncbi.nlm.nih.gov/
3. Andreianov, B.P., Coclite, G.M., Donadello, C.: Well-posedness for vanishing viscosity solutions of scalar conservation laws on a network. Discrete Contin. Dyn. Syst. A **37**(11), 5913–5942 (2017). https://doi.org/10.3934/dcds.2017257
4. Arenas-López, C., et al.: The genetic basis of 3-hydroxypropanoate metabolism in Cupriavidus necator h16. Biotechnol. Biofuels **12**(1) (2019). https://doi.org/10.1186/s13068-019-1489-5
5. Asimakopoulos, K., Gavala, H.N., Skiadas, I.V.: Reactor systems for syngas fermentation processes: a review. Chem. Eng. J. **348**, 732–744 (2018). https://doi.org/10.1016/j.cej.2018.05.003. http://www.sciencedirect.com/science/article/pii/S1385894718307903
6. Averesch, N.J.H., Kracke, F.: Metabolic network analysis of microbial methane utilization for biomass formation and upgrading to bio-fuels. Front. Energy Res. **6** (2018). https://doi.org/10.3389/fenrg.2018.00106
7. Bardos, C., Titi, E.S., Wiedemann, E.: The vanishing viscosity as a selection principle for the Euler equations: the case of 3D shear flow. Comptes Rendus Mathematique **350**(15), 757–760 (2012). https://doi.org/10.1016/j.crma.2012.09.005. http://www.sciencedirect.com/science/article/pii/S1631073X1200235X
8. Bianchini, S., Bressan, A.: Vanishing viscosity solutions of nonlinear hyperbolic systems. Ann. Math. **161**(1), 223–342 (2005). http://www.jstor.org/stable/3597345
9. Bressan, A., Guerra, G., Shen, W.: Vanishing viscosity solutions for conservation laws with regulated flux. J. Differ. Equ. **266**(1), 312–351 (2019). https://doi.org/10.1016/j.jde.2018.07.044
10. Chatzis, M.N., Chatzi, E.N., Smyth, A.W.: On the observability and identifiability of nonlinear structural and mechanical systems. Struct. Control Health Monit. **22**(3), 574–593 (2014). https://doi.org/10.1002/stc.1690
11. Chen, G.Q.G., Perepelitsa, M.: Vanishing viscosity limit of the Navier-stokes equations to the Euler equations for compressible fluid flow. Commun. Pure Appl. Math. **63**(11), 1469–1504 (2010). https://doi.org/10.1002/cpa.20332

12. Chen, J., Daniell, J., Grin, D., Li, X., Henson, M.A.: Experimental testing of a spatiotemporal metabolic model for carbon monoxide fermentation with clostridium autoethanogenum. Biochem. Eng. J. **129**, 64–73 (2018). https://doi.org/10.1016/j.bej.2017.10.018. http://www.sciencedirect.com/science/article/pii/S1369703X17303029

13. Chen, J., Gomez, J.A., Höffner, K., Barton, P.I., Henson, M.A.: Metabolic modeling of synthesis gas fermentation in bubble column reactors. Biotechnol. Biofuels **8**(1) (2015). https://doi.org/10.1186/s13068-015-0272-5

14. Chen, J., Gomez, J.A., Höffner, K., Phalak, P., Barton, P.I., Henson, M.A.: Spatiotemporal modeling of microbial metabolism. BMC Syst. Biol. **10**(1) (2016). https://doi.org/10.1186/s12918-016-0259-2

15. Constantin, P., Kukavica, I., Vicol, V.: On the inviscid limit of the Navier-stokes equations. Proc. Am. Math. Soc. **143**(7), 3075–3090 (2015). https://doi.org/10.1090/s0002-9939-2015-12638-x

16. Donea, J.: Finite Element Methods for Flow Problems. Wiley (2003). https://www.xarg.org/ref/a/0471496669/

17. Gomez, J.A., Höffner, K., Barton, P.I.: DFBAlab: a fast and reliable MATLAB code for dynamic flux balance analysis. BMC Bioinform. **15**(1) (2014). https://doi.org/10.1186/s12859-014-0409-8

18. Grubben, N.L., Keesman, K.J.: Controllability and observability of 2D thermal flow in bulk storage facilities using sensitivity fields. Int. J. Control **91**(7), 1554–1566 (2017). https://doi.org/10.1080/00207179.2017.1321782

19. IAPWS: Release on the IAPWS formulation 2008 for the viscosity of ordinary water substance (2018). http://www.iapws.org/relguide/viscosity.html

20. Khalil, H.K.: Nonlinear Systems. PEARSON - SUPERPEDIDO, March 2018. https://www.xarg.org/ref/a/B00A2KG8B8/

21. Kopke, M., et al.: Clostridium ljungdahlii represents a microbial production platform based on syngas. Proc. Natl. Acad. Sci. **107**(29), 13087–13092 (2010). https://doi.org/10.1073/pnas.1004716107

22. Kou, S.R., Elliott, D.L., Tarn, T.J.: Observability of nonlinear systems. Inf. Control **22**(1), 89–99 (1973). https://doi.org/10.1016/S0019-9958(73)90508-1. http://www.sciencedirect.com/science/article/pii/S0019995873905081

23. Lecca, P.: CIBB 2019 - LNBI extension (2020). http://www.mediafire.com/folder/z2lwnr8ucvgvw/CIBB%202019%20-%20LNBI%20extension

24. Lecca, P., Re, A.: Identifying necessary and sufficient conditions for the observability of models of biochemical processes. Biophys. Chem. **254**, 106257 (2019). https://doi.org/10.1016/j.bpc.2019.106257. http://www.sciencedirect.com/science/article/pii/S0301462219302807

25. Liew, F., Henstra, A.M., Köpke, M., Winzer, K., Simpson, S.D., Minton, N.P.: Metabolic engineering of clostridium autoethanogenum for selective alcohol production. Metab. Eng. **40**, 104–114 (2017). https://doi.org/10.1016/j.ymben.2017.01.007. http://www.sciencedirect.com/science/article/pii/S1096717617300319

26. Liew, F., Henstra, A.M., Winzer, K., Köpke, M., Simpson, S.D., Minton, N.P.: Insights into CO_2 fixation pathway of clostridium autoethanogenum by targeted mutagenesis. mBio **7**(3) (2016). https://doi.org/10.1128/mbio.00427-16

27. Lularevic, M., Racher, A.J., Jaques, C., Kiparissides, A.: Improving the accuracy of flux balance analysis through the implementation of carbon availability constraints for intracellular reactions. Biotechnol. Bioeng. **116**(9), 2339–2352 (2019). https://doi.org/10.1002/bit.27025

28. Maes, K., Chatzis, M., Lombaert, G.: Observability of nonlinear systems with unmeasured inputs. Mech. Syst. Signal Process. **130**, 378–394 (2019). https://doi.org/10.1016/j.ymssp.2019.05.010.   http://www.sciencedirect.com/science/article/pii/S0888327019303115

29. Mohammadi, M., Mohamed, A.R., Najafpour, G.D., Younesi, H., Uzir, M.H.: Kinetic studies on fermentative production of biofuel from synthesis gas Using-Clostridium ljungdahlii. Sci. World J. **2014**, 1–8 (2014). https://doi.org/10.1155/2014/910590

30. Mojtabi, A., Deville, M.O.: One-dimensional linear advection-diffusion equation: analytical and finite element solutions. Comput. Fluids **107**, 189–195 (2015). https://doi.org/10.1016/j.compfluid.2014.11.006.   http://www.sciencedirect.com/science/article/pii/S0045793014004289

31. Müller, V.: New horizons in acetogenic conversion of one-carbon substrates and biological hydrogen storage. Trends Biotechnol. **37**(12), 1344–1354 (2019). https://doi.org/10.1016/j.tibtech.2019.05.008.   http://www.sciencedirect.com/science/article/pii/S0167779919301155

32. Norman, R.O., et al.: Genome-scale model of C. autoethanogenum reveals optimal bioprocess conditions for high-value chemical production from carbon monoxide. Eng. Biol. **3**(2), 32–40 (2019). https://doi.org/10.1049/enb.2018.5003

33. Norman, R.O., Millat, T., Winzer, K., Minton, N.P., Hodgman, C.: Progress towards platform chemical production using clostridium autoethanogenum. Biochem. Soc. Trans. **46**(3), 523–535 (2018). https://doi.org/10.1042/bst20170259

34. Raman, K., Chandra, N.: Flux balance analysis of biological systems: applications and challenges. Brief. Bioinform. **10**(4), 435–449 (2009). https://doi.org/10.1093/bib/bbp011

35. Schuchmann, K., Müller, V.: Autotrophy at the thermodynamic limit of life: a model for energy conservation in acetogenic bacteria. Nat. Rev. Microbiol. **12**(12), 809–821 (2014). https://doi.org/10.1038/nrmicro3365

36. Slotine, J.J.: Applied Nonlinear Control. Pearson (1991). https://www.xarg.org/ref/a/0130408905/

37. Slyusarchuk, V.Y.: Necessary and sufficient conditions for the invertibility of nonlinear differentiable maps. Ukr. Math. J. **68**(4), 638–652 (2016). https://doi.org/10.1007/s11253-016-1247-9

38. Stigter, J.D., Joubert, D., Molenaar, J.: Observability of complex systems: finding the gap. Sci. Rep. **7**(1) (2017). https://doi.org/10.1038/s41598-017-16682-x

39. Stoll, I.K., Boukis, N., Sauer, J.: Syngas fermentation to alcohols: reactor technology and application perspective. Chemie Ingenieur Technik **92**(1–2), 125–136 (2019). https://doi.org/10.1002/cite.201900118

40. Sun, X., Atiyeh, H.K., Huhnke, R.L., Tanner, R.S.: Syngas fermentation process development for production of biofuels and chemicals: a review. Bioresour. Technol. Rep. **7**, 100279 (2019). https://doi.org/10.1016/j.biteb.2019.100279. http://www.sciencedirect.com/science/article/pii/S2589014X19301690

41. Valgepea, K., et al.: Maintenance of ATP homeostasis triggers metabolic shifts in gas-fermenting acetogens. Cell Syst. **4**(5), 505–515.e5 (2017). https://doi.org/10.1016/j.cels.2017.04.008

42. Villaverde, A.F.: Observability and structural identifiability of nonlinear biological systems. Complexity **2019**, 1–12 (2019). https://doi.org/10.1155/2019/8497093

43. Villaverde, A.F., Tsiantis, N., Banga, J.R.: Full observability and estimation of unknown inputs, states and parameters of nonlinear biological models. J. R. Soc. Interface **16**(156), 20190043 (2019). https://doi.org/10.1098/rsif.2019.0043

On the Simulation and Automatic Parametrization of Metabolic Networks Through Electronic Design Automation

Nicola Bombieri[1], Antonio Mastrandrea[1], Silvia Scaffeo[1], Simone Caligola[1],
Franco Fummi[1], Carlo Laudanna[2], Gabriela Constantin[2],
and Rosalba Giugno[1(✉)]

[1] Department of Computer Science, University of Verona, Verona, Italy
{Nicola.Bombieri,Antonio.Mastrandrea,Silvia.Scaffeo,Simone.Caligola,
Franco.Fummi,Rosalba.Giugno}@univr.it
[2] Department of Medicine, University of Verona, Verona, Italy
{Carlo.Laudanna,Gabriela.Constantin}@univr.it

Abstract. This work presents a platform for the modelling, simulation and automatic parametrization of semi-quantitative metabolic networks. Starting from a network modelled through Petri Nets (PN) and represented in SBML, the platform converts the model into an internal representation implemented through an Electronic Design Automation (EDA) description language. It applies techniques and tools well established in the EDA field to simulate the model and to automate the network parametrization. We present the validation of the model simulation and of the parameters automatically extrapolated by the platform with the state of art modelling and simulation tools for PNs. The validation uses a real metabolic network and shows the platform opportunities and limitations in reproducing the experimental results, simulating the models in different conditions, and facilitating the analysis of the dynamics that regulate the network.

Keywords: Metabolic network · Modelling · Simulation · Parametrization · EDA

1 Scientific Background

Model development and analysis of metabolic networks is recognized as a key requirement for integrating *in vitro* and *in-vivo* experimental data. *In silico* simulation of a biochemical model allows one to test different experimental conditions and simulating biochemical systems for both quantitative and qualitative analysis.

Quantitative models, which can be deterministic or stochastic [18], are dynamic models based on mathematical formalism like ordinary differential equations (ODEs). In deterministic models, the parameter values and initial conditions are well-known and they always produce the same output. In contrast, stochastic models are characterized by random properties.

© Springer Nature Switzerland AG 2020
P. Cazzaniga et al. (Eds.): CIBB 2019, LNBI 12313, pp. 323–334, 2020.
https://doi.org/10.1007/978-3-030-63061-4_28

Obtaining and analysing mathematical models can be difficult or even prohibitive, due to the large number of independent variables and the lack of quantitative information [10,20]. Differently, the goal of qualitative models is to discover particular properties of the network and to employ a high level of abstraction in reproducing the system behaviour [2,12,13,16]. Both qualitative and quantitative models can be effectively represented by Petri Nets (PNs) [14]. PNs are directed bipartite graphs, used to mathematically represent the structure of biological systems. Systems are described in SBML (Systems Biology Markup Language), which is, at the moment, the standard for representing computational models in Systems Biology.

Different software applications have been developed for the simulation of complex biochemical systems including metabolic pathways, signal transduction pathways, and gene expression networks. Examples are COPASI [11], for quantitative analysis of general complex biochemical systems, Monalisa [1], and Snoopy [19], which are both focused on the simulation of PN models.

Despite their difference in supporting different levels of abstraction, SBML descriptions and simulation paradigms (i.e., deterministic, stochastic, etc.), all these tools allow for system simulation starting from kinetic and semi-quantitative data parameters.

Since the value of parameters are rarely available, the most important limitation of these tools concerns the discovery of quantitative information, referred as the problem of *parameterization*. Very often it is necessary to explore *the solution space* of the network parameters to identify which network configurations lead the model to satisfy certain biological properties. The parameter setting, which must be done manually in these tools, makes their applicability not suitable for the analysis of actual complex dynamic systems.

To solve this limitation, our intuition is that such a reverse engineering and design space exploration, like several other characteristics such as concurrency, reactivity, and abstraction levels to model biological systems are common to the modelling of electronics systems. We have shown that using languages, techniques, and tools well established in the context of EDA (Electronic Design Automation) can introduce automation and efficiency to model, simulate metabolic pathways [5].

In this work, we present a platform that converts semi-quantitative models of metabolic networks into an internal representation. In particular, the platform implements a front-end tool that converts PN models from SBML to the EDA standard SystemC language. The platform then requires, as input data, the system properties to be observed and the known experimental data. As a result, the platform automatically extrapolates the system parametrization to reproduce the experimental results and to simulate the model under different system dynamics.

We present a comparison between the simulation results obtained with the proposed platform and those provided by one of the most widespread software at the state of art (Snoopy [19]). To evaluate the applicability and potential of the platform, we applied both Snoopy and our system to model the purine metabolism pathway. Our team members have studied it thoroughly as biological experts in its dynamics. The aim of the study was to reproduce the metabolomics data obtained from naive lymphocytes and autoreactive T cells implicated in the induction of experimental autoimmune disorders. Immunometabolism represents a novel field of investigation linking immune cell function and intracellular metabolic networks. We focused on purine metabolism related to nucleotide moieties, which are constantly de novo synthesized intracellular metabolites required for a wide variety of biological processes. During activation and proliferation, T cells require increased nucleotide synthesis for DNA replication and RNA production to support protein synthesis at different stages of the cell cycle [15]. Furthermore, nucleotide concentrations cannot fall below critical levels in quiescent cells in which there is a considerable turnover of RNA, involved in cell maintenance, repair, and regulation [15]. Purine metabolism is regulated at transcriptional level by a set of master transcription factors, but also at the enzymatic level by allosteric regulation and feedback inhibition, making this pathway suitable for our simulation study taking advantage of the mass action law. The flux through a metabolic pathway depends on the supply of the initial metabolites, the concentration and activities of all the enzymes in the pathway, and the presence of feedback or other regulatory controls [15].

In this work, our goal is to compare Snoopy and the proposed platform to decipher how much control each enzyme in a pathway exerts on the net flux in order to guarantee a stabilized intracellular concentration of initial, intermediate or final metabolites.

2   Materials and Methods

Section 2.1 introduces the Snoopy software and its use [19] to understand how it can be compared to the proposed platform. Both software applications apply to metabolic networks modelled as stochastic Petri nets (SPNs) [6] and simulate the model through the Gillespie's algorithm [9]. Then, the section presents the proposed platform in detail.

2.1   Snoopy: Description and Usage

Snoopy [19] is a software application that allows for the analysis of biomolecular networks by supporting their representation in different variants of PN models. These include qualitative PNs, which contain only information about the structure of the network, quantitative PNs, which include information about kinetics, and semi-quantitative PNs, which include *partial* information about kinetics.

Besides representing PNs in a standard way, with arcs and nodes such as places and transitions, an extended form of PN can also be considered. Snoopy supports different types of arcs designed to represent the behaviour of several biological reactions. It is possible to build a network from scratch and to export it into several formats, or import its description. Whether it is built or imported, the final network can be visible in the GUI, making each element information (such as number of tokens or kinetic function) easily accessible and adjustable.

The simulation can be performed by setting the type of simulator, the time interval and some visualization options.

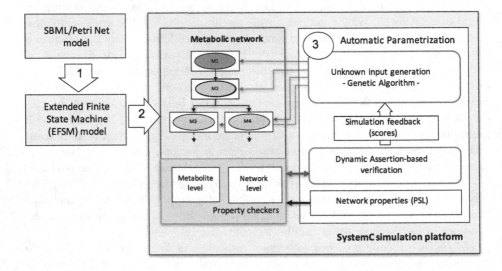

Fig. 1. The platform overview.

As an example, the stochastic simulator may follow algorithms such as the Gillespie's exact method to estimate the exact sequence of reaction firings, or the Tau-leaping method to specify an interval in which more than one reaction firing can happen. Starting from the SBML description of a network, the user imports the network in Snoopy according to the target simulation model (e.g., deterministic, stochastic, etc.) and the software displays the visual representation of the dynamic simulation. Having all the correct kinetic information, a simulation dialogue can be opened in order to set the preferred simulator algorithm and choose the simulation total time. The resulting trends can be plotted and, eventually, exported as a CSV file.

2.2 EDA Based Modeling Platform

The proposed platform relies on the SystemC library to implement Petri nets (PNs) and Stochastic Petri nets (SPNs) formalism.

Starting from the PN model of the metabolic network, the platform implements three main phases (see Fig. 1):

1. Representation of the PN through Extended Finite State machines (EFSM).
2. Synthesis of EFSM into SystemC with the support of a kernel for stochastic simulation.
3. Automatic parametrization of the network through a genetic algorithm based input generation guided by dynamic assertion-based verification.

Modeling PN Through EFSM. To perform the event-driven simulation of the PN through an EDA simulator, we represent each node and each reaction of the PN through Extended Finite State Machines (EFSMs). The EFSM model is widely used for modelling complex systems in the context of Electronic Design Automation (EDA), and allows for discrete-event simulation.

Formally, an EFSM is defined as a 5-tuple $M = \langle S, I, O, D, T \rangle$ in which: S is a set of states, I is a set of input data, O is a set of output data, D is a n-dimensional linear space $D_1 \times \ldots \times D_n$, T is a transition relation so that $T : S \times D \times I \rightarrow S \times D \times O$. An n-tuple $x = (x_1, \ldots, x_n)$ represents a point in D that models the state variables of the model. A configuration of M is defined as a pair $\langle s, x \rangle \in S \times D$, while a step in a EFSM $M = \langle S, I, O, D, T \rangle$ is defined as follows: If M is in a configuration $\langle s, x \rangle$ and it receives an input $i \in I$, it moves to the configuration $\langle t, y \rangle$ iff $((s, x, i), (t, y, o)) \in T$ for $o \in O$.

The EFSM model allow us to modularly synthesize the whole PN model into executable SystemC code.

EFSM-SystemC Synthesis with Stochastic Simulation Support. The framework implements an adapted version of the First Reaction Method (FRM) [21]. FRM is a variant of the classic Gillespie's Stochastic Simulation Algorithm (SSA). In the PN model of a metabolic network, each reaction R from the reactant M_i to the product M_j can fire if the concentration of M_i is sufficient to execute R, based on its stoichiometry. According to the FRM method, at each step and for each reaction R, the system generates a possible reaction delay τ_R that depends on the concentration of the reactant and on the mass-action rate constant. We defined the reaction delay as follows:

$$\tau_R = \left\lfloor \left(\frac{1}{a_R(m_i)} \right) \cdot ln \left(\frac{1}{rand_k} \right) \right\rfloor + 1$$

where $rand_k \in (rand_1, \ldots, rand_n)$ is a random number from the uniform distribution $U(0, 1)$. $a_R(m_i) = c_i m_i$ is the mass-action propensity function where c_i is the reaction rate constant and m_i is the number of tokens of the reactant M_i. We consider a discrete formulation in which a lower-bound value is associated to each reaction delay (i.e., 1) in our discrete-event simulation.

According to the FRM model, the stochastic simulation of the system evolves along discrete steps, where each next time step corresponds to the expiration time of the immediately next reaction. In particular, considering t_i as the current simulation time of the system, $t_{i+1} = t_i + \tau_{R_i}$, where $\tau_{R_i} = min(\tau_{R_1}, \ldots, \tau_{R_k})$.

At each simulation step, all the reaction delays are updated, and the updating considers the elapsed time and a random component. The updating phase at each simulation step of all the reaction delays $\tau_{R_i} \forall i$, which include a random component $(rand_k)$, is the necessary condition to perform the stochastic simulation of the PN model.

Parameter Estimation Through Dynamic Assertion-Based Verification (ABV). In EDA, functional verification based on *assertions* represents one of the main applied and investigated techniques that combines simulation-based (i.e., dynamic) and formal (i.e., static) verification [4,7]. The proposed methodology applies simulation-based ABV, by which assertions are defined in a formal language (i.e., Property Specification Language - PSL), they are automatically synthesized into *checkers*[1], and plugged to the SystemC model representing the network [3]. In our context, checkers aim at monitoring the concentration of the metabolites and to give a *score* (i.e., fitness) to the input generation module, which implements a genetic algorithm to generate configurations of the kinetic parameters (see Fig. 1). The module generates a configuration of parameters and runs a dynamic simulation of the network for such a set of input values for a given simulation time. Then, the module generates a new different configuration for a new simulation. The fitness function evaluates the goodness of each potential solution estimated through simulation. The run ends when the module finds the parameter configuration that allows the system properties to be satisfied. The definition of the fitness function depends on the property to be checked and it is formulated as the distance between *state vectors* representing the *simulation trend* $s_i(t)$ in comparison to a defined *reference trend* $r_i(t)$. In our platform, a trend is a formulation of the system state that represents the behaviour of the system until a certain simulation point and the target behaviour of the system to be reached.

The ABV checks the simulation trend $s_i(t)$ and, eventually, it stops the simulation to provide a score compared to $r_i(t)$. Given the state vectors $S = [s_1, s_2, \ldots, s_n]$ and $R = [r_1, r_2, \ldots, r_n]$ representing, respectively, the simulation and the reference trend, the score is defined as follows:

$$score_c(t) = d(S, R)$$

where d is a distance function (e.g.., Euclidean distance but any distance function suitable for the domain of application can be applied, including normalized distances) between the state vectors. The formulation of state vectors is general and allows us to model biological behaviours such as the stability in concentration and the change of concentrations between time points.

3   Results

We applied our methodology to understand how the dynamics of the purine pathway change between normal and proliferating conditions. To do this, we first constructed the purine pathway from metabolomic data, represented it in SBML format, and then modeled the system in each experimental condition to observe how the flows and the products of the pathway change. We also evaluated the simulation trends of the analyzed network obtained with the proposed SystemC framework by considering the Snoopy results as ground truth [19].

---

[1] The framework relies on the IBM FoCs synthesizer for the automatic synthesis of assertions.

3.1   Network Construction from Metabolomics Data

We analyzed the intracellular metabolic profile of resting, naive CD4+ T cells isolated from lymph nodes and spleens of healthy SJL mice by magnetic cell sorting (all reagents from Miltenyi Biotech). Production of actively proliferating PLP139-151-specific encephalitogenic T cell lines was previously described in [17]. In brief, SJL mice were immunized subcutaneously with 300 mg of proteolypid protein (PLP)139-151 peptide. 10–12 days later, draining lymph nodes were removed and total cells were *in vitro* stimulated with PLP139-151 peptide for 4 d. T cell lines were obtained by re-stimulation of these cultures every 14 d for at least 3 times in the presence of irradiated splenocytes as antigen presenting cells at a ratio of 1:8 (T cell vs irradiated spleen cells). For the metabolomics analysis, actively proliferating PLP-specific encephalitogenic T cells were collected after two days of *in vitro* re-stimulation. The levels of purine metabolites were determined by performing metabolomics analysis in lysates from naive lymphocytes and PLP-specific T cells (http://www.metabolon.com/). Pathway and metabolism ontology analysis were performed using Cytoscape and MetScape plugin. We built a network containing the most important features of ex novo purine synthesis and catabolism, consisting of rounded and squared nodes, related to metabolites and metabolic reactions respectively (see Fig. 2(a) to extrapolate the structure of the network; node colors in the network will be explained below). The network was represented according to SBML language for subsequent implementation in our simulation platform and for the comparison study.

3.2   Integrating Metabolomics Observations with Simulation Analysis

The purine metabolic pathway was summarized in our network starting from PRPP transformation to IMP and ending to the production of dAMP, dGMP and urate as a waste product of purine synthesis and catabolism. In metabolic pathways, the concept of steady state refers to the stability of the levels of metabolites and it is crucial to keep homeostasis inside the cell. We simulated our model to find the mass-action parameters of all network reactions which keep stable the concentrations of each metabolite except dAMP, dGMP and Urate. Due to the lack of experimental data in the transformation rate from PRPP to IMP we considered two different flow speeds, slow and fast. *Our simulations led to interesting differences in the regulation of the entire purine network, suggesting that some chemical reactions may be favored in PLP-specific cells versus naive lymphocytes* as shown in Fig. 2, where gradients of colors are associate to reaction speeds from fast (antique rose) to slow (blue).

In our model the inhibition mechanisms are represented through the inhibition arcs, which are an extension of the classical Petri nets to represent particular biological interactions. We assumed that a metabolite can inhibit a reaction when it grows up by 10% from its initial concentration. The pathway is considered at steady state if the concentration of each element does not differ by more than

±15% from the initial concentration and it is maintained stable throughout the simulation time. We specified this property formally through ABV by imposing that the concentration of all metabolites except the terminal ones could not exceed the above threshold. For each parameter configuration, the network simulation required around 7 s to simulate 1 million time instants (see time t in Fig. 3) All the simulations were run on a machine equipped with an Intel(R) Core(R) i7-4720HQ CPU clocked at 2.6 Ghz and 16 GBs RAM, and the Ubuntu 16.04 operating system. We empirically set such a time limit to correctly verify the steady state property of the pathway. Automatic parameter estimation required from 1 to 12 min for each network version in our laptop. The genetic algorithm used by the automatic input generator has been configured with a population size of 200 individuals, a mutation probability of 0.05 and a crossover probability of 0.1. The selection method used to pick an individual from the population is *rank-based*, meaning reproduction is always done by taking individuals with better fitness. This set-up of parameters has empirically proved to be a good compromise between *exploration* and *exploitation* of the parameter configurations. *The simulation results suggest that the strength of the initial flow may impact the final network output.*

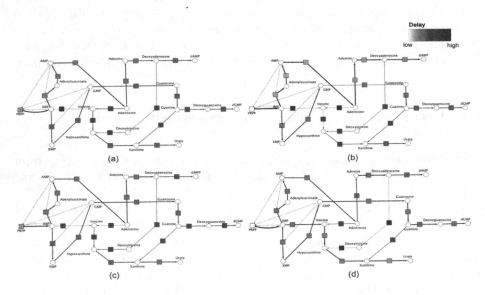

Fig. 2. Pathway activity based on PRPP→IMP flux: slow naive (a), fast naive (b), slow PLP-specific (c) and fast PLP-specific (d). (Color figure online)

The simulation of the naive T cell network resulted in the production of comparable amounts of dAMP, dGMP and urate by setting an initial slow rate of PRPP transformation (Fig. 3(a)). By increasing the initial flow we observed an increment in dAMP production but no changes in dGMP and urate production in naive T cells (Fig. 3(b)). However, when we simulated the PLP-specific network,

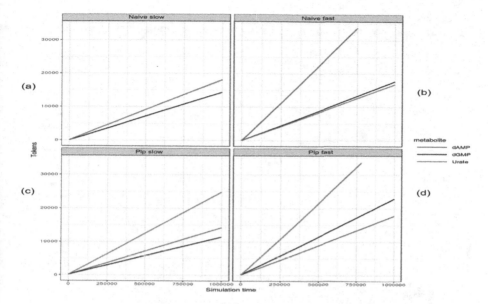

Fig. 3. Production of dAMP, dGMP and Urate: initial slow rate in naive cells (a), initial fast rate in naive cells (b), initial slow rate in PLP-specific cells (c) and initial fast rate in PLP-specific cells (d).

we observed a higher urate production than dAMP and dGMP production under initial slow flow (Fig. 3(c)). Moreover, by setting a high initial flow of PRPP transformation, we further boosted the production of all the three products urate, dAMP and dGMP with urate production being higher than dAMP and dGMP (Fig. 3(d)). Overall these results indicate that, although the networks had the same topology, metabolites, and chemical reactions, our simulations displayed two different metabolic outputs between naive and PLP-specific T cells. The results also suggest that the metabolic flux of each chemical reaction may potentially impact the final metabolic output of the purine metabolism network. Notably, the increased urate and dGMP production in PLP-specific network reflects a well-known metabolic feature of proliferating lymphocytes [8] and validates the potentiality of our platform in simulating metabolic processes.

3.3   Comparing Our Platform with Snoopy

Among the already existing software applications available in literature, we chose to focus on Snoopy as it is, in our opinion, one of the most complete and well-suited for our purpose, as well as being implemented in C++ language. Since also our framework is based on a C++ extension, this will allow a fairer comparison in the future on different properties such as the execution time and memory usage. Figure 4 shows, as an example, the results obtained by simulating the purine network without any parameter estimation, in order to have a comparison with Snoopy software based only on network topologies. The plots show how the

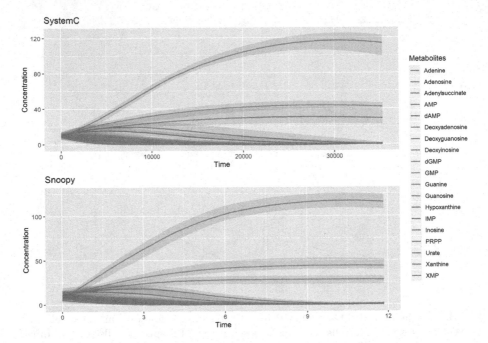

Fig. 4. Comparing of simulation trend of the purine network with our model implemented in SystemC (top) and with Snoopy [19] (bottom).

concentration of the metabolites change on average over time, as well as the range in which the minimum and maximum concentration of each represent the boundaries, for each time instant. The average, minimum and maximum concentration were calculated on 10 simulation runs for both software. Although stochastic simulations were performed, as it can be seen from the results, the general trends are similar.

Our current simulation is based on the FRM method, and the implementation of other simulation models is part of our future work. In addition, the support for SBO (Systems Biology Ontology) terms, which are used in Systems Biology for computational modelling (and that is missing in Snoopy) is part of our current work.

4   Conclusion

This paper presented an EDA-based platform for modelling and simulation of metabolic pathways. We applied the platform to model the purine metabolism and to reproduce the metabolomics data obtained from naive lymphocytes and autoreactive T cells implicated in the induction of experimental autoimmune disorders. We first validated the obtained simulation results with state-of-the-art alternatives. Then, we showed that the model configured with the obtained parameters led to interesting differences in the regulation of the entire purine network, suggesting that some chemical reactions may be favored in PLP-specific cells versus naive lymphocytes and that the strength of the initial flow may impact the final network output.

Acknowledgments. G.C. was supported by the European Research Council (ERC) grants IMMUNO ALZHEIMER (nr. 695714, ERC advanced grant).
R. G. is supported by GNCS-INDAM and JPND 2019-466-037.

References

1. Balazki, P., Lindauer, K., Einloft, J., Ackermann, J., Koch, I.: MONALISA for stochastic simulations of Petri Net models of biochemical systems. BMC Bioinform. **16**(1), 215 (2015). https://doi.org/10.1186/s12859-015-0596-y
2. Barabasi, A.L., Oltvai, Z.N.: Network biology: understanding the cell's functional organization. Nat. Rev. Genet **5**(2), 101 (2004)
3. Bombieri, N., et al.: Reusing RTL assertion checkers for verification of SystemC TLM models. J. Electron. Test. **31**(2), 167–180 (2015). https://doi.org/10.1007/s10836-015-5514-8
4. Bombieri, N., Fummi, F., Pravadelli, G.: On the evaluation of transactor-based verification for reusing TLM assertions and testbenches at RTL. In: Proceedings of ACM/IEEE DATE, pp. 1–6 (2006)
5. Caligola, S., et al.: Efficient simulation and parametrization of stochastic Petri Nets in systemc: A case study from systems biology. In: 2019 Forum for Specification and Design Languages (FDL), pp. 1–7 (2019)
6. Chaouiya, C.: Petri Net modelling of biological networks. Brief. Bioinform. **8**(4), 210–219 (2007)
7. Coelho, C.N., Foster, H.D.: Assertion-based verification. In: Drechsler, R. (ed.) Advanced Formal Verification, vol. 4. Springer, Boston (2004). https://doi.org/10.1007/1-4020-2530-0_5
8. Eleftheriadis, T., et al.: Uric acid induces caspase-1 activation, il-1β secretion and p2x7 receptor dependent proliferation in primary human lymphocytes. Hippokratia **17**(2), 141 (2013)
9. Gillespie, D.T.: A general method for numerically simulating the stochastic time evolution of coupled chemical reactions. J. Comput. Phys. **22**(4), 403–434 (1976)
10. Hoffmann, A., Levchenko, A., Scott, M.L., Baltimore, D.: The iκb-nf-κb signaling module: temporal control and selective gene activation. Science **298**(5596), 1241–1245 (2002)
11. Hoops, S., et al.: COPASI - A COmplex PAthway SImulator. Bioinformatics **22**(24), 3067–3074 (2006)
12. Jeong, H., Mason, S.P., Barabási, A.L., Oltvai, Z.N.: Lethality and centrality in protein networks. Nature **411**(6833), 41 (2001)
13. Klamt, S., Saez-Rodriguez, J., Lindquist, J.A., Simeoni, L., Gilles, E.D.: A methodology for the structural and functional analysis of signaling and regulatory networks. BMC bioinformatics **7**(1), 56 (2006). https://doi.org/10.1186/1471-2105-7-56
14. Koch, I.: Petri Nets in systems biology. Softw. Syst. Model. **14**(2), 703–710 (2014). https://doi.org/10.1007/s10270-014-0421-5
15. Lane, A.N., Fan, T.W.M.: Regulation of mammalian nucleotide metabolism and biosynthesis. Nucleic Acids Res. **43**(4), 2466–2485 (2015)

16. Morris, M.K., Saez-Rodriguez, J., Sorger, P.K., Lauffenburger, D.A.: Logic-based models for the analysis of cell signaling networks. Biochemistry **49**(15), 3216–3224 (2010)
17. Piccio, L., et al.: Molecular mechanisms involved in lymphocyte recruitment in inflamed brain microvessels: critical roles for P-selectin glycoprotein ligand-1 and heterotrimeric Gi-linked receptors. J. Immunol. **168**(4), 1940–1949 (2002)
18. Puchałka, J., Kierzek, A.M.: Bridging the gap between stochastic and deterministic regimes in the kinetic simulations of the biochemical reaction networks. Biophys. J. **86**(3), 1357–1372 (2004)
19. Rohr, C., Marwan, W., Heiner, M.: Snoopy-a unifying Petri net framework to investigate biomolecular networks. Bioinformatics **26**(7), 974–975 (2010)
20. Steuer, R., Junker, B.H.: Computational models of metabolism: stability and regulation in metabolic networks. Adv. Chem. Phys. **142**, 105 (2009)
21. Thanh, V.H., Priami, C.: Simulation of biochemical reactions with time-dependent rates by the rejection-based algorithm. J. Chem. Phys. **143**(5), 08B601_1 (2015)

Deep Clustering for Metagenomics

Isis Bonet[1](✉), Alejandro Pena[1], Christian Lochmuller[1], Alejandro Patino[1], and Mario Gongora[2]

[1] EIA University, km 2 + 200 Vía al Aeropuerto José María Córdova, Antioquia, Envigado, Colombia
ibonetc@gmail.com, {fjapena,christian.lochmuller, hector.patino}@eia.edu.co
[2] De Montfort University, The Gateway, Leicester LE1 9BH, UK
mgongora@dmu.ac.uk

Abstract. Metagenomics is an area that is supported by modern next generation sequencing technology, which investigates microorganisms obtained directly from environmental samples, without the need to isolate them. This type of sequencing results in a large number of DNA fragments from different organisms. Thus, the challenge consists in identifying groups of DNA sequences that belong to the same organism. The use of supervised methods for solving this problem is limited, despite the fact that large databases of species sequences are available, by the small number of species that are known. Additionally, by the required computational processing time to analyse segments against species sequences. In order to overcome these problems, a binning process can be used for the reconstruction and identification of a set of metagenomic fragments. The binning process serves as a step of pre-processing to join fragments into groups of the same taxonomic levels. In this work, we propose the application of a clustering model, with a feature extraction process that uses an autoencoder neural network. For the clustering a k-means is used that begins with a k-value which is large enough to obtain very pure clusters. These are reduced through a process of combining various distance functions. The results show that the proposed method outperforms the k-means and other classical methods of feature extraction such as PCA, obtaining 90% of purity.

Keywords: Metagenomic · Clustering · Deep learning · Autoencoder · k-means

1 Introduction

Microorganisms are a main focus of research. Although they are abundant living beings that dominate our planet, we are still far from knowing them in detail. In 2016 a study concluded that there could be approximately 1 trillion microorganisms on earth, of which only 0.01% is known [1]. Microorganisms are relevant in different areas, which are considered as important by society, such as agriculture and medicine. Studying microorganisms leads to a better understanding

© Springer Nature Switzerland AG 2020
P. Cazzaniga et al. (Eds.): CIBB 2019, LNBI 12313, pp. 335–347, 2020.
https://doi.org/10.1007/978-3-030-63061-4_29

of global cycles that keep the biosphere in balance. In addition, knowing their functions is important for the development of antimicrobial therapies. Also, for providing solutions to current environmental problems. This is why, recently, we can observe an increasing interest in knowing more species that can provide solutions for problems in our daily life.

The growth of the amount of studies about microorganisms has led to searching for more effective alternatives to identify them. However, a major obstacle in this area of research consists in the low survival rate of some microorganisms in artificial environments, which prevents that they can grow isolated in a laboratory. Metagenomics pretend to face and resolve this limitation. That can be done, if we manage to sequence the genome of microbes from samples that are taken directly from the environment, without the need to isolate them [2]. Thus, metagenomics allow the discovery of new species of microbes.

However, metagenomics are not the immediate solution to this problem. Analyzing the results of a metagenomic process entails significant challenges, as the environmental sample contains a set of different microorganisms. The result is a cluster of DNA sequences that have no distinction, in terms to which organism they belong. Taking into account that in one sample there might be up to 10000 species, the solution of this "puzzle", to assemble all species, all of sudden appears to be an impossible endeavour [2]. This process, called binning, consists of identifying which groups of DNA fragments belong to a single organism. In order to improve the results of the binning process it is common to make a partial assembly. This is done to obtain the largest fragments that are called contigs.

One solution to identify the organisms in this puzzle consists in using large existing databases with information about the sequences of known microorganisms. Even though these databases contain large volumes of sequences, as mentioned above, this amount is very small compared to what actually exists. Even so, the volume of these bases and the quantity of fragments that must be analyzed make the process slow and sometimes little effective.

With regard to the binning process the focus is on two methodologies: methods based on composition and similarity. Where the binning based on similarity is a supervised method, which uses similarity techniques, such as alignment, comparing metagenomic sequences with genes or known proteins of the available databases. With the application of grouping algorithms and artificial intelligence, based on the similarity of the sequences, it is possible to generate groups of sequences that belong to the same organism. Whereas composition-based binning uses a representation based on the characteristics of the sequence. Among the most commonly used characteristics are the content of GC and the use of codons or the frequencies of oligonucleotides. This binning process can be implemented as supervised or unsupervised, depending on the use of a training set.

Although supervised methods are more accurate than unsupervised methods, available examples of all possible species do not exist, which has led to the use of unsupervised methods or a combination of both methods.

Previous research applied different clustering algorithms to perform a binning process. Where different distance measurements and features to characterize the DNA fragments are emphasized. One of the first works in this regard is on the development of the named TETRA [3], where the novelty consists in the use of the frequencies of tetranucleotides (k-mers, with k = 4) as features to represent the sequences. Another model, which also uses this feature, is MetaCAA [4].

There are some issues that can arise from a binning process: The databases are large and heterogeneous, the number of species in a sample is unknown, fragments vary in size and the number of fragments, derived from each species, is different, which results in an unbalanced database. These problems increase the difficulties for unsupervised binning, and require better attributes to represent the DNA fragments to be determined and also improved algorithms that can handle large amounts of complex data. Hence, in metagenomics the process of clustering has to consider these aspects. With respect to the application of these methods, it is important that a method is not only accurate, but also fast, in terms of computer processing. By using distributed computing and machine learning platforms like Apache Spark and TensorFlow, it is possible to implement an optimization of clustering algorithms. These implementations manage to accelerate the processing of large volumes of data for data analytics. In this context, cloud computing solutions can also generate benefits in terms of reliability, processing time and costs. In addition, deep learning methods have increased the effectiveness, not only in prediction processes, but also in data clustering. Authors such as [5], due to the complexity of image processing, have developed deep clustering methods for image clustering. In our present case of metagenomics it is similar because metagenomics are also characterized by a large number of features and separating them is a complex process. The use of deep learning can help to reduce the dimensionality of the data without losing information. This way, improving the effectiveness of data clustering.

In this paper, we focus on unsupervised composition-based methodologies. Our paper is based on a previous work [6], which applies a variant of k-means (iterative k-means) with k-mers as features to represent the fragments. We propose a deep clustering model to improve the computational execution and processing time. This model applies autoencoder neural networks combined with a k-means++ and different distance measurements. We use k-mers frequencies for the representation of features and compare the results of the new method with the results obtained from a simple clustering algorithm. In addition, we compare the extraction of features using the autoencoder and PCA.

The remainder of this paper is structured as follows. Section 2 describes the data used to create the metagenomic database, the features selected to describe the sequences and the proposed clustering method. Section 4 discusses the obtained results. The paper ends with the conclusions of our research.

2   Methodology, Materials and Methods

2.1   Data and Features

The data were extracted from a publicly accessible sequenced DNA database, which is available on the FTP site of the Sanger institute (ftp://ftp.sanger.ac. uk). The database was built simulating the result of sequencing a set of organisms that are described below. This way, obtaining contigs of different sizes. We use assembled genomic sequences at a contig level of 16 organisms in total that are divided into three domains, as it is shown in Table 1.

Table 1. Species in the database

Domain	Specie
Viruses	HIV
	Chikungunya
	Ebola
	Influenza
	Dengue
Bacteria	Bacteroides dorei
	Bifidobacterium longum
Eukaryotes	Ascaris suum
	Aspergillus fumigatus
	Bos taurus
	Candida parasilopsis
	Glossina morsitans
	Malus domestica
	Manihot esculenta
	Pantholops hodgsonii
	Zea mays

The database contains 872576 contigs, where the sizes range from 50 to 30000 nucleotides. It is important to note that one of the important problems, when it comes to finding clusters in metagenomics, is that the dataset is very unbalanced, and our dataset is no exception in this respect. This is due to the large difference in the size of the DNA of the different organisms. There are organisms, such as viruses like Chikungunya, Ebola and HIV that have only one DNA fragment. Whereas other Eukaryote like Bos taurus have 315841 DNA fragments.

Taking into account the features that are most commonly used in literature, we use in our work k-mers, which is a k-combination of the 4 nucleotides (i.e. A,G,C,T). That means words or subsequences of 4 letters of length k. Here, we use 4-mers or tetranucleotides. These are computed as the number of each tetranucleotide and normalized, considering the total of tetranucleotides in the contig. This means that we have 256 (4^4) features to describe the contigs.

2.2 Clustering Method

K-means is one of the most used clustering methods [7]. K-means++ is a variation of the k-means, which achieves more efficient centroids [8]. The k centroids are based on a weighted probability distribution, selecting a point x with a probability that is proportional to a function of distance. This selection provides centroids that are further apart. The only difference, compared to the k-means, is how the centroids are selected.

2.3   Research Methodology

Figure 1 shows the steps of the proposed methodology for carrying out the metagenomic data clustering process.

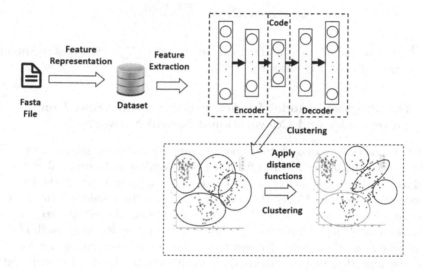

Fig. 1. Research methodology for the proposed model

The first step is to find the appropriate features to represent the contigs that are naturally different in size. In this work, as previously stated, k-mers will be used for this representation.

The second step is to reduce the dimensionality of the data, based on a feature extraction process. For this process we propose a deep learning autoencoder model. This network has two processes, one of data compression and another of data decompression, to bring them to the original dimension. The data compression or coding process can be used as a form of trait extraction, since it produces a reduction in dimensionality, corresponding to the number of neurons in the central coding layer. This process will be explained in more detail in the next section.

The third step refers to the clustering process. This process is based on the database, which has been built with the output of the last layer of the autoencoder coding stage. In this work, the tests were performed with k-means++ as a clustering method. The clustering process is carried out in two stages, the first with a number of k clusters. After obtaining the clusters, intra and inter-cluster distances are calculated. By using a certain threshold, the clusters that are less compact and closer to each other are selected, in order to be clustered again. The two clustering processes perform different distance functions. This allows that different patterns of closeness can be found in the data.

2.4   Validation and Clustering Evaluation Metrics

To assess the final quality of clustering methods we use a labeled database, intra and inter-cluster distances, and purity measure Eq. 1 [6].

$$Purity\,(C_j) = \frac{\max\,(n_{ij})}{n_j} \tag{1}$$

Where n_j is the number of organisms in cluster j (C_j) and n_{ij} is the number of organisms of class i in cluster j.

2.5   The Proposed Model for Feature Extraction Based on Autoencoder and Convolutional Neural Networks

Feature extraction is one of the most common pre-processing processes. The objective is to find better features from the combination of the original features or to reduce the dimensionality of data. One of the most known ways is the Principal Component Approach (PCA) [9]. Here, we compare the results of the proposed model for feature extraction with PCA. This section describes two proposed methods for performing feature extraction based on deep learning methods. The first is based on an autoencoder and the second on a convolution network. For the implementation of these methods we rely on the algorithms proposed by [5]. In both cases, they are used as the start of the data pre-processing and the output is used to initialize the centroids with a k-means++.

An Autoencoder Model for Feature Extraction. An autoencoder is an unsupervised compression neural network that can be divided in two components: encoder and decoder. Encoder is a set of layers, which is responsible for data compression by reducing the dimensionality. This can be represented as a function, which results from the combination of a combination of layers, which receives as input data a set of characteristics X, and as output a set of features Y of low dimensionality. This means that the length of Y is shorter than the length of X (Eq. 2).

$$Encoder(X) = Y \tag{2}$$

While the decoder is responsible for reconstructing the compressed data Y to restore the data into a new x data, where X' is expected to be as close to the original data as possible (Eq. 3).

$$Decoder(Y) = X' \tag{3}$$

Autoencoders are typically used for dimensionality reduction, noise reduction and outlier detection.

In [5] a method that is called Deep Embedded Clustering is proposed, which uses deep neural networks in order to learn feature representations and cluster assignments. This method learns a map, starting from the data space to a lower-dimensional feature space to optimize the clustering process.

Such networks have been used for image pre-processing, and the way these networks can reduce the dimensionality of data can be useful in different circumstances of feature extraction. This is why we apply these networks for feature extraction with regard to the first encoding process.

We use the autoencoder model as a feature extraction process before applying a clustering method. Although the first part of the model is the one that performs the compression of the data, we are going to use the complete model since the second part serves to validate how effective the compression is. However, for the second part of the model, the clustering part, the encoder output will be used as input.

After several experiments, we selected a topology with 3 layers, each with 128, 64 and 32 neurons respectively. Thus, the k-means will receive 32 input features. In this example, 16 output classes were used, although experiments with a larger number of classes were carried out.

The Clustering Process. As already mentioned above, k-means will be used as a clustering method. This method applies the value of k as an adjustment parameter. That is, k represents the number of clusters to build, which is precisely the number we do not know beforehand. Hence, two strategies will be used to select the value of k that best fits the problem. The first consists in applying a range of k-values and select the value with the highest data compression, based on the calculation of the intra and inter-cluster distance.

The second strategy is based on using a k-value which is large enough to subsequently reduce it, according to the clusters that can be put together. K-means will be trained with a defined distance function. After obtaining the clusters these will be analyzed. Those clusters that show a distance between the centroids below a certain threshold and will be joined.

The next step is to use the combination of other distance functions to see which other clusters may be close. A threshold will also be used to join the closest clusters. Although this strategy can led to a larger number of clusters, it is expected to form clusters that are purer.

In this work we start with k = 2000. The k-means uses the cosine distance for training and the Euclidean distance for the unification of the clusters.

3   Results

A metagenomic database composed of 16 different organisms is used to evaluate the method. The descriptive features of the sequence are tetranucleotides.

Here, we start with 256 features, which represent the repetition of four subsequent nucleotides in the contigs. Many of the features that are part of the database have various zero values, even more of these zero values are present, when the contigs are of short length. An extraction of the traits may improve the results, when we try to combine several of them and stay with the traits that have less zero values (Fig. 2).

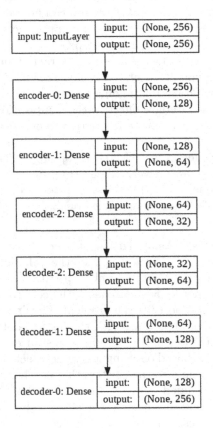

Fig. 2. The final topology of the autoencoder.

The application of deep learning techniques, allows to work with large volumes of data. This facilitates the treatment of these bases, which in general are large. As mentioned above, work has been performed that applies an autoencoder to reduce the dimensionality for different problems, outside image processing. In this context, it has been demonstrated, for example, that an encoder can be built, where the dimensionality of data is decreased first, and then the data is reconstruct from this state. This shows that, despite of dimensionality reduction, sufficient information is embedded in the data that allows for returning to the original state.

With regard to this problem, the use of clustering, before making the taxonomic classification, can be very useful as this means, not to classify contigs by contigs, but an entire group of contigs would be passed, which can help to find a corresponding species easier, if it exists.

Table 2. Results of purity in 25 clusters with lengths greater than 5000.

	k-means	PCA+k-means	Encoder+k-means	Encoder+k-means (length > 1000)
Purity-original	56.4%	62.1%	65.5%	75%
Purity-combination	64.8%	70.3%	75.1%	90.2%

It is best to create as many clusters as real groups exist in the database, in this case, 16 by the 16 species found. However, this is not the objective of this work. What we pretend to do is creating groups, which are composed of contigs that belong to the same species, although they are not all of this species. When we have a group of contigs of a species, this allows for the identification of the species in the future. That is in fact the aim of clustering in the first place, to achieve groups of contigs that can be used for the taxonomic identification of the species that are in the metagenomic data set. To obtain these clusters, as pure as possible, the grouping error is minimized. To start, an even larger number of groups is chosen than an expert would expect to obtain in the sample being analyzed. With this in mind, several clusters can be obtained that represent the same species, but may have less dispersal than other species in the group.

The results show that the use of encoder layers, for a pre-processing of the data, when reducing the dimensionality, can improve the clustering results.

The topologies were adapted to use a four-layer encoder and $256 \times 128 \times 64 \times 32$ neurons, with an output of 32 features as input to the k-means. They were generated with different sizes of clusters, beginning with a k-value of 15 and increasing this value stepwise to 40.

As mentioned before, more clusters than expected will be used. So experiments were performed varying the number of clusters in a range between 20 and 40, where the best results were obtained with k = 32. This means 32 clusters that are presented in Fig. 3 and which shows that the difference between the total of contigs of each cluster and the maximum of contigs that belong to the predominant species in this cluster is very small.

Considering that the number of contigs is very large and there are clusters with more than 160,000 contigs, we will still observe some clusters with contigs that belong to different species.

Table 2 shows the percentage of purity of the clusters, that is, the percentage of the element with the greatest occurrence in the cluster. As can be observed, the proposed methods outperform the classic k-means. Feature extraction improves outcomes with both, using PCA and the autoencoder. Where the latter achieves a higher percentage. Despite the observed improvement, the elements in the different groups are still widely dispersed. The second row of the table is related to the last strategy, which began with a large number of clusters that was decreased by combining different distance measures.

Another strategy for improving the results consists in filtering the base for longer contigs. Very small contigs can generate noise because it might be a piece of DNA that belongs to different species. By using lengths greater than 5000, the algorithm reaches 80% purity for all groups and 70% when the lengths are shorter. Despite of the fact that each species does not belong to a single group, if a separation is enforced that generates more than the expected amount of groups, good purity is obtained.

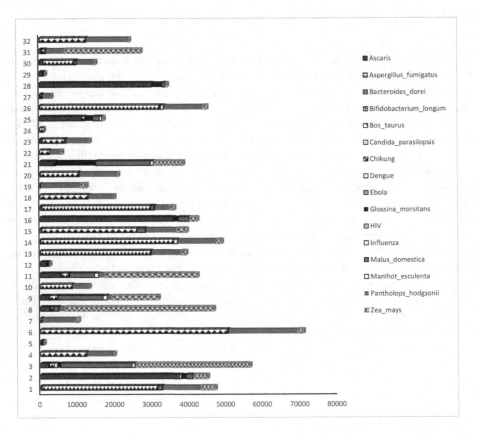

Fig. 3. Result of the difference between the total of contigs (Total) and the total of the predominant species in the cluster (MaxSpecie).

So, we can conclude that there are still some dispersed fragments, but these are those that may create noise. This can be corroborated, as the percent of purity turned out to be 90% when filtering the data with a contigs size greater than 10000.

Another way to increase the degree of purity is to create much more clusters and then reduce their amount by eliminating clusters with a very small cluster distance. For this test, a k-means with k = 2000 was trained.

A combination of Euclidean and Cosine distance was used, where the average of the two distances with respect to the obtained centroids was the decision criteria to join the clusters or to leave them separated.

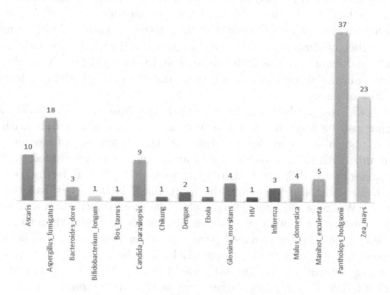

Fig. 4. Result related to the difference between the total of contigs (Total) and the total of the predominant species in the cluster (MaxSpecie).

The average between the Euclidean and Cosine distance was used to compare the distances between the centroids of the obtained clusters. The closest clusters were grouped together, based on a threshold of 0.5. The number of clusters was reduced from 2000 to 117, having at least one cluster for each specie, which yielded 99% of purity. The number of clusters oscillated between 37 and 1 clusters by specie, as can be observed in Fig. 4. In this figure the number of clusters for each specie is shown. Ascaris for example, is divided into 10 clusters.

The species that cannot be completely separated from the rest are Bifidobacterium, Influenza and Aspergilus. In most of cases we observe sequences with lengths lower than 10000.

Summing up, it can be said that the obtained results with feature extraction, an autoencoder neural network and k-means with a combination of distance measures outperforms the results of the classic k-means. By taking into account the lengths of the sequences we can divide the problem and create models that focus on short sequences and models based on large sequences. While the combination of different distances can generate a significant change with respect to the separation of the space.

4   Conclusion

A deep clustering method of two parts is used. The first is responsible for pre-processing the data, achieving a data dimensionality reduction. The second takes the reduced data and uses k-means to perform the clustering. Feature extraction was applied with a neural network autoencoder and compared with the classical PCA algorithm. Several experiments were performed with a different number of clusters. The best results were obtained with 32 clusters.

Another strategy, that obtained the best result, consists in beginning with a large number of clusters. At the end, a cluster unification process is applied, based on the distances between the clusters and with a defined threshold. After this process all models improve their results, but the model with an autoencoder reaches 90% of purity.

This model was applied to a metagenomic database with 16 species. It is a model that can be applied, as shown, not only to images and data with more dimensions, but also to complex problems like the present one of metagenomics. In accordance with the results presented above, it can be concluded that small DNA fragments are difficult to classify, which in biological terms, is logical, because the smaller the fragments the more likely is that they can belong to any species. Whereas the longer fragments have more information and produce a better clustering. Although the number of clusters is greater than the number of organisms, the method provided pure groups for organisms, and achieved a high percentage of purity in all groups. Although the experiments were carried out by using data of a only one database, the use of deep learning proves to be sufficiently effective to group larger sequences in the taxonomic classification process.

References

1. Locey, K.J., Lennon, J.T.: Scaling laws predict global microbial diversity. In: Proceedings of the National Academy of Sciences, vol. 11, issue 21, pp. 5970–5975 (2016)
2. Wooley, J.C., Godzik, A., Friedberg, I.: A primer on metagenomics. PLOS Comput. Biol. **6**(2), 1–13 (2010)
3. Teeling, H., Waldmann, J., Lombardot, T., Bauer, M., Glöckner, F.O.: Tetra: a web-service and a stand-alone program for the analysis and comparison of tetranucleotide usage patterns in dna sequences. BMC Bioinform. **5**(1), 163 (2004)
4. Reddy, R.M., Mohammed, M.H., Mande, S.S.: Metacaa: a clustering-aided methodology for efficient assembly of metagenomic datasets. Genomics **10**(2), 161–168 (2014)
5. Xie, J., Girshick, R. B., Farhadi, A.: Unsupervised deep embedding for clustering analysis. CoRR, abs/1511.06335, 2015
6. Bonet, I., Escobar, A., Mesa-Múnera, A., Alzate, J.F.: Clustering of metagenomic data by combining different distance function. Acta Polytech. Hung. **14**(3), 223–236 (2017)

7. MacQueen, J.: Some methods for classification and analysis of multivariate observations. In Proceedings of the Fifth Berkeley Symposium on Mathematical Statistics and Probability, Volume 1: Statistics, Berkeley, Calif. University of California Press (1967)

8. Arthur, D., Vassilvitskii, S.: K-means++: the advantages of careful seeding. In: Proceedings of the Eighteenth Annual ACM-SIAM Symposium on Discrete Algorithms, SODA 2007, Philadelphia, PA, USA. Society for Industrial and Applied Mathematics (2007)

9. Abdi, H., Williams, L.J.: WIREs principal component analysis. Comput. Stat. **2**(4), 433–459 (2010)

Author Index

Printed in the United States
By Bookmasters